BOTANICAL DIETARY SUPPLEMENTS:
QUALITY, SAFETY AND EFFICACY

# BOTANICAL DIETARY SUPPLEMENTS: QUALITY, SAFETY AND EFFICACY

GAIL B. MAHADY, PH.D., R.PH.[1,2]

[1]UIC/NIH Center for Botanical Dietary Supplements Research
Program for Collaborative Research in the Pharmaceutical Sciences
College of Pharmacy
PAHO/WHO Collaborating Centre for Traditional Medicine
University of Illinois at Chicago, Chicago IL

[2]Department of Pharmacy Practice,
College of Pharmacy, University of Illinois at Chicago

HARRY H.S. FONG, PH.D.[1]
NORMAN R. FARNSWORTH, PH.D.[1]

SWETS & ZEITLINGER
PUBLISHERS

LISSE          ABINGDON          EXTON (PA)          TOKYO

*Library of Congress Cataloging-in-Publication Data*

Mahady, Gail B., 1957-
  Botanical dietary supplements : quality, safety, and efficacy / Gail B. Mahady, Harry H.S. Fong, Norman R. Farnsworth.
    p. cm.
  Includes bibliographical references and index.
  ISBN 9026518552
    1. Materia medica, Vegetable--Quality control. 2. Materia medica, Vegetable--Standards. I. Fong, Harry Hong Sang, 1935- II. Farnsworth, Norman R. III. Title.

RS164.M235 2001
615'.32--dc21

                                                    2001031370

Cover design: Brian West, London, England
Typesetting: Red Barn Publishing, Skeagh, Skibbereen, Co. Cork, Ireland
Printed by Grafisch Produktiebedrijf Gorter, Steenwijk, The Netherlands

*Acknowledgement*
This research was supported in part by NIH grant P50-AT00155, the Office of Dietary Supplements, the National Institute of General Medicine, the Office for Research on Women's Health, and the National Center for Complementary and Alternative Medicine.

*Disclaimer*
This book is intended to serve as an information resource and not a final authority for all botanical products. While great care has been taken to ensure the accuracy of the information presented, the reader is advised that the authors, editors, and publishers cannot be responsible for any errors, omissions, or the application of this information, and for any consequences arising therefrom.

ISBN 90 265 1855 2 Hardback

# Contents

# 1

# Introduction to the Botanical Market

## Introduction

Over the past decade botanical dietary supplements, also known as herbal medicinal products, have become a topic of increasing global importance, with both medical and economic implications.[1-4] In the United States, annual retail sales of botanical products rose from a meager $200 million in 1988, to an estimated $5.1 billion in 1997, and consumer use of these products has increased by 380% in the past ten years.[1,2,5] In developing countries such as Africa, China, and India, botanicals have always played a central role in healthcare. Data from the World Health Organization suggests that 65 to 80% of the populations in these countries depend on traditional and botanical medicines as the primary source of healthcare.[6]

In contrast to developing countries, where botanicals have been used in traditional systems of medicine for thousands of years, the use of plant-based medicines in North America dates back for only 300–400 years. In 1620 when the pilgrims first landed in North America, they brought with them plants to use both as food and medicine. Eventually, the settlers incorporated some indigenous American plants into their medical practice, especially *Echinacea*, which was used extensively by the Native American Indians to treat everything from the common cold to rattlesnake bites.[7] Thus, historically, all medical systems, including the American medical system, were once botanically based.

In fact, botanical medicine was practiced in North America relatively successfully for over 300 years, from approximately 1620 to 1930. It was only during the 1940–50s, with the advent of chemical synthesis and new synthetic drugs, coupled with a lack of clinical data to establish the safety and efficacy of botanicals, that herbal usage declined in the United States. However, what is not generally recognized is that up to 25% of all prescription drugs dispensed in the U.S. are derived either directly or indirectly from natural sources such as plants, bacteria, and fungi.[8] Considering that the global prescription medicines market is estimated at over $350 billion USD annually, and the sales in North America and Europe represent almost $210 billion, then the prescription market for natural medicines represents approximately $50–75 billion dollars annually.

*The Herbal Renaissance*

Over the past ten years, botanical remedies have enjoyed a revival in many western countries, including Australia, Canada, Europe and the United States.[2,5,8] In the U.S. alone, it is estimated that herbal usage increased by 380%, just in the period between 1990 and 1997.[2] This "herbal renaissance" has been fueled by strong consumer interest in natural therapies, preventative medicine, coupled with a disappointment with allopathic medicine, and the perception that botanicals are safe and free from side effects.[1] Although, data to accurately calculate the entire global market for botanical medicines is sparse, it can be conservatively estimated that worldwide sales of herbal products are in the range of $25.0 billion USD per year.

Within the European Community, herbal medicinal products represent an important share of the pharmaceutical market, with annual sales in the range of $7 billion US. In order to market herbal medicines in Europe, manufacturers must obtain a "marketing authorization" from the regulatory authorities, which requires the submission of a formal dossier providing scientific proof of safety and efficacy for each product.[9,10] Thus, in many respects, botanicals in Europe are regulated more like drugs. This is in sharp contrast to the U.S. where they are regulated more like foods, under the Dietary Supplements Health and Education Act of 1994. Another difference between the U.S. and Europe is recent market growth. In Europe, the sale of herbal medicinal products has remained stable over the past number of years increasing at a rate of approximately seven percent annually.[11] However, in the U.S. the botanical dietary supplements market increased dramatically over the past ten years. Increasing an average of 15 to 55% per year, and is now estimated at $5.1 billion USD.[2]

*Botanicals and Traditional Medicine*

In developing countries, traditional systems of medicine, such as traditional Chinese Medicine, Ayurvedic medicine, and Kampo medicine have always played an important role in healthcare.[6] It is estimated that up to 65–80% of the population in developing nations rely on traditional medicine as a primary source of healthcare.[6] If one assumes that the current world population is approximately six billion, then as many as three billion people may rely on traditional medicine as their only source of health care. A common characteristic of all traditional systems of medicine is its focus on herbal medicines (botanicals). In many circumstances, botanical drugs may be the only medicines available to populations in developing nations, due to the fact that modern pharmaceutical drugs are in great demand and costly. Low accessibility and the high price of prescription medications, make them unavailable to the general population. Unfortunately, the use of botanicals in developing countries is based primarily on empirical knowledge, and only a small percentage of these plants have been scientifically evaluated for their safety and efficacy.[12]

Furthermore, there are significant concerns about the lack of adequate regulation of botanical products in many countries, and the encouragement of the sale of unregistered products that are not controlled by any regulatory authority.[13] The legal status of botanicals varies widely from country to country, and the regulation of these products has not evolved in any systematic manner.[10] Most countries have different ways of defining medicinal plants, herbs or the products derived from them, and have adopted various approaches to licensing, dispensing, manufacturing and trading to ensure their quality, safety and efficacy.[10] For example, in Europe, herbal medicinal products play an essential role in rational drug therapy, and many herbal medicinal products have already been integrated into conventional medical practice. For years herbal medicinal products have maintained an important share of the European pharmaceutical market, with current annual sales in excess of $7 billion USD.[9–11]

## Safety and Efficacy Questions

The widespread use of botanicals throughout the world has raised serious questions as to the quality, safety and efficacy of these products.[1,14] The assessment of the safety and efficacy of herbal medicines is a complicated and controversial subject that is often hampered by the conflicting philosophies of modern versus complementary and alternative medicine (CAM).[14] In the United States, modern medicine is defined as evidence-based medical science.[15,16] In contrast, CAM, which includes botanical medicines, is defined as those medical practices not normally taught as part of the core curriculum in medical schools (although most schools now offer such courses), and are generally viewed as practices that have not been scientifically validated.[2,5] Historically, modern medicine has provided a scientific framework in which to develop and test medical theories, to ensure that specific therapeutic modalities are both safe and effective.[16] Thus, the credibility of any new and untested therapeutic modality is generally questioned, until it is either, validated and accepted, or invalidated and rejected.[16]

In the case of botanicals, there is much concern by medical professionals that the proponents of these therapies are opposed to scientific validation of such treatments.[16,17] There is also a general misconception that some advocates of alternative medicine believe that these therapies cannot, or should not be subjected to robust scientific investigation.[15] Moreover, some physicians believe that it is the scientific rigor by which medical treatments are evaluated, that actually separates alternative from conventional medicine.[16] Since we are supposed to practice evidence-based medicine, it is difficult for any pharmacist or physician to accept a medical treatment that lacked sound scientific data to support its safety and efficacy claims. Botanicals are no exception.[14]

The first step toward the acceptance of botanicals by healthcare professionals must therefore rely on an accurate assessment of the quality, safety and efficacy, based on the existing scientific and clinical literature.[14] While, many healthcare professionals believe that there is very little evidence to support the safety and efficacy of botanicals, for some of the most commonly used herbal medicines such as garlic, ginkgo, valerian, saw palmetto, *Echinacea* and St. John's wort, this is not the case. In reality, there is a wealth of scientific information available globally, published in peer-reviewed scientific journals throughout the world.[14] However, the amount of experimental and clinical data published in the peer-reviewed U.S. medical and pharmaceutical journals is only a small fraction of what is available globally.[14] Thus, if healthcare professionals restrict their reading to only U.S. based pharmaceutical and medical journals, they would not recognize how much information is actually available.

Much of the clinical data supporting safety and efficacy of herbal medicines have been published in prominent, peer-reviewed European, Chinese, Korean, Russian and Indian journals over the past 30–40 years.[14] The reasons for this are simple, unlike the U.S., where botanicals are regulated as dietary supplements, in most nations, botanical medicine has been an integral part of the healthcare system for hundreds or thousands of years, and as a consequence, there has been a continuum of research in this field. While some U.S. clinicians would prefer to ignore scientific data published in other countries, the truth is that there is currently not enough scientific data published in the U.S. literature to adequately evaluate botanicals with any degree of scientific accuracy. Therefore, until the results of new U.S. clinical trials are published, assessment of the global scientific literature is currently the only way to validate safety and efficacy claims of these products.[14]

Such a task, however, is no small undertaking. In fact, just acquiring copies of scientific articles from foreign journals can be a difficult and time-consuming job. Moreover, since many of the trials are published in languages other than English, it is very difficult to accurately assess this literature, unless it is translated. Thus, the approach to addressing the botanical literature must be one of collaborative teamwork. Along with trained professionals having clinical experience, some members of the team must have expertise in the field of pharmacognosy (natural products research), otherwise key pieces of data may be overlooked (i.e., was the correct plant species, plant part or extract used), and erroneous conclusions reached. In fact, the assessment of the global botanical literature requires a joint collaboration between scientific experts in the fields of pharmacognosy, pharmacology, toxicology and clinical trials.[14]

### Safety and Efficacy Initiatives

Various groups have already begun to address the issues of quality, safety and efficacy. For example, from the period of 1998–2000 the United States Pharmacopoeial Convention convened an *Ad Hoc Advisory Panel on Botanicals* to help revise the USP's drug information botanical monographs. In 2000, the USP convention now has an official panel of experts on botanicals to address safety and efficacy issues. Furthermore, the American Pharmaceutical Association in collaboration with the American Dietetic Association will be publishing an official document describing how to assess safety and efficacy of botanicals.

In Europe, the *European Scientific Cooperative on Phytotherapy* (ESCOP) has formulated herbal monographs for the European market, and the German Commission E monographs are used extensively in Germany and other parts of Europe. For the global botanical market, the World Health Organization's Traditional Medicine Programme

(WHO-TRM) has assessed the quality, safety and efficacy of 58 widely used botanicals, which will be published in a series of books entitled *"WHO Monographs on Selected Medicinal Plants"* (Volumes I, II & III). The monographs were prepared according to the *"Guidelines for the Assessment of Herbal Medicines"* published by WHO-TRM. Over 220 pharmacists, physicians, regulatory authorities, scientists, and world experts in herbal medicines collaborated in the drafting and reviewing of these books. The importance and impact of these published assessments is already apparent. In 1998, the *Ad Hoc Working Group on Herbal Medicinal Products*, within the European Agency for the Evaluation of Medicinal Products, recommended that the scientific monographs drafted by both ESCOP and WHO-TRM be used in the support of the demonstration of the safety and efficacy of herbal medicines.[18]

This book was written as an update of the previous assessments of the quality, safety and efficacy of the top selling botanical dietary supplements. The UIC/NIH Center for Botanical Dietary Supplements Research, within the Program for Collaborative Research in the Pharmaceutical Sciences at the College of Pharmacy, University of Illinois at Chicago has been involved in assessing botanical medicines for the World Health Organization's Traditional Medicine Program since 1995. The goal of this book is to provide healthcare professionals with an overview of the scientific and clinical information available for botanical dietary supplements. Also, to provide a quick reference with regard to the contraindications, precautions, adverse reactions and drug interactions associated with botanicals. As consumers continue to self-medicate with these products, they will increasingly seek council from pharmacists, physicians, nurses, dietitians and other healthcare professionals as to the safe and judicious use of these products.

## References

1   Mahady, GB. Herbal medicine and pharmacy education. *J APhA*. 1998;38:274.
2   Eisenberg DM, Davis RB, Ettner SL et al. Trends in alternative medicine uses in the United States, 1990–1997. Results of a follow-up National survey. *JAMA*. 1998;20:1569–1575.
3   Tyler V. The new age of herbals: a pharmacognosy renaissance *J APhA*. 1999;39:11–12.
4   Eliason BC, Kruger J, Mark D et al. Dietary supplement users: demographics, product use and medical system interaction. *J Am Board Fam Pract*. 1997;10:265–271.
5   Eisenberg DM, Kessler RC, Foster CF et al. Unconventional medicine in the United States – prevalence, costs, and patterns of use. *N Engl J Med*. 1993;328:246–252.
6   Bannerman R, Burton J, Chen WC (eds.). *Traditional Medicine and Health Care Coverage*, World Health Organization, Geneva, Switzerland, 1983.
7   Hobbs C. *Echinacea*, a literature review. *Herbal Gram*. 1994;30(Suppl):33–46.
8   Farnsworth NR, Morris RW. Higher plants: the sleeping giant of drug discovery. *Amer J Pharm* 1976;40:115–121.
9   Association Européenne des Spécialités Pharmaceutiques Grand Public. *Herbal Medicinal Products in the European Union*, Final Report, 1998.
10  WHO-Traditional Medicine Programme. *Regulatory situation of herbal medicines*. A worldwide review. World Health Organization, Geneva, Switzerland. 1998:pp 1–5.

11    Wagner H. Phytotherapy in Europe. *J Natl Inst Environ Toxicol.* 1999.
12    Akerele O. Medicinal plants and primary health care: an agenda for action. *Fitoterapia.* 1988;355–363.
13    DeSmet PAGM. Should herbal medicine-like products be licensed as medicines. *Brit Med J.* 1995;310:1023–1024.
14    Mahady GB. Botanicals: The complexities associated with assessing the clinical literature on efficacy. *Pharmacy and Therapeutics.* 2000;25:127–132.
15    Fontanarosa PB, Lundberg GD. Alternative medicine meets science. *JAMA.* 1998;280:1618–1619.
16    Ember LS. Alternative medicine goes mainstream. *Chemical and Engineering News.* 1998; December 7, 14–19.
17    Angell M, Kassirer JP. Alternative medicine – the risk of untested and unregulated remedies. *N Engl J Med.* 1998;339:839–841.
18    *The European Agency for the Evaluation of Medicinal Products*, Ad Hoc Working Group on Herbal Medicinal Products, Executive Summary, January 28, 1998.

# 2

# The Regulatory Status of Botanical Dietary Supplements

For thousands of years humans have used plants (botanicals) as both as source of food and medicine. In fact, plants (botanicals) represent a broad range of products, in that they are used as conventional foods and culinary adjuvants, as well as drugs. In the United States, the Federal Food and Drug Administration regulations lists approximately 250 botanical ingredients (and their essential oils and extracts) that are generally recognized as safe (GRAS) for use in food as spices and flavorings, essential oils, and natural extracts.[1] In addition, more than 100 botanical ingredients are listed as approved flavoring agents for the use as natural flavorings in foods and beverages.[2] Other botanical products are regulated as dietary supplements under the Dietary Supplements Health and Education Act. Furthermore, numerous naturally occurring chemical compounds have laid the foundation for many of today's well-known pharmaceutical drugs. For example, foxglove (*Digitalis lanata*) is the source of the cardiotonic drug, digitalis, and the bark of the Pacific yew tree yields the drug known as paclitaxel, a potent chemotherapeutic agent.

In most countries around the world, botanicals (herbal medicinal products) constitute major portions of the drugs listed in their pharmacopoeias. A number of countries including China, France, Germany, Japan, Switzerland and the United Kingdom have also established herbal pharmacopoeias. Systems of regulation applicable to the therapeutic use of botanicals have also been established in many developed and developing countries, including Australia, Canada, China, Germany, and France. Twelve of eighteen countries, for which there is information available, have formal mechanisms by which therapeutic claims may be made for botanicals, based on historical and scientific evidence. In some countries, clinical evidence is required to support recommended uses, while in other countries, traditional use is sufficient. Some countries have established lists of ingredients that are permitted and also lists of permitted therapeutic claims. In fact, the World Health Organization has established guidelines for the regulation of herbal medicines, and has published books on the regulatory status of botanicals worldwide and also a series of volumes of botanical monographs.[3–5] A complete review of the global regulatory situation for botanicals is beyond the scope of this chapter. Instead the focus is the regulatory situation in the United States, including how botanicals became dietary supplements and the impact of these regulations on health care.

## Background of U.S. Regulation

Historically, there was little regulation of botanicals or herbal medicinal products, the primary medications of the 18th century. However, in 1848, Congress attempted to regulate the quality of drugs in the U.S. primarily as a consequence of the provision of adulterated quinine to American troops. The profession of pharmacy first united around the need to establish standards for the safety of the medicines of the time. The first attempt to develop standards and guidelines for drugs was initiated after the founding of the American Pharmaceutical Association, the national professional society of pharmacists, in 1852.[6]

With the development of drug products on a large scale for the general population, instead of individualized preparations, there was increasing concern about the safety of such proprietary products, and the government increased its oversight of medications. In 1906, Congress passed the Pure Food and Drug Act, which dealt with unsafe foods, unregulated elixirs and misbranded products.[2] This Act addressed the purity and quality of drugs and foods, and the accuracy of branding.[7] However, the 1906 Act did not require pre-market screening of drugs, which may have accidentally contributed to the 1937 tragedy in which least seventy-three people died after ingesting a batch of Elixir Sulfanilamide, which was contaminated with diethylene glycol, commonly known as antifreeze. This tragedy prompted the passing of the Federal Food, Drug and Cosmetic Act (FDC Act) of 1938, which exerted more legal control over the safety and efficacy of drugs and also required pre-marketing approval.[7,8] The 1938 FDC Act also established a category of foods for special dietary use and required that the labels of such food products provide information to the consumer regarding the vitamin and mineral content as well as other dietary properties.[2] In 1941 the Food and Drug Administration established regulations governing the labeling of vitamin and mineral supplements and other foods for special dietary use containing added vitamins and minerals.

Federal control was again increased with the Drug Amendments of 1962, adopted in response to the thalidomide-induced fetal deformities in Europe, although thalidomide was never approved for the U.S. market. New amendments requiring clinical testing for approval and marketing of new drugs were established, and the burden of proving safety and efficacy was placed on the manufacturers.[2] Together, these legislative initiatives yielded a comprehensive regulatory structure for the introduction, production, and marketing of prescription and over-the-counter medications. The Congressionally mandated oversight of the drug manufacturing and approval process is based on a proportionate focus on science, and the validation of safety and efficacy of drugs through clinical trials. This oversight is necessary and appropriate as prescription and over-the-counter medications form the cornerstone of consumer health. Both consumers and health care professionals alike rely on the assurances provided by a stringent regulatory process: prescription and over-the-counter medications are available because the FDA has reviewed the data showing the product's safety and efficacy and have "approved" the product for marketing and sale. This comprehensive regulation of drugs supported the shift in the health care system to reliance on "scientifically-valid" treatments as the primary component of medical practice.

In 1976, the Congress passed the Rogers/Proxmire amendment that prohibited the FDA from classifying vitamin and mineral supplements as drugs based only on their combinations or potency (unless drug claims were made). The amendment also prohibited the establishment of standards for identifying these products, and from limiting the quantity or combination of nutrients in them, except for reasons of safety.[2] This amendment also incorporated the FDA's 1941 definition of special dietary use into the FDC Act. During the period between 1973 and 1994, no formal labeling requirements were in effect for dietary supplements. It was not until early in 1994 that the FDA finalized nutrition labeling regulations for supplement products. However, the Dietary Supplements Health and Education Act (DSHEA) later amended the labeling provisions in 1994. Prior to DSHEA, Congress passed the Nutrition Labeling Education Act (NLEA) in 1990 that affected nutrition labeling of food and dietary supplements. The NLEA mandated that all food labels must contain specific information on nutrient content and "health claims could be made relating specific nutrients to diseases and disorders."[2] Such claims however, had to be based on significant scientific agreement on the validity of the claimed relationship between the nutrient and the disease. The FDA established standards for the types and levels of evidence necessary to meet the criteria for approval of health claims. In 1992, Congress passed the Dietary Supplement Act that prohibited the implementation of NLEA with respect to dietary supplements except for approved health claims. In 1993 the FDA published a statement with regard to the labeling of foods and dietary supplements in response to the outbreak of a number of cases of eosinophilia myalgia syndrome due to the ingestion of contaminated L-tryptophan, and other reports of adverse reactions to botanical supplements. This statement suggested that botanical products were inherently drugs and not dietary supplements, and that many dietary supplements, including amino acids, were unapproved food additives.[2] This reasoning met with very strong opposition from both American consumers and the dietary supplement industry. The overwhelming opposition was in response to reports that FDA was attempting to remove vitamins and other "natural products" from the market. For consumers with a renewed interest in unconventional therapy, the threat of government intervention to regulate something as safe as "vitamins" seemed ridiculous. The result of this public opposition to the FDA statement eventually led to the passing of the Dietary Supplements Health and Education Act (DSHEA) in 1994.[2,9]

Following public debate about the importance of dietary supplements in health promotion, and the need for American consumers to have access to accurate information about dietary supplements, DSHEA was passed on October 25, 1994.

The Act amends the Federal Food, Drug and Cosmetic Act of 1938, and alters the way dietary supplements are regulated and labeled. DSHEA also codified a new category of products for consumer use for health promotion, maintenance, and chronic disease prevention. DSHEA was enacted by Congress based on 15 findings, one of which was the finding that "improving the health status of United States citizens ranked at the top of the national priorities of the Federal government."[2] Other Congressional findings supporting the adoption of DSHEA included the increasing reliance of the use of "unconventional" health care providers to avoid the increasing costs of conventional medical treatments, and the belief that consumers should be able to make choices about preventative health programs based on information obtained from scientific studies of health benefits related to particular dietary supplements. In the absence of harm from vitamins, most American consumers preferred that science-based health care allow the use of unconventional therapy, specifically dietary supplements, without the cost and delay of safety and efficacy testing required of prescription and over-the-counter medications.

In addition to defining a dietary supplement, DSHEA influences the FDA's regulatory practice in four primary areas by: (1) placing the burden of proof of safety on the FDA in actions against a dietary supplement manufacturer; (2) limiting the FDA's ability to require pre-marketing information about product safety and efficacy; (3) establishing a minimal pre-marketing notification system for certain new dietary ingredients not marketed in the United States before October 15, 1994, and (4) specifying acceptable claims that could be made without triggering requirements for classification as drugs.

DSHEA defines a dietary supplement as "a product (other than tobacco) intended to supplement the diet that bears or contains one or more of the following ingredients: (A) a vitamin; (B) a mineral; (C) a herb or other botanical; (D) an amino acid; (E) a dietary supplement used by man to supplement the diet by increasing the total dietary intake; or (F) a concentrate, metabolite, constituent, extract, or combination of any ingredient described in clause (A, B, C, D or E)". Also according to DSHEA, a dietary supplement is a product that is labeled as a dietary supplement and is not represented for use as a conventional food or as a sole item of a meal or the diet. The Act also describes the variety of forms-capsules, powders, softgel, tablet, liquid, or other forms in which these products may be ingested (except in the form of a conventional food). It also removed dietary supplements from the category of food additives.

DSHEA established standards for the safety of dietary supplements by describing conditions under which supplements may be designated as adulterated (unsafe). The burden of proving that a dietary supplement is unsafe rests with the FDA. Products that were not on the U.S. market prior to October 15, 1994 are considered new dietary supplements and may be deemed unsafe if there is inadequate information to provide reasonable assurance that such an ingredient does not present a significant or unreasonable risk of illness or injury.

## Impact of DSHEA on Regulatory Practice

Under DSHEA, the burden of proof of safety is placed on the FDA, rather than on the manufacturer.[2] Thus, in contrast to the system required of prescription and over-the-counter medications, dietary supplements under DSHEA are first presumed safe. The FDA must investigate perceived problems and pursue action to remove dangerous products from the market. This limitation may present problems in enforcing DSHEA in that dietary supplements should meet the identity and strength of the product indicated on the label and should have quality guidelines related to chemical purity and physical properties, such as tablet/capsule dissolution and disintegration. These dissolution and disintegration requirements will ensure that the tablet or capsule will actually release a consistent, specific amount of product when swallowed. Failure to disintegrate within specified guidelines may mean the consumer gets less of the supplement released than expected. Conversely, if the product disintegrates too quickly, it may produce high blood levels of the supplement faster than expected, risking a possible adverse reaction. In addition, without the ability to require pre-marketing information, the FDA may not be aware of the presence in products of ingredients the FDA that has already determined to be unsafe. For example, the FDA issued a proposed rule to limit high-dose ephedrine and caffeine products after numerous reports of adverse reactions and approximately thirty deaths.[10] The attempt to limit the availability of these products with documented risks met intense public opposition, and the FDA has yet to finalize a ruling. To address some of the quality issues, the FDA intends to establish guidelines for Good Manufacturing Practices (GMP) for dietary supplements and has issued a proposed rule making on current GMP, pertaining to the maintenance of buildings and facilities, requirements for product handlers, cleanliness of equipment and procedural requirements for maintaining safety during the production and processing of products. All GMPs established by the FDA are specific to the product category under consideration, i.e., "food" or "drug.".[11]

DSHEA also limits the FDA's ability to require pre-marketing information about a product's safety and efficacy. If a dietary supplement includes ingredients on the U.S. market prior to October 1994, the manufacturer is not required to provide the FDA with any information about its production, and can wait until thirty days after marketing the product before providing the FDA with information. With the large number of botanical dietary supplements in the marketplace, the ability of the FDA to identify and pursue those companies who fail to comply with this minimal requirement is questionable. Short of frequently reviewing the shelves of all pharmacies, grocery stores, health food stores and a large number of mail order and other types of outlets, the detection of a product marketed in violation of the 30 days notice is very difficult. The lack of clarity in FDA's enforcement policy further compounds this problem. If a potential violation is identified, the steps the FDA will follow in establishing harm and recommending corrective action are specific, although the current consequences of marketing violations by means of "courtesy letters" may be insufficient to encourage the removal of the products from the market.

Another area in which DSHEA affects the authority of the FDA involves the marketing of "new" dietary supplement ingredients, which are those not already on the market prior to October 15, 1994. For marketing these dietary supplement ingredients, manufacturers must only provide the FDA with a "history of use" or other evidence of safety establishing that the dietary ingredient will reasonably be expected to be safe. Thus, the manufacturer may rely on previously published scientific data, and need not conduct any scientific research of its own to demonstrate safety of its product. The manufacturer is required to provide safety information to the FDA seventy-five days before introducing the product into the market. It has not yet been clearly defined if specific dietary supplement ingredients may be legally exempt from the seventy-five day notification requirement. As a result, some manufacturers have failed to evaluate the status of ingredients in their products at all and simply assume exemption under DSHEA, marketing products without providing the information to the FDA. This regulatory dilemma is in sharp contrast to the regulations concerning prescription and over-the counter medications. In general, to bring a new pharmaceutical drug to market, manufacturers must perform their own research and conduct double-blind, placebo-controlled clinical trials. To bring even one new drug to market requires substantial evaluation of product safety, with an average price tag of $350 million and an average of 12 to 15 years from the laboratory to its use in general medical practice.[2]

DSHEA also mandated the establishment of an Office of Dietary Supplements (ODS) within the National Institutes of Health. According to the Act, the purpose of the ODS is to explore the potential role of dietary supplements as a significant part of the efforts of the United States to improve health care and promote scientific study of the benefits of dietary supplements in maintaining health and preventing chronic disease and other health-related conditions. DSHEA also established a Commission on Dietary Supplements Labels to develop recommendations for the regulation of label claims and statements for dietary supplements.[2] The Commission was appointed by the White House on October 2, 1995, and began work in 1996. In November 1997 the Commission delivered its final report to the President, Congress and the Department of Health and Human Services. The FDA reviewed the Commission's report and recommendations and has drawn up some final rules with regard to the labeling of dietary supplements and claims and statements to be made about the use.

The initial proposed rules in DSHEA regulated the ability of the manufacturer to make certain claims about its products without triggering requirements for classifying the product as a drug.[8] Prior to 1994 dietary supplement products were allowed to be marketed as long as no health or drug claims were made. Under the rules of DSHEA, dietary supplement labels may not make therapeutic drug claims, only statements of nutritional support and specific structure/function claims are allowed without requiring pre-marketing authorization from the FDA. No claims that state: "intended for use in the diagnosis, cure, mitigation, treatment or prevention of diseases" are allowed, as these are considered drug claims. On January 6, 2000 the FDA issued its final rules on structure function claims for dietary supplements under DSHEA.[12] The regulations are intended to clarify the types of claims that may be made for dietary supplements without prior review by the FDA, as well as the types of claims that require prior authorization through the establishment of criteria for determining when a statement about a dietary supplement is a disease claim. The new rules expand the structure/function claims to now include many conditions previously allowed for over-the-counter medications. As some drug claims currently permitted in the OTC drug monographs are not disease claims, such as antacid, digestive aid, short-term laxative, the FDA will allow these claims to be made for dietary supplements as well. Thus, the claim "for occasional relief of constipation" will be allowed for dietary supplements, but the label of these products should provide clear information that it is not meant for the treatment of chronic constipation, which may be the symptom of a serious disease. Under the new rules, standard medical texts will be used to determine whether a label implies treatment or prevention of a disease by listing the characteristic signs and symptoms of a disease or class of disease. Disease is defined as "damage to an organ, part, structure, or systems of the body such that it does not function properly, or a state of health leading to a disease (i.e., hypertension), except that diseases resulting from essential nutrient deficiencies are not included in this definition". Thus statements such as "decreases joint pain" would not be allowed, as this is a symptom of rheumatoid arthritis, while "helps support joint function" is acceptable. Mild conditions commonly associated with aging or normal physiological processes will not be considered diseases. For example claims can be made for "absent mindedness and memory problems associated with aging, hair loss associated with aging, hot flashes, and premenstrual syndrome.[12] Alzheimer's disease and senile dementia, benign prostatic hyperplasia, osteoporosis, hyperemesis gravidarum are all disease claims under the new ruling. The new rules clarify that expressed and implied disease claims made through the name of a product (i.e., Carpaltum); through a statement about the formulation of the product; or through the use or pictures, vignettes or symbols can be made. The new rules will also use the definition of disease that was issued as part of the Nutrition Labeling and Health Act of 1990.

Under the final ruling, the FDA agreed that by enacting DSHEA, the U.S. Congress intended to encourage the dissemination of scientific research and truthful, nonmisleading information on dietary supplements. Therefore, publications and other promotional material containing a title referring to a disease will be deemed a disease claim if, in the context of the labeling as a whole, the citation implies treatment or prevention of a disease.[12]

Under DSHEA and existing regulations, dietary supplement manufacturers are required to maintain documentation substantiating structure/function claims and must include a disclaimer on their labels that their products are not drugs and receive no prior FDA approval. The content of the substantiation files for such claims including the 30-day notification letter to the FDA should contain the identification of the product ingredients, evidence to substantiate the claim, evidence to substantiate safety, assurances of GMP's, as well as the qualifications of the persons who reviewed the data on safety and efficacy.[12]

## Impact of DSHEA on Health Care

Both prescription and over-the-counter drugs are subject to strict control by the FDA, including the regulation of manufacturing processes, specific requirements for the demonstration of safety and efficacy, as well as well-defined limits on advertising and labeling claims. Such controls provide assurances to consumers and health care professionals about the quality of the products and contribute to their acceptance as "legitimate" treatments. The exemption of dietary supplements from these specific regulatory controls may impact their consideration as "legitimate". If health care professionals feel that the quality of dietary supplement products is lacking, and if they consider dietary supplements outside the scope of "prevailing" medical or pharmacy practice, then physicians and pharmacists will have a low level of confidence in recommending these products to their patients for fear of litigation. Physicians and pharmacists cannot, however, separate themselves from the use of dietary supplements. With more than 29,000 dietary supplements on the market, consumers have broad access to and are using these products. In fact, a survey of consumers has suggested that in 1998, approximately 42% of American consumers were using complementary and alternative therapies, with 24% of consumers using botanical dietary supplements on a regular basis.[13]

## Conclusion

Since physicians, pharmacists and other health care professionals are responsible for ensuring the accurate distribution and appropriate use of medications, including the use of quality products, it is important that they have some basic understanding of botanical dietary supplements. This will become increasingly more important as the lines between distinguishing dietary supplements from over-the-counter medications becomes more difficult. As the self-medication practices made by American consumers will ultimately have a significant impact on the treatment alternatives in the health care system, physicians and pharmacists must ask about OTC and dietary supplement use prior to recommending appropriate therapy. Some botanical dietary supplements, such as St. John's wort are known to interact with prescription drug products. Therefore the healthcare provider should be able to accurately assess the safety of supplements for each individual patient, counsel them on the judicious use of these products. To do anything less would go against the professional ethics and our health care practice responsibilities.

## References

1   Food and Drug Administration. *Generally regarded as safe list.* 1998.
2   Report on the Presidential Commission of Dietary Supplements Labels, Office of Disease Prevention and Health Promotion, Washington, DC, November 1997.
3   WHO *"Guidelines to the assessment of herbal medicines"*, WHO-TRM, Geneva, Switzerland, 1991.
4   WHO "Worldwide Regulatory Situation of Herbal Medicines", WHO-TRM, Geneva, Switzerland, 1998.
5   Anon. *WHO Monographs on Selected Medicinal Plants*, Volume I, WHO Publications, Geneva, Switzerland, 1999.
6   APhA began to issue a corollary to the United States Pharmacopoeia (the National Formulary) in 1888, focusing on products not found in the USP. United States Pharmacopeia 23 – National Formulary 18, Rockville: The United States Pharmacopoeial Convention Inc; xiv 1995.
7   Pure Food and Drug Act of 1906, ch. 3915, 34 Stat. 768 (repealed 1938).
8   Dietary Supplement Health and Education Act of 1994, Pub. L. No. 103–417, 108 Stat. 4325 (1994) § 4, 21 U.S.C. § 342(f)(1).
9   Dietary Supplements Containing Ephedrine Alkaloids, *Fed Reg.* 1997;62;30677.
10  Food Drug and Cosmetic Act of 1938, 21 U.S.C. § 351 (1999). 49 Supra note 18, 19.
11  Current Good Manufacturing Practice in Manufacturing, Packing, or Holding Dietary Supplements, *Fed Reg.* 1997;62;5699.
12  FDA. Regulations on Statements made for dietary supplements concerning the effect of the product on the structure or function of the body; final rules. *Fed Reg.* 2000;65:999–1050.
13  Eisenberg DM, Davis RB, Ettner SL et al. Trends in alternative medicine uses in the United States, 1990–1997. Results of a follow-up National survey. *JAMA.* 1998;20:1569–1575.

# 3

# Standardization of Botanicals and Botanical Products

## Introduction

In the U.S.A., botanical products are available as prescription and over-the-counter (OTC) drugs, and dietary supplements for healthcare needs. By law, good manufacturing practices (GMP) are required in the production of prescription and OTC drugs. On the other hand, the regulatory provisions under the Dietary Supplement Health and Education Act (DSHEA) of 1994 which allows botanical products to be marketed as dietary supplements, provides little assurance of identity, quality or purity. The problem is that although DSHEA places the responsibility for product safety on manufacturers, it places the burden of proof on the FDA. In turn, botanical dietary supplement products have not been subjected to mandated quality assurance (QA)/quality control (QC) standards as in the case of prescription and OTC drugs. Consequently, product quality may differ from brand to brand, and even from lot to lot (within the same brand); post-marketing analysis of certain botanical supplements in recent years confirm this situation.

*Current Botanical Product Quality*
In a recently completed Ginseng Evaluation Program study[1,2] conducted at the University of Illinois at Chicago and at the University of Ottawa, in collaboration with the American Botanical Council, selected commercial ginseng products prepared from *Panax ginseng, Panax quinquefolius* and *Eleutherococcus senticosus* marketed as botanical supplements in North America in the 1995–1998 period were subjected to qualitative profiling (finger-printing) and quantitative analysis of ginsenosides and/or eleutherosides. A total of 232 Asian ginseng (*P. ginseng*) products in 15 different formulations, and a total of 81 North American ginseng (*P. quinquefolius*) products in 8 different formulations were evaluated for their ginsenoside content. At the same time, 111 eleuthero *(E. senticosus)* products in 10 formulation types were analyzed for eleutherosides B and E content. The qualitative high-performance liquid (HPLC) chromatograms of the majority of the products showed ginsenoside and eleutheroside

profiles consistent with those obtained for reference root materials. Although analysis showed that *ca* 74% of the products derived from the *Panax* species met product label claim for ginsenosides, there are considerable variations. The quantitative ginsenoside contents of the 232 Asian ginseng products ranged from 0.00 to 13.54% in the various dosage forms. The lack of consistency in the ginsenoside contents of these products is evident not only among different products of all types, but also among the products from within a given formulation type. Capsules composed of dried extracts contained the highest quantity of ginsenosides (average > 3%). Tablets and liquid dosage forms, on the other hand, showed the lowest ginsenoside content (average < 0.7%). The total ginsenoside content in 81 North American ginseng products ranges from 0.009–8.00%. As observed for the Asian ginseng products these materials, as represented by the powdered root capsules, showed extensive variation. As in the case of Asian ginseng, the lowest concentration of ginsenosides was found in the liquid and tablet formulations of the North American ginseng products. The eleutherosides B and E content of eleuthero root powder and other formulated extract products also showed large variation.[1,2] Studies on the quality of St. John's Wort (*Hypericum perforatum*) products showed similar results. The hypericin content of ten products was examined in a study commissioned by the *Los Angeles Times*.[3] The results of the study, which employed the DAC-91 (German Pharmacopeial Codex, 1991) spectrophotometric procedure, showed "hypericin" content ranging from 22–140% of label claim. It should be noted that this assay assesses total hypericins, including pseudohypericin and structurally related compounds, not just hypericin used as the reference standard. High-performance liquid chromatographic (HPLC) procedures, on the other hand, can measure the individual hypericins, including pseudohypericin and structurally related species. Nevertheless, the large variation in the "hypericin" content reported in the *Los Angeles Times* study attests to the variability of product quality. The findings of this study were confirmed by a study of eight brands of commercial St. John's Wort products employing an HPLC procedure, in which "hypericin" content varied from 47–165% of label claim.[4]

Aside from the variation in the chemical content of dietary supplements, there can also be pharmaceutical quality differences in these products. In a dissolution/disintegration study, tablets from two of nine brands of melatonin tablets did not disintegrate after more than 20 hours of testing.[5]

*Factors Contributing to Chemical Variability and Other Quality Features in Botanicals and Botanical Products*

Unlike prescription or OTC drugs, botanical supplements do present some unique problems in their quality standardization. These variables are caused by intrinsic and extrinsic factors such as species difference, organ specificity, seasonal variation, cultivation, harvest, storage, transportation, adulteration, substitution, contamination; post harvest treatment and manufacturing practices.[6–10]

Intrinsically, botanicals are derived from dynamic living organisms, each of which is capable of being slightly different in its physical and chemical characters due to genetic influence, climate, geography, etc. This phenomenon has been well documented. A case in point concerns the accumulation of hypericin in *Hypericum perforatum* (St. John's Wort), which showed that narrow leafed populations have greater concentrations than the broader leafed variety.[11,12] In general, both qualitative and quantitative variations of phytochemicals are greater in wild than in domesticated populations of the same species. Recent studies on the content of artemisinin, the antimalarial agent, in *Artemisia annua*;[7] on michellamine B, a compound with *in vitro* anti-HIV activity, in *Ancistrocladus korupensis*;[7] and on the essential oil composition of *Ocimum basilicum*[7] showed greater variations in the chemical content of the wild than of cultivated populations.

It is also well established that the secondary chemical constituents of medicinal plants can, and do, differ from species to species as demonstrated by the presence of structurally different alkylamides in the roots of *Echinacea angustifolia and E. purpurea*, but by their total absence in *E. pallida*.[13,14] To insure chemical uniformity, it is necessary that the starting plant material for the manufacture of botanicals be accurately identified and authenticated by their scientific names (Latin binomial). The use of common names is inadequate as they often refer to more than one species. Ginseng, for example, is not only the common name for members of the genus *Panax*, but has also been accepted in North American commerce to include *Eleutherococcus senticosus* ("eleuthero", "Siberian", "Russian", "Manchurian" ginseng). As indicated above, the ginsenoside profiles of the two *Panax* species are not identical and *E. senticosus* contains entirely different classes of chemical compounds. Consequently, products purported to be prepared from one species should not be made from another plant known by the same common name. Medicinal plants and their derived products must therefore be identified by their scientific names.

In regards to plant organ specificity, the site of biosynthesis and the site of accumulation and storage are normally different. Chemical biosynthesis usually takes place in the leaves, and then transported through the stems to the roots for storage, with the chemical profiles in these organs being different from each other. Accumulation and storage can also take place in the leaves, but to a much lower extent, and very infrequently in the stems. An example of site-specific accumulation, as well as species specificity, is that of the compounds considered responsible for the immunostimulant effect of *Echinacea* species. These compounds encompass five groups of chemicals: caffeic acid derivatives, alkylamides, polyacetylenes (ketodialkenes and ketodialkynes), glycoproteins and polysaccharides. As indicated above, alkylamides are found in the roots of *Echinacea angustifolia* and *E. purpurea*, but they are structurally different; and are totally absent in *E. pallida* roots. Polyacetylenes, on the other hand, are present abundantly in the roots of *E. pallida*, but absent in *E. angustifolia* and *E. purpurea* roots. While the glycoproteins and polysaccharides are present in the fresh juices and aerial parts of all three species, they occur only in minute quantities in the roots.[13,14]

Seasonal variation is another intrinsic factor affecting chemical accumulation in both wild and cultivated plants. Depending on the plant, the accumulation of chemical constituents can occur at any time during the various stages of their growth. In the majority of cases, maximum chemical accumulation occurs at the time of flowering, followed by a decline beginning at the fruiting stage.

Compounding the problem in botanical standardization is that in the case of prescription and OTC drugs, each product has a single defined "active" chemical constituent, which is used to measure quality and determine shelf-life. The active principle(s) of botanical dietary supplements, on the other hand, are largely unknown. For example, there is no evidence that the marker compound, 27-deoxyactein is the active principle in black cohosh (*Cimicifuga racemosa*), nor that eleutherosides B and E are the prime actives in eleuthero (*Eleutherococcus senticosus*). There is also presently considerable disagreement on the biological roles of hypericin and hyperforin in St. John's Wort, and of the ginsenosides in Asian and North American ginsengs (*Panax ginseng* and *P. quinquefolius*). Consequently, major constituent(s) of the source plant, whether biologically active or not, are employed as marker compounds for the standardization of most of the botanical dietary supplements products currently in the market.

There are many extrinsic factors affecting qualities of medicinal plants. It has been well established that factors such as soil, light, water, temperature and nutrients can, and do, affect phytochemical accumulation in plants, as exemplified by alkaloid concentrations of 1.3 and 0.3%, respectively, in *Atropa belladonna* grown in the Caucasus and those cultivated in Sweden;[8] essential oil content in shade-grown (1.09%) and normal light-grown (1.43%) *Mentha* × *piperita* plants;[8] and by the silymarin content being highest in the fruits of plants grown under 60% water/field capacity (1.39%) and nitrogen level of 100 (1.46%) and 150 kg (1.42%) per feddan.[15]

The methods employed in the field collection from the wild, as well as in commercial cultivation, harvest, post-harvest processing, shipping and storage can also influence the physical appearance and chemical quality of the botanical source materials. Contaminations by microbial and chemical agents (pesticides, herbicides, heavy metals), as well as by insect, animal, animal parts and animal excreta during any of the stages of source plant material production can lead to lower quality and/or unsafe materials.[7–10]

Botanicals collected in the wild often include non-targeted species either by accidental substitution or by intentional adulteration. However, adulteration and/or substitution of cultivated botanicals have also been documented. Substitution of *Periploca sepium* for eleuthero (*Eleutherococcus senticosus*) has been widely documented, and is regarded as responsible for the "hairy baby" case involving maternal/neonatal androgenization.[16] More recently, the case of a previously healthy patient, who presented with a toxic serum level of digoxin after ingestion of a botanical dietary supplement led FDA investigators to determine that the plantain (*Plantago ovata*) consumed was contaminated by *Digitalis lanata* at the foreign supplier end; and that the primary USA distributor had dispatched the tainted source material to a number of secondary distributors, and subsequently to manufacturers, wholesalers and retail distributors.[17] Other examples of adulteration/substitution of botanicals include *Echinacea angustifolia* roots being contaminated with *E. atrorubens*, *E. pallida*, *E. paradoxa*, *E. simulata*, *Lespedeza capitata* and *Parthenium integrifolium*; and ginseng being adulterated with *Mirabilis jalapa*, *Phytolacca acinosa*, *Platycodon grandiforum*, and *Talinum paniculatum*.[18]

Although multi-component herbal mixtures are not in the main stream of botanical supplements in the U.S.A. market, there are, nevertheless, a number of such products available, especially in ethnic stores. They are primarily imports and a number of them are adulterated with therapeutic chemical substances. Foremost among these herbal mixtures are multi-component Chinese herbal remedies. Chemical analysis of some arthritis remedies led to the discovery that synthetic anti-inflammatory drugs such as phenylbutazone, indomethacin and/or corticoid steroids have been added.[19] In a recent study of chemical adulteration of traditional medicine in Taiwan, 23.7% (618 of 2,609) of samples collected by eight major hospitals were found to contain one or more synthetic therapeutic agents. These compounds included caffeine, acetaminophen, indomethacin, hydrochlorothiazide, prednisolone, ethoxybenzamide, phenylbutazone, betamethasone, theophylline, dexamethasone, diazepam, bucetin, chlorpheniramine maleate, prednisone, oxyphenbutazone, diclofenac sodium, ibuprofen, cortisone, ketoprofen, phenobarbital, hydrocortisone acetate, niflumic acid, triamcinolone, diethylpropion, mefenamic acid, prioxicam and salicylamide.[20] The most frequent adulterants were caffeine (213), acetaminophen (167), indomethacin (152), hydrochlorothiazide (127), prednislone (91), and chlorzoxazone (87 cases).

Heavy metal contamination can occur at the cultivation, post-harvest treatment or product manufacturing stages. Lead and thallium contamination has been reported in multi-component herbal mixtures. Besides the unintentional in-process adulteration, it is well established that Ayurvedic and Traditional Chinese Medicine sometimes employ complex mixtures of plant, animal and mineral substances, including heavy metals. It is not uncommon to find appreciable quantities of heavy metals such as lead, mercury, cadmium, arsenic and gold in certain formulations. Cases of lead, thallium, mercury, arsenic, gold, and cadmium poisoning from the consumption of such products have been documented.[19]

*Quality Assurance and Quality Control of Botanicals*
The assurance of the quality of prescription and OTC drugs is affected through production under good manufacturing practices (GMP) mandated by FDA and/or FTC regulations. In the case of botanicals, GMP requirement is not yet mandated. Further, GMP regulation at the manufacturing end alone will not be sufficient to assure quality. For botanicals, quality control measures must be taken from the point of medicinal plant procurement, whether by field collection from the wild, or by cultivation, as the quality of the finished botanical products is obviously directly related to the quality of the raw materials. Whether field collected or produced by cultivation, authentication of plant species by a taxonomic botanist is paramount to insure that the correct source material is acquired. It is essential that the plant materials are identified by their binomial Latin names, and a description of the macroscopic,

microscopic and organoleptic (sensory) characters, be provided along with herbarium specimens, drawings or photographs.[6,21–24] In the field collection of medicinal plants, care must be exercised to avoid the acquisition of non-targeted species, freed of undesirable plant parts, soil, rock, insects, animals, animal excreta and other contaminants. Post collection treatments should mirror those accorded cultivated plant materials. Due to their genetic and chemical content variations, the site and date should be recorded for each collection. The production of raw materials by cultivation should normally lead to more uniform botanical products due to greater genetic uniformity. The production of quality raw materials can only be assured by employing good agricultural practices (GAP) as carried out in the commercial cultivation of *Ginkgo biloba*[25] and of *Echinacea* species.[26] GAP[27] requires as the first step, the botanical identification of the starting seeds or cuttings to the plant species, variety, cultivar, chemotype and origin. Whether cultivation is by conventional or by organic methods, care must be taken to avoid environmental disturbance; the soil should be freed of sludge, heavy metal, pesticide/herbicide or other unnatural chemical contaminants; fertilizers should be applied sparingly and indirectly; animal manures must be well composted prior to application to minimize contamination by pathogenic microbes. Human manure should never be used due to its contained pathogenic microbes. Irrigation should be applied in accordance with the need of the target plant, and the water used must be freed of faecal matter, unnatural chemicals, heavy metal and other toxic substances. Pesticides and herbicides are to be applied only when absolutely necessary using the minimum level of materials approved by the target country of commerce, and be fully documented.

Harvesting of the target plant part should be carried out when the plants are at the growth stage of maximal chemical accumulation, and desired physical appearance. It should be accomplished under optimal environmental conditions to minimize the effect of moisture, and employing clean equipment and implements to minimize microbial contamination. The harvested materials must not touch bare soil, and washing in contaminated water must be avoided; they must be promptly collected and transported under dry condition using clean sacks, baskets, containers and trailers. Post harvest drying can be by open air under the sun, in the shade, in well-aerated buildings or by artificial heat. Ideally, the materials for drying should be spread out in a thin layer on wire-mesh frames off the ground, with frequent turning to assure uniform drying and to avoid mold formation. The dried plant materials must be stored in clean sacks, bags, or cardboard containers protected from insects, rodents, and other animals in well aerated buildings prior to packaging and shipment to the processor and manufacturers of the finished products under GMP. All personnel involved in the GAP process must be adequately trained and all production steps be documented.

GMP procedures[6] employed for the manufacture of botanical products are similar to those employed for the manufacture of conventional drugs employing various QA/QC analytical techniques and instruments. The World Health Organization has published some general guidelines for the GMP production of botanical products.[28]

*QA/QC Methods of Analysis*

The methods and procedures employed for the QA/QC of botanicals involve taxonomic, chemical, spectroscopic and microbial protocols. At the raw material production end, botanical taxonomic identification must be performed to assure species identification. Macroscopic, microscopic and organoleptic analysis should be undertaken at the processing and manufacturing stage to assure quality and purity by appropriate protocols.[23,24] Microscopic and organoleptic examinations will help assure botanical identity and purity as each plant species possesses characteristic microscopic cellular characters, and may have distinct sensory properties. Macroscopical examination will reveal the presence of deterioration and signs of contamination by molds, insects, rodent and other animals, as well as by other plants.

As in the case of conventional drugs, microbiological evaluations are required to monitor the presence and level of microbial contamination in botanical products.[23] This aspect can be performed at the processing stage, and should be part of the GMP protocol in the manufacturing process.

Procedures for the QA/QC analysis of active and/or marker chemical compounds in botanical products can be accomplished by colorimetric, spectroscopic or chromatographic methods. Colorimetric and spectroscopic methods are older analytical procedures quantifying the absorption of structurally related compounds at a specific wavelength of light, and expressed as concentration of a reference compound, which is normally the active or major chemical constituent in that plant material. Since other unrelated plant constituents absorbing at the same wavelength will also be included in the measurement, a higher concentration can be erroneously ascribed to the test material as in the case of hypericins cited above. There has been a decline in the use these procedures in recent years.

In recent years, chromatographic procedures have become the method of choice for the analysis of secondary chemical constituents. Thin-layer chromatographic (TLC) procedures have the advantage of being simple, rapid, can provide useful characteristic profile patterns, and are inexpensive to use. However, their resolving power is limited. Gas liquid chromatography (GLC) can provide high resolution of the more volatile complex mixtures, but is of limited value in the case of non-volatile polar compounds, especially the polar polyhydroxylated and glycosidic compounds. High-performance liquid chromatography (HPLC) is capable of resolving complex mixtures of polar and non-polar compounds, and has become the method of choice for the qualitative and quantitative analysis of botanical extracts and products. The literature is replete with HPLC methods for the analysis of more than 95% of the botanical extracts or products on the market. The Ginseng Evaluation Program described above is but one representative of this methodology. Combined high performance liquid chromatography – mass spectrometry (LC-MS) and liquid chromatography – tandem mass spectrometry (LC-MS-MS) are analytical methods coming on line and will be the analytical methods of choice in the near future. The advantage of these methods is that as each compound is being eluted, it is captured by the mass spectrometer and provides an immediate molecular ion and/or major mass fragment, which allows for positive identification of the eluting "peak". This technique has been used for the identification of ginsenosides in the aforementioned ginseng evaluation program.[1,29]

## Conclusion

The current lack of uniform quality in botanical dietary supplement products undermines consumer confidence. Fortunately, there is a growing awareness and acceptance by the dietary supplements industry of the absolute need for standardization of botanical extracts to ensure batch to batch consistency.[30] A major challenge to the industry is the harmonization of standards and methods of standardization of the same product to ensure consistent quality not only batch to batch within a given company, but industry-wide. The final methods adopted should take into consideration the need for quality control monitoring over the entire process, from field collection and the selection of the germplasm and cultivation to post harvest processing and manufacture of the final botanical product. The literature is replete with the required analytical methods.

## References

1    Fitzloff J, Yat P, Lu Z et al. Perspectives on the quality assurance of ginseng products in North America. In: *Advances in Ginseng Research – Proceedings of the 7th International Symposium on Ginseng*. H. Huh, KJ Choi, YC Kim, (eds.). Seoul, Korea, The Korean Society of Ginseng, 1998;138–145.

2    Awang DVC, Yat PN, Arnason JT, Lu ZZ, Fitzloff JF, Fong H, Hall T, Blumenthal M. The American Botanical Council's program of evaluation of 'ginseng' products on the North American market. *International Ginseng Conference '99. Ginseng: Its Science and Its Markets. Advances in Biotechnology, Medicinal Applications and Marketing*, 8–11 July 1999, Hong Kong, China, Abstract No. A17.

3    Monmaney T. Remedy's U.S. Sales zoom, but quality control lags. *The Los Angeles Times*, 1998, August 31:A1–A10.

4    Constantine GH, Karchesy J. Variations in hypericin concentrations in *Hypericum perforatum* L. and commercial products. *American Society of Pharmacognosy Interim Meeting*, Tunica, Mississippi, April 29–May 1, 1999;Abstract No. P-16.

5    Hahm H, Kujawa J, Augsburger L. Comparison of melatonin products against USP's nutritional supplements standards and other criteria. *J APhA*. 1999;39:27–30.

6    Reichling J, Saller R. Quality control in the manufacturing of modern herbal remedies. *Quarterly Rev Nat Med*. 1998;Spring:21–28.

7    Simon JE. Domestication and production considerations in quality control of botanicals. In: *Botanical Medicine – Efficacy, Quality Assurance and Regulation*. D. Eskinazi, M. Blumenthal, N. Farnsworth, CW Riggins, (eds.). Larchmont, N.Y.: Mary Ann Liebert, Inc. Publishers, 1999;133–137.

8    McChesney JD. Quality of botanical preparations: Environmental issues and methodology for detecting environmental contaminants. In: *Botanical Medicine – Efficacy, Quality Assurance and Regulation*. D. Eskinazi, M. Blumenthal, N. Farnsworth, CW Riggins, (eds.). Larchmont, N.Y.: Mary Ann Liebert, Inc. Publishers, 1999;127–131.

9    Flaster T. Shipping, handling, receipt, and short-term storage of raw plant materials. In: *Botanical Medicine – Efficacy, Quality Assurance and Regulation*. D. Eskinazi, M.

Blumenthal, N. Farnsworth, CW Riggins, (eds.). Larchmont, N.Y.: Mary Ann Liebert, Inc. Publishers, 1999;139–142.

10    Busse W. The processing of botanicals. In: *Botanical Medicine – Efficacy, Quality Assurance and Regulation*. D. Eskinazi, M. Blumenthal, N. Farnsworth, CW Riggins, (eds.). Larchmont, N.Y.: Mary Ann Liebert, Inc. Publishers, 1999;143–145.

11    Southwell IA, Campbell MH. Hypericin content variation in *Hypericum perforatum* in Australia. *Phytochemistry*. 1991;30:475–478.

12    Campbell MH, May CE, Southwell IA, Tomlinson JD, Michael PW. Variation in *Hypericum perforatum* L. (St. John's wort) in New South Wales. *Plant Protection Quarterly*. 1997;12:64–66.

13    Bauer R. *Echinacea*: Biological effects and active principles. In: *Phytomedicines of Europe. Chemistry and Biological Activity*. LD Lawson and R Bauer, (eds.). Washington, DC, American Chemical Society, 1988;140–157.

14    Bauer R, Wagner H. *Echinacea* species as potential immunostimulatory drugs. In: *Economic and Medicinal Plant Research*. Vol. 5, H Wagner, NR Farnsworth, (eds.). New York, Academic Press, 1991;253–321.

15    Hammouda FM, Ismail SI, Hassan NM, Zaki AK, Kamel A. Evaluation of the silymarin content in *Silybum marianum* (L.) Gaertn. cultivated under different agricultural conditions. *Phytother Res*. 1993;7:90–91.

16    Awang DVC. Quality control and good manufacturing practices: Safety and efficacy of commercial herbs. *Food and Drug Law J*. 1997;52:341–344.

17    Slifman NR, Obermeyer WR, Aloi BK et al. Contamination of botanical dietary supplements by *Digitalis lanata*. *N Engl J Med*. 1998;339:806–811.

18    Brevoort P et al. Role of ethnomedicine in selecting the level of purification of botanical products. *The Drug Information Association Workshop on Botanical Quality: Identification and Characterization*, April 9–10, 1996. Thru: Ref. 16.

19    Farnsworth NR. Relative safety of herbal medicines. *Herbal Gram*. 1993;29:36A–367H.

20    Huang WF, Wen K-C, Hsiao M-L. Adulteration by synthetic therapeutic substances of Traditional Chinese Medicine in Taiwan. *J Clin Pharmacol*. 1997;37:344–350.

21    Anon. *Research Guidelines for Evaluating the Safety and Efficacy of Herbal Medicines*, Manila, World Health Organization, Regional Office for the Western Pacific, 1993.

22    Anon. Annex 11. Guidelines for the assessment of herbal medicines. *WHO Technical Report Series No. 863*, Geneva, World Health Organization, 1996:178–184.

23    Anon. *Quality Control Methods for Medicinal Plant Materials*, Geneva, World Health Organization, 1998.

24    Houghton PJ. Establishing identification criteria for botanicals. *Drug Inform J*. 1998;32:461–469.

25    Balz J-P. Agronomic aspects of pharmaceutical plants production on the example of *Ginkgo biloba*. *Phytopharmaka in Forschung und Klinischer Anwendung*. Vol. 4. Paris, Beaufour-IPSEN, 1998;45–50.

26    Li TSC. *Echinacea*: Cultivation and medicinal value. *Herbology*. 1998;8(2):122–129.

27    Anon. Guidelines for good agricultural practice (GAP) of medicinal and aromatic plants. *Zeitschrift Arzneimittel Gew Pflanzen*. 1998;3:166–173.

28    Anon. Annex 8. Good manufacturing practices: Supplementary guidelines for the manufacture of herbal medicinal products. *WHO Technical Report Series No. 863*, Geneva, World Health Organization, 1996;109–113.

29    van Breemen RB, Huang C-R, Lu Z-Z, Rimando A, Fong HHS, Fitzloff JF. Electrospray liquid chromatography/mass spectrometry of ginsenosides. *Anal Chem*. 1995;67:3985–3989.

30    Schutt E. Hot herbs. *Nutraceuticals World*, 1998;November/December:44–54.

# 4

# Black Cohosh

## Synopsis

Clinical evidence supports the use of black cohosh (*Cimicifuga racemosa*) for the symptomatic treatment of menopausal symptoms, such as hot flashes, profuse sweating, insomnia and anxiety. However, there is little clinical evidence to support the use of black cohosh for the treatment of premenstrual syndrome, amenorrhea or dysmenorrhea. The recommended dose is a 40 to 60% ethanol or isopropyl extract of the rhizome of *Cimicifuga racemosa* (black cohosh) corresponding to 40 mg per day. Four to twelve weeks of treatment may be required before the full therapeutic effects apparent. A few minor adverse reactions such as nausea, vomiting, headaches and dizziness have been reported in clinical trials. No drug interactions have been reported in the medical literature. Administration of black cohosh preparations to children, or during pregnancy and lactation is contraindicated due to potential hormonal effects.

## Introduction

*Cimicifuga racemosa* is a perennial woodland herb whose underground portions consist of a thick, knotted rhizome system.[1] The plant is native to North America and was routinely used as a medicine by the Native American Indians for the treatment of female ailments, and hence it was commonly known as squawroot.[2] The Native American Indians also used infusions (teas) of the root to treat coughs, colds, constipation, fatigue, and rheumatism and to increase milk production.[2,3] In 1832, Dr. John King, an Eclectic physician, began using a tincture of black cohosh root for the treatment of pain and inflammation associated with endometriosis, rheumatism, neuralgia and dysmenorrhea.[2] In fact, a fluidextract of black cohosh was listed in the United States National Formulary for over 100 years from 1840 until 1946. During the early part of the 19th century, numerous case reports appeared in the medical literature describing the use of black cohosh for a variety of ailments. By 1849 the use of black cohosh began to focus specifically on the treatment of various female ailments such as amenorrhea and dysmenorrhea.[3]

The first pharmacological studies on black cohosh were published in 1944, and suggested that the plant had estrogen-like effects in rodents.[3] At least 14 clinical studies and case reports were reported prior to 1962, and at least 10 clinical trials assessing the effect of black cohosh on menopausal symptoms have been published since 1982.[1,3] Today, standardized extracts and other commercial products of black cohosh are prepared from the dried rhizomes and roots of the plant.

**Quality Information**

- The correct Latin name for black cohosh is *Cimicifuga racemosa* (L.) Nutt.[1] Botanical synonyms that may appear in the scientific literature include: *Actaea gyrostachya* Wender, *A. orthostachya* Wender, *A. monogyna* Walt., *A. racemosa* L., *Bortrophis actaeoides* Raf., *B. serpentaria* Raf., *Christophoriana canadensis racemosa* Gouan, *Cimicifuga racemosa* (Torr) Bart., *C. serpentaria* Pursh, *Macrotis racemosa* Sweet, *M. serpentaria* Raf, *Macrotrys actaeiodes* Raf.[1] Common names for the plant include: actée à grappes, black cohosh, black snakeroot, black root, bugbane, bugwort, bugwort rattleroot, cimicifuga, cohosh bugbane, Frauen Wurzel, herbe aux punaises, macrotnys, macroty's, macrotys, natsushirogiku, Qatil el baq, racine d'actée à grappes, rattle root, rattle top, rattle snake root, rattleweed, rich weed, schwarze Schlangenwurzel, Traubensilberkerze, squaw root, squawroot, Wanzenkraut and zilberkaars.[1]
- Standardized extracts and other commercial products of black cohosh are prepared from the dried rhizomes and roots of *Cimicifuga racemosa* (L.) Nutt. (Ranunculaceae).[1]
- The major and characteristic constituents of the roots and rhizomes include cycloartanol based triterpenes such as acetol, acetylacteol, 27-deoxyacteol, cimigenol, actein, 27-deoxyactein, cimifugoside, and isoferulic acid.[1] Although, the estrogenic isoflavone, formononetin was reported to be a constituent of the root, its presence was not detected in alcohol extracts of the root and rhizomes.[4]

**Medical Uses**

Extracts of black cohosh are used for the symptomatic treatment of climacteric symptoms such as anxiety, hot flushes, profuse sweating, insomnia and vaginal atrophy.[5–14] Extracts of black cohosh have also been used for the symptomatic treatment of premenstrual syndrome, amenorrhea and dysmenorrhea.[9,15] While there are published case reports to support these latter claims, evidence from controlled clinical trials is lacking.

**Summary of Clinical Evidence**

Ten clinical trials (6 controlled, 4 uncontrolled) have assessed the effects of black cohosh for the symptomatic treatment of menopause.[5–14] All ten of the trials were performed with either a 40% isopropyl alcohol or a 60% ethanol extract of black cohosh roots and rhizomes. A 12-week double-blind, placebo-controlled comparison trial compared the efficacy of a black cohosh with that of conjugated estrogens or placebo for the treatment of climacteric symptoms and vaginal atrophy.[11] Eighty women (ages 45–58) were treated with 8 mg of the extract, or 0.625 mg of conjugated estrogens, or placebo. A significant reduction in climacteric symptoms was observed in the black cohosh treated group as compared with the groups treated with either conjugated estrogens or placebo. The group treated with black cohosh extract had a reduction in the Kupperman index (decreased from 34 to 14), a decrease in the Hamilton Anxiety Scale and the proliferative status of the vaginal epithelium (p < 0.001).[11]

A randomized comparison trial assessed the efficacy of black cohosh for the treatment of climacteric symptoms, induced by hysterectomy.[8] Sixty women under the age of 40, who had undergone a hysterectomy but retained one ovary, were treated with estriol (1 mg/day), conjugated estrogens (1.25 mg/day), an estrogen-progesterone sequence therapy or an isopropyl alcohol extract of the rhizome (8 mg/day). The results of each treatment were determined at 4, 8, 12, and 24 weeks and outcomes were measured using a modified Kupperman index. The results of this trial demonstrated a statistically significant decrease in climacteric symptoms in all treatment groups (p < 0.01), and verified by reductions in a modified Kupperman index. Conjugated estrogens or estrogen-progesterone combinations appeared to be slightly more effective than black cohosh extracts; however no statistically significant differences between the three treatment groups were observed.[8] Serum levels of LH and FSH did not change in any of the groups during treatment (p > 0.05).[8]

A controlled comparison trial, involving 60 women between the ages of 45 and 60 years, assessed the efficacy of an extract of the rhizome with hormonal replacement therapy for the treatment of climacteric symptoms.[14] The outcomes measured included a menopause index comprised of the following symptoms: hot flushes, nocturnal sweating, nervousness, headache and palpitations. Psychological symptoms were measured using the Hamilton Anxiety Scale (HAMA) and Self-assessment Depression Scale (SDS). The patients were treated with 40 drops of an ethanol extract of the rhizome, or 0.625 mg conjugated estrogens or 2 mg diazepam for a period of 12 weeks. No placebo group was used. All three forms of therapy reduced the menopause index, HAMA and SDS. The rhizome extract and conjugated estrogens also reduced atrophic changes in the vaginal mucosa.[14]

A placebo-controlled clinical trial involving 110 women with climacteric symptoms assessed the effect black cohosh extract on the serum levels of LH and FSH.[6] The subjects were treated with the extract (8 mg/day) or matching placebo for two months. Although a significant reduction in serum LH levels (p < 0.01) was seen in the treated group as compared with placebo, no reduction in serum FSH levels was observed.[6] A 6-month randomized, double-blind clinical trial involving 152 women with climacteric symptoms compared the effects of two different doses of an isopropyl alcohol extract (corresponding to 40 mg drug versus 127 mg drug/day).[9,16,17]

A decrease in the Kupperman-Menopause Index (beginning value 31) was observed after two weeks in both treatment groups. Both dosage levels had similar therapeutic safety and efficacy. After 6 months of treatment, the number of responders (Kupperman index < 15) was approximately 90%. No effects on the levels of LH, FSH, sex-hormone binding globulin, prolactin, estradiol or vaginal cytology parameters were observed.[9,16,17]

A placebo-controlled clinical trial assessed the efficacy of a black cohosh extract for the treatment of 41 women with climacteric symptoms.[7] In the treatment group, 31 women reported a decrease in symptoms as compared with placebo, while 10 women with severe climacteric symptoms showed no improvement.[7]

In an uncontrolled clinical trial involving 50 women with climacteric complaints, oral administration of a black cohosh extract (2 × 40 drops daily) reduced moderate symptoms to "requiring no therapy" after 12 weeks of treatment.[13] Another uncontrolled trial 36 women with climacteric symptoms were treated with a 60% ethanol extract of the rhizome (40 drops twice daily) for 12 weeks.[5] A statistically significant decrease in the average values of the Kupperman menopause index was reported, and an increase in the Clinical Global Impressions scale was observed.[5] In an open study, 50 women with climacteric symptoms, who had previously been treated with intramuscular injections of estradiol valerate 4 mg and prasteronenantate (200 mg, every 4–6 weeks), were alternatively treated with a 40% isopropyl alcohol rhizome extract for 6 months (2 tablets twice daily).[10] The therapeutic results were rated as good to very good in 41 of the patients. Twenty-eight patients (56%) required no further injections, twenty-one patients (44%) required one injection in 6 months and 1 patient required two injections. The Kupperman Index decreased below 15 points (p < 0.001) indicating successful symptomatic treatment.[10]

A multicenter open drug-monitoring study of 629 patients with menopausal symptoms assessed the efficacy of an ethanol extract of the rhizome (2 × 40 drops daily) over 8 weeks of treatment.[12] Symptoms such as hot flushes, profuse sweating, headache, vertigo, nervousness and depression were improved in over 80% of all cases after 6 to 8 weeks of treatment with the extract.[12]

Numerous case reports have described the use of a 60% ethanol or a 40% isopropyl alcohol extract of black cohosh for the treatment of over 833 women with climacteric symptoms, as well as menstrual disorders such as primary or secondary amenorrhea, and premenstrual disorders.[18–22]

## Mechanism of Action

The mechanism of action of black cohosh is not well understood, although estrogenic-like effects have been described. However, reports on the estrogenic activity of black cohosh is controversial, and both *in vitro* and *in vivo* investigations have been attempted to clarify this issue. The proliferation of human mammary carcinoma cells was measured after treatment with an isopropyl alcohol extract of the rhizome *in vitro*.[23] Treatment of the cells with extract concentrations below 2.5

μg/ml did not appear to enhance the growth of the carcinoma cells. However, concentrations including, and above 2.5 μg/ml caused a significant inhibition of cell proliferation.[23] Similar results were reported in an investigation employing estrogen-receptor-positive human mammary cancer cell line MCF-7.[24] Mammary cancer cells were treated with a 40% isopropyl alcohol extract of the rhizome in concentrations ranging from 1 ng/ml to 100 μg/ml. The extract induced a dose-dependent inhibition of cell proliferation, and further augmented the antiproliferative effects of tamoxifen on this cell line.[24]

Administration of a black cohosh extract (type of extract not described) to female rats had estrogenic effects.[25] The extract was added to a standard liquid diet and fed to ovariectomized female rats daily for 3 weeks. An increase in uterine weight was observed, along with an increase in serum ceruloplasmin levels, suggesting estrogenic activity.[25] Conversely, intragastric or subcutaneous administration of a 50% ethanol extract of black cohosh (30, 300, or 3000 mg/kg) to immature mice for 3 days did not produce estrogenic effects as assessed by changes in uterus weight and vaginal smears.[26] The difference in the results observed is likely due to the length of treatment in these studies (3 days as compared to 3 weeks), and the type of extract used in the investigations.

A chloroform fraction, isolated from a methanol extract of the rhizome, was shown to bind to the estrogen receptors of rat uteri in vitro.[27] Formononetin, a minor constituent of the extract, showed a low relative molar binding affinity to the estrogen receptor ($1.15 \times 10^{-2}$).[27] The effects of formononetin, and a dichloromethane (DCM) extract of the rhizome, on luteinizing hormone (LH) secretion was tested in vivo.[28] Ovarectomized rats received nine intraperitoneal injections, receiving a total dose of 10 mg of formononetin or 108 mg of the DCM extract. While the DCM extract inhibited LH secretion, formononetin did not exhibit estrogenic activity in vivo, and it did not reduce the serum concentrations of LH.[27,28] Intraperitoneal administration of a chloroform (140 mg), a 60% ethanol (0.3 ml) or a dichloromethane soluble extract (27 mg) of black cohosh reduced the serum concentration of pituitary LH in ovariectomized rats after 3–3.5 days of treatment.[6,27,29] Neither serum FSH or prolactin levels were affected.[27] The effects of estradiol on estrogen-dependent uterine parameters were compared with that of a DCM fraction of a hydroalcoholic extract of black cohosh.[30] Daily injection of the DCM fraction (60 mg/rat) had no significant effect on uterine weight, but reduced serum LH levels, whereas estradiol increased uterine weight and reduced serum LH levels. Up-regulation of estrogen receptor-α gene expression was observed in MCF-7 mammary carcinoma cells treated with either the extract (35 μg/ml) or estradiol.[30] The results suggest that the organic fraction of an extract of black cohosh may act as a selective estrogen receptor modulator.[30]

## Pharmacokinetics

Due to a lack of knowledge concerning the active constituents of black cohosh, no pharmacokinetics studies have been reported in the scientific literature.

## Safety Information

### A.  Adverse Reactions
In the clinical trials, minor cases of nausea, vomiting, dizziness and headaches have been reported.[7,11,12,13]

### B.  Contraindications
The use of black cohosh use during pregnancy or lactation is contraindicated, due to a lack of safety data. In addition, there is no therapeutic rationale for the administration of black cohosh to children.[1]

### C.  Drug Interactions
No drug interactions have been reported.

### D.  Toxicology
A fluid extract of black cohosh (445 mg/oz) has been used for over 100 years (1882–1982) at a total daily dose of 890 mg per day without report of serious adverse reactions.[31]

Acute toxicity: In mice, intragastric administration $LD_{50}$ 7.73 g/kg, intravenous administration $LD_{50}$ 1.1 g/kg.[32]

Chronic toxicity: No evidence of toxicity was observed in rats at a dose of 500 mg/kg/day for 27 weeks, or in dogs at 400 mg/kg/day for 26 weeks.[32]

A 40% isopropyl alcohol extract of the rhizome was not mutagenic in the Ames test using *Salmonella typhimurium* strains TA98 or TA100 *in vitro*.[32] Intragastric administration of 100–1600 mg/kg/d to rats or 100–900 mg/kg/d to rabbits did not show any teratogenic effects or affects on reproduction.[33,34]

### E.  Dose and Dosage Forms
Daily dosage: a 40–60% ethanol or isopropyl alcohol extract of the rhizome (v/v),[5–14] corresponding to 40 mg of herbal drug daily.[15]

## References

1    Anon. Rhizoma Cimicifugae Racemosae. *WHO Monographs on Selected Medicinal Plants*. Volume II, WHO, Geneva, Switzerland, WHO Publications, 2001.
2    Brinker, F. Review of *Macrotys* (black cohosh). *Eclectic Med J.* 1996;2(1):2–4.
3    Foster S. Black cohosh: *Cimicifuga racemosa*. A literature review. *Herbal Gram.* 1999;45:35–49.
4    Struck D, Tegtmeier M, Harnischfeger G. Flavones in extracts of *Cimicifuga racemosa*. *Planta Med.* 1997;63:289.
5    Daiber W. Klimakterische Beschwerden: Ohne Hormone zum Erfolg. *Ärztliche Praxis.* 1983;35:1946–1947.
6    Düker EM et al. Effects of extracts from *Cimicifuga racemosa* on gonadotropin release in menopausal women and ovariectomized rats. *Planta Med.* 1991;57:424–427.

7     Földes J. Die Wirkungen eines Extraktes aus *Cimicifuga racemosa*. *Ärztliche Forschung*. 1959;13:623–624.

8     Lehmann-Willenbrock E, Riedel HH. Klinische und endokrinologische Untersuchungen zur Therapie ovarieller Ausfallerscheinungen nach Hysterektomie unter Belassung der Adnexe. *Zentralblatt Gynäkologie*. 1988;110:611–618.

9     Liske E. Therapeutic efficacy and safety of *Cimicifuga racemosa* for gynecological disorders. *Advances in Therapy*. 1998;15:45–53.

10    Pethö A. Klimakterische Beschwerden. Umstellung einer Hormonbehandlung auf ein pflanzliches Gynakologikum möglich? *Ärztliche Praxis*. 1987;38:1551–1553.

11    Stoll W. Phytopharmakon beeinflußt atrophisches Vaginalepithel Doppelblindversuch *Cimicifuga* vs. Östrogenpraparat. *Therapeutikon*. 1987;1:7–15.

12    Stolze H. Der andere Weg, klimakterische Beschwerden zu behandeln. *Gyne*. 1982;1:14–16.

13    Vorberg G. Therapie klimakterischer Beschwerden. *Zeitschrift für Allgemeinmedizin*. 1984;60(13):626–629.

14    Warnecke G. Influencing menopausal symptoms with a phytotherapeutic agent. *Die Medizinische Welt*. 1985;36:871–874.

15    German Commission E Monograph: Cimicifugae racemosae rhizoma. *Bundesanzeiger*. 02.03.1989.

16    Liske E et al. Human-pharmacological investigations during treatment of climacteric complaints with *Cimicifuga racemosa* (Remifemin®): No estrogen-like effects. In: *Proceedings of the 5th International ESCOP Symposium*, London, October 1998.

17    Liske E, Wüstenberg P. Therapy of climacteric complaints with *Cimicifuga racemosa*: A herbal medicine with clinical proven evidence. In: *Proceedings of the 9th Annual Meeting of the North American Menopause Society*, Toronto, September 1998.

18    Görlich N. Behandlung ovarieller Storungen in der Allgemeinpraxis. *Ärztliche Praxis*. 1962;14:1742–1743.

19    Heizer H. Kritisches zur *Cimicifuga*-Therapie bei hormonalen Storungen der Frau. *Medizinsche Klinik*. 1960;55:232–233.

20    Schotten EW. Erfahrungen mit dem *Cimicifuga*-Präparat Remifemin. *Der Landarzt*. 1958;11:353–354.

21    Starfinger W. Therapie mit östrogen wirksamen Pflanzenextrakten. *Medizin Heute*. 1960;9:173–174.

22    Stiehler K. Über die Anwendung eines standardisierten *Cimicifuga*-Auszuges in der Gynäkologie. *Ärztliche Praxis*. 1959;26:916–917.

23    Neßlhut T et al. Untersuchungen zur proliferativen Potenz von Phytopharmaka mit östrogenähnlicher Wirkung bei Mammakarzinomzellen. *Arch Gynecol Obstet*. 1993;817–818.

24    Freudenstein J, Bodinet C. Influence of an isopropanolic aqueous extract *of Cimicifugae racemosae* rhizoma on the proliferation of MCF-7 cells. *Twenty-third International LOF-Symposium of Phyto-estrogens*, Gent, Belgium, January 15, 1999.

25    Elm CL et al. Medicinal botanicals: estrogenicity in rat uterus and liver. In: *Proceedings of the American Association for Cancer Research*, 1997;38:293.

26    Einer-Jensen N et al. *Cimicifuga* and *Melbrosia* lack oestrogenic effects in mice and rats. *Maturitas*. 1996;25:149–153.

27    Jarry H et al. Studies on the endocrine effects of the contents of *Cimicifuga racemosa* 2. *In vitro* binding of compounds to estrogen receptors. *Planta Med*. 1985;4:316–319.

28    Jarry H et al. Treatment of menopausal symptoms with extracts of *Cimicifuga racemosa*: *in vivo* and *in vitro* evidence for estrogenic activity. In: *Phytopharmaka in Forschung und klinischer Anwendung*. D. Loew and N. Rietbrock, (eds.). Darmstadt, Steinkopff, 1995:99–112.

29    Jarry H, Harnischfeger G. Studies on the endocrine effects of the contents of *Cimicifuga racemosa*: 1. Influence on the serum concentration of pituitary hormones in ovariectomized rats. *Planta Med.* 1985;(1):46–49.

30    Jarry H et al. Organ-specific effects of *Cimicifuga racemosa* (CR) in brain and uterus. *Twenty-third International LOF-Symposium of Phyto-estrogens*, Gent, Belgium, January 15, 1999.

31    Anon. *Federal Register*, 1982;47:55092

32    Beuscher N. *Cimicifuga racemosa* L. Black cohosh. *Zeitschrift für Phytotherapie.* 1995;16:301–310.

33    Fukunishi K et al. Teratology study of hochu-ekki-to in rats. *Pharmacometrics.* 1997;53:293–297.

34    Sakaguchi Y et al. Teratology study of otsuji-to in rats. *Pharmacometrics.* 1997;53:287–292.

# 5

# Chaparral

## Synopsis

In spite of the fact that chaparral has been used in traditional medicine for the treatment of various ailments including bronchitis, the common cold, rheumatic pain, stomach pain, chickenpox, snakebites, tuberculosis and venereal disease, there are no clinical or scientific data to support these claims. Furthermore, numerous reports from the medical and scientific literature have demonstrated that chaparral is hepatotoxic, and infection causes hepatitis, cholestasis and hepatocellular injury. The dose and duration of exposure to chaparral, before the onset of hepatotoxicity varies depending on the individual. Hepatotoxicity has been observed in persons exposed to chaparral for as little as 6 weeks or as long as 15 months. Although the chemical constituents responsible for the toxic effects of chaparral have not been identified, similarities between several lignan components of chaparral, and the compound diethylstilbesterol, have been described. Based on the lack of scientific evidence for efficacy of chaparral for any medical use, and the potential for serious adverse reactions, the administration of chaparral for any therapeutic indication is not justifiable. In 1992, the FDA Center for Food Safety and Applied Nutrition issued a press release warning that the ingestion of chaparral posed a potential health risk to the public and that individuals with hepatic dysfunction or other chronic diseases may be at higher risk for irreversible liver damage. Unfortunately, products containing chaparral are still available to the general public, and still present a potentially serious threat to public safety.

## Introduction

Chaparral (*Larrea tridentata*) commonly referred to as "creosote bush" or "greasewood", is a low evergreen desert shrub native to the arid regions of southwestern United States and Mexico.[1] Traditionally, a tea prepared from the plant was used by the Native American Indians for the treatment of various ailments including bronchitis, the common cold, rheumatic pain, stomach pain, chickenpox, and snakebites, tuberculosis and venereal disease.[1] In recent years, chaparral has been used as an ingredient in dietary supplements used for the treatment of burns, cancer, obesity, liver ailments, and dermatological disorders.[1,2,3] The leaflets of the plant have a strong odor, and contain up to 20% of a resin with antioxidant activity, due to the high concentration of nordihydroguaiaretic acid (NDGA).[1] Although NDGA is a potent antioxidant, and is used to inhibit the oxidation of lard and other animal shortenings, it has been shown to produce cysts and kidney damage in rats and hypoplasia of the testes in guinea pigs.[4] Since 1990, there have been numerous reports in the medical literature indicating that ingestion of chaparral causes liver damage. In 1992 alone, more than 10 cases of acute nonviral hepatitis associated with chaparral ingestion were reported to the U.S. Food and Drug Administration and the Centers for Disease Control.[5] In December of 1992, the Food and Drug Administration issued a public warning against the consumption of chaparral due to numerous reports of acute, toxic hepatitis associated with the ingestion of chaparral in dietary supplements.[6] As a consequence products containing chaparral were removed from store shelves. However, from the period between 1994 and 1997 there were at least nine new reports of chaparral-induced liver toxicity,[3,7] indicating that the products are still generally available to the public.

## Quality Information

- The correct Latin name for chaparral is *Larrea tridentata* (Sessé & Moc.) Coville (Zygophyllaceae).[1] Botanical synonyms that may appear in the scientific literature include *Covillea tridentata*, *Larrea divaricata* Cav. and *Larrea mexicana* Moric.[8] Various vernacular (common) names for this plant include chaparral, creosote bush, el gobernadora, falsa alcaparra, greasewood, guamis, hedion dilla, palo ondo, sonora, tasajo, and yah temp.[1,8]
- Commercial products of chaparral are prepared from the leaves and twigs of *Larrea tridentata*.[1]
- Chaparral is native to the arid regions of southwestern United States and Mexico.[1]
- The chemical constituents of chaparral include the flavonoid glycosides and aglycones, triterpenes, a ligno-naphthaquinone, and nordihydroguaiaretic acid and related lignans 3-O-methyl nordihydroguaiaretic acid, 3'-demethoxy-6-*O*-demethylisoguaiacin(nor-3'-demethoxyisoguaiacin), 3"-hydroxy-4-epi-larrea tricin, 4-*epi*-larreatricin, 3'-hydroxynorisoguaiacin, isoguaiacin, enterolactone, enterodiol, guaiaretic acid, guiaretic acid diquinone, secoisolariciresinol, matairesinol and dihydrolarreatricin. Nordihydroguaiaretic acid is the most abundant lignan in the plant, and is present in the leaves at concentrations up to 10% of the dry weight.[2,9,10]
- The use of products containing chaparral should be prohibited, therefore no dosage recommendations can be made.

## Medical Uses

Traditionally, preparations containing chaparral were used for the treatment of various ailments including bronchitis, the common cold, rheumatic pain, stomach pain, chickenpox, snakebites, tuberculosis and venereal disease.[1,2,8] However, there are no clinical or scientific data to support these claims.

## Summary of Clinical Evidence

No clinical trials were found in the medical or scientific literature.

## Pharmacokinetics

No pharmacokinetics studies were found in the medical or scientific literature.

## Safety Information

### A. Adverse Reactions

Although there are no clinical trials assessing the safety and efficacy of chaparral for any therapeutic indication, there are numerous case reports in the scientific and medical literature of hepatotoxicity associated with the ingestion of chaparral.[2,5,7,11–14] A review of 18 reports of adverse reactions associated with the ingestion of chaparral, reported to the FDA between 1992 and 1994, showed evidence of hepatotoxicity in 13 cases.[2] Patients presented as jaundiced with a marked increase in serum liver enzymes occurring 3 to 52 weeks after the ingestion of chaparral. Resolution of symptoms occurred 1 to 17 weeks after intake was stopped. The predominant pattern of liver injury was characterized as toxic or drug-induced cholestatic hepatitis, with progression to cirrhosis in 4 patients and acute fulminant liver failure requiring liver transplantation in 2 patients.[2] In six other cases, the period of exposure before the onset of symptoms was 6 weeks to 15 months. In one of the cases, administration of 160 mg of chaparral for 2 months resulted in cholangiolithic hepatitis, characterized by severe cholestasis and hepatocellular injury.[14] In another patient who had taken a chaparral preparation for 40 weeks, hepatitis was complicated by encephalopathy and progressed to fulminant hepatic failure requiring liver transplantation.[7]

In addition to hepatotoxicity, some renal and skin adverse effects have been associated with the ingestion of chaparral.[15,16] Cystic renal cell carcinoma and acquired renal cystic disease was reported in one patient after the consumption of chaparral tea.[15] At least 16 cases of contact dermatitis have been attributed to chaparral or one of its chemical constituent's nordihydroguaiaretic acid.[16]

### B. Contraindications and Warnings

Use of chaparral containing preparations internally or externally on broken skin may cause serious hepatotoxic effects. In individuals with chronic liver diseases, the ingestion of chaparral may cause irreversible liver damage.[6]

## C. Drug Interactions

None reported.

## D. Toxicology

Neither the pathophysiology, nor the chemical constituents responsible for the hepatotoxicity induced by chaparral ingestion are known. However, a variety of mechanisms have been proposed.[16] It has been suggested that since chaparral has estrogenic activity, due to the similarity between the lignan components of chaparral and diethylstilbesterol, and estrogenic-like mechanism may be involved.[2,9] Most cases of hepatotoxicity have been associated with the ingestion of capsules and tablets, as compared to the traditional method of administering the crude drug as an infusion (tea). A recent analysis of the lignan composition of an infusion and a methanol extract of chaparral demonstrated that the infusion contained only traces of lignans, while the methanol extract contained most of the lignan components. These data suggest that exposure to the toxic lignan components may be much greater after ingestion of capsules or tablets containing chaparral.[9]

Preliminary studies with brine shrimp and cultured rat hepatocytes indicated that a methanol extract of the plant was cytotoxic *in vitro*, but the chemical constituents responsible for this activity have not been identified.[16] However, in other *in vitro* studies, nordihydroguaiaretic acid was cytotoxic in both Hep 2 and Vero cells lines,[17] and inhibited cytochrome P450-dependent arachidonic acid metabolism in liver cells *in vitro*.[18] The ligno-naphthaquinone, larreantin was cytotoxic to the P-388 leukemia in cell culture. In one animal study, the administration of a hydroalcoholic extract of the leaves to male hamsters, at a dose of 4% of their diet, caused a marked reduction of growth, pronounced irritability and aggressiveness and hypoplasia of the testes, but no effects on liver damage were reported.[19] 3'-demethoxy-6-*O*-demethylisoguaiacin exhibited anti implantation effect on timed pregnant rats at the oral dose of 15mg/kg.[2]

## E. Dose and Dosage Forms

Due to the lack of scientific data supporting the use of chaparral, and the numerous reports of hepatotoxicity, no dose of the crude drug should be recommended.

## References

1     Foster S, Tyler V. *The Honest Herbal*, Hawthorne Herbal Press, 1999.

2     Konno C, Lu ZZ, Xue HZ et al. Furanoid lignans from *Larrea tridentata. J Nat Prod.* 1990;53:396–406.

3     Sheikh NM, Philen RM, Love LA. Chaparral-associated hepatotoxicity. *Arch Int Med.* 1997;157:913–919.

4     Grice HC, Becking G, Goodman T. Toxic properties of nordihydroguaiaretic acid. *Fd Cosmet Toxicol.* 1968;6:155–161.

5     Clark F, Reed R. Chaparral-induced toxic hepatitis – California and Texas. *Morb Mortal Week Rep.* 1992;41:812–814.

6     Anon. Public warning about herbal product "chaparral". From the Food and Drug Administration. *J Amer Med Assoc.* 1993;269:328.

7    Gordon DW et al. Chaparral ingestion. The broadening spectrum of liver injury caused by herbal medications. *J Amer Med Assoc.* 1995;273:489–490.

8    Farnsworth NR, ed. *Larrea tridentata.* Napralert[sm] database, copyright the Board of Trustees, University of Illinois at Chicago, College of Pharmacy, 1999.

9    Obermeyer WR, Musser SM, Betz JM et al. Chemical studies of phytoestrogens and related compounds in dietary supplements: flax and chaparral. *Proc Soc Ex Biol Med.* 1995;6–12.

10   Luo ZY et al. Larreantin, a novel, cytotoxic naphthoquinone from *Larrea tridentata.* *J Org Chem.* 1998;53:2183.

11   Katz M, Saibil F. Herbal hepatitis: subacute hepatic necrosis secondary to chaparral leaf. *J Clin Gastroenterol.* 1990;12:203–206.

12   Batchelor WB, Heathcote J, Wanless IR. Chaparral-induced hepatic injury. *Amer J Gastroenterol.* 1995;90:831–833.

13   Smith BC, Desmond PV. Acute hepatitis induced by ingestion of the herbal medication chaparral. *Aust NZ J Med.* 1993;23:526.

14   Alderman S et al. Cholestatic hepatitis after ingestion of chaparral leaf: confirmation by endoscopic retrograde cholangiopancreatography and liver biopsy. *J Clin Gastroenterol.* 1994;19:242–247.

15   Smith AY et al. Cystic renal cell carcinoma and aquired renal cystic disease associated with consumption of chaparral tea: a case report. *J Urol.* 1994;152:2089–2091.

16   DeSmet PAGM. *Larrea tridentata.* Adverse Effects of Herbal Drugs, Volume II, pp 233.

17   Prichard D et al. Primary rat hepatocyte cultures aid in the chemical identification of toxic chaparral (*Larrea tridentata*) fractions. *In vitro Cell Develop Biol.* 1994;30a:91–92.

18   Zamora JM. Cytotoxic, antimicrobial and phytochemical properties of *Larrea tridentata* Cav. *Dissertation Abstracts.* 1985;45:3809–3810.

19   Capdevilla J et al. Inhibitors of cytochrome P-450-dependent arachidonic acid metaboleim. *Arch Biochem Biophys.* 1988;261:257–263.

20   Granados H, Cardenas R. Cálculos Biliares en el Jámster Dorado. XXXVII. Accion reventiva de "Gobernadora" (*Larrea tridentata*) en la Colelitiasis Pigmentaria Producida por la Vitamina A. *Rev Gastroenterol Mex.* 1994;59:31–35.

# 6

# Comfrey

## Synopsis

Comfrey-containing preparations have been used traditionally for the treatment of bruises and sprains, pharyngitis, periodontal disease, rheumatism, gastritis and peptic ulcers. However, due to the serious potential for toxicity, and the lack of strong scientific data supporting its use, there are no justifiable therapeutic applications for this herb. Chronic ingestion of comfrey is a potential health hazard due to the presence of the pyrrolizidine alkaloids, which occur naturally in all parts of the plant. Scientific evidence has shown that the pyrrolizidine alkaloids are hepatotoxic in both animals and humans, and in high doses may cause liver cancer in rodents. While there are no data from human studies, the evidence from rodents suggests that comfrey should be regarded as a potential human carcinogen, with a small risk even at low levels of exposure. At least three cases of liver toxicity, in the form of veno-occlusive disease, have been reported in humans after the ingestion of comfrey containing preparations. While the dermal absorption of the pyrrolizidine alkaloids appears to be limited, the use of topical preparations, such as creams and ointments, containing comfrey are not recommended due to the lack of scientific evidence. Despite the knowledge of potentially fatal hepatoxicity associated with the ingestion of comfrey, some products still remain commercially available as dietary supplements in the United States.

## Introduction

*Symphytum* species, the genus of comfrey, originated in Ancient Greece and was used by Dioscordies and Pliny in the 1st century AD for the treatment of wounds.[1] Comfrey, known scientifically as *Symphytum officinale*, is a perennial herb used in traditional medicine for the treatment of arthritis, hematomas, rheumatic diseases, sprains, ulcerations, and wounds.[2] The herb is now commonly used as a salad vegetable, an herbal tea or in the form of poultice for the treatment of wounds.[3] While there is some evidence to support the traditional uses of the plant, comfrey contains at least eight pyrrolizidine alkaloids, which have been shown to cause hepatotoxicity and liver cancer in rats.[4] Ingestion of a tea prepared from comfrey roots is especially dangerous as the roots contain much higher concentrations of the pyrrolizidine alkaloids than the aerial parts (leaves and stems) of the plant. A single cup of tea prepared from comfrey roots may contain 12 to 36 mg of the pyrrolizidine alkaloids, depending on their original concentrations in the plant.[5] Currently, it is difficult to define the oral dose at which liver toxicity will occur and lead to progressive damage in humans. The external use of comfrey containing preparations may not be hazardous since the alkaloids are converted to toxic metabolites by liver enzymes after ingestion.[6] However, in light of the lack of evidence for efficacy and the potential for serious toxicity if applied to broken skin, the risk-benefit ratio is low. Comfrey root capsules and other products are still available as ingredients of dietary supplements in the U.S. and phytomedicines in Europe, thereby explaining the reports of adverse reactions in the medical literature.

### Quality Information

- The correct Latin name for comfrey is *Symphytum officinale* L. (Boraginaceae). No botanical synonyms appear in the scientific literature. Numerous vernacular (common) names for the plant include: ass ear, black root, boneset, bruisewort, common comfrey, consuelda, consuelda mayor, grand consoude, gum plant, healing herb, Herba Consolidae, Herba Symphyti, knitbone, knitback Radix Symphyti, slippery root and wallwort.[1–3]
- Commercial products of comfrey are prepared from the fresh or dried aerial parts, or leaves, or roots of *Symphytum officinale*.[7,8]
- The chemical constituents present in comfrey leaves include allantoin, and the pyrrolizidine alkaloids including echinatine, lycopsamine, 7-acetyllycopsamine, echimidine, lasiocarpine, symphytine, and intermedine. Other constituents include silicic acid, and tannins.[7]
- Daily dose: External use only on intact skin. Ointments containing 5 to 20% of the dried crude drug. The total daily dose should not exceed 1 mcg of pyrrolizidine alkaloids with a 1,2 unsaturated necine structure, including their *N*-oxides. External applications should not exceed 4–6 weeks of treatment per year.[8] NOT FOR INTERNAL USE!

### Medical Uses

Comfrey-containing preparations have been used traditionally for the treatment of bruises and sprains, pharyngitis, periodontal disease, rheumatism, gastritis and peptic ulcers, and wounds.[2,7,8] However, due to the serious potential for toxicity, and the lack of strong scientific information supporting its efficacy, there are no justifiable therapeutic applications.

## Summary of Clinical Evidence

No clinical studies were found in the published medical or scientific literature.

## Pharmacokinetics

No pharmacokinetic studies were found in the published medical or scientific literature.

## Mechanism of Action

Two *in vivo* studies have investigated the effect of comfrey extracts on wound healing in cats and rats.[9,10] External application of the extract to wounds increased epithelialization, cell proliferation, and neovascularization.[9,10] The wound healing effects of the extract were attributed to one of its chemical constituents, allantoin.[11] Antiphlogistic activities have also been reported. Application of extracts of comfrey roots to UV-induced erythema of human skin resulted in a reduction of pain and redness.[12,13] The chemical constituent responsible for the antiphlogistic effects of the extract was established as $\alpha$-hydroxycaffeic acid.[13] An aqueous extract of the leaves stimulated the synthesis of prostaglandin F2 $\alpha$ in isolated rat stomach.[14] However, prostaglandin synthesis was inhibited by rosmarinic acid, another constituent of comfrey, through the inhibition of cyclo-oxygenase activity.[15]

## Safety Information

### A. Adverse Reactions

At least five cases of hepatotoxicity, primarily in the form of veno-occulsive disease (VOD) has been reported in human subjects and associated with ingestion of comfrey containing preparations.[16–21] In the first case, a female patient was diagnosed as having veno-occlusive disease on the basis of a liver biopsy.[21] Her consumption of pyrrolizidine alkaloids was estimated at 15 $\mu$g/kg per day over a period of several months (85 mg total ingestion) due to the ingestion of comfrey containing teas and other preparations.[21] The second case involved a 13 year-old male patient diagnosed with veno-occlusive disease after regular administration for several years of teas containing comfrey.[18] Hepatotoxicity in two female and one male patient, presenting as hepatic (VOD), was directly associated with ingestion of comfrey tea over various time periods ranging from two weeks to two years.[20]

Hepatic VOD is a progressive form of portal hypertension characterized by non-cirrhotic ascites and often progressing to hepatic failure. While the disease is relatively rare in humans, it is common in livestock, and almost always associated with the ingestion of plants containing pyrrolizidine alkaloids (PA). In fact, there are numerous reports of liver toxicity, ranging from veno-occulsive disease to cirrhosis, from various countries all associated with the consumption of PA from medicinal herbs and food plants.[20]

## B.  Contraindications

Due to the potential for serious liver toxicity, the use of comfrey containing teas or other preparations are not recommended.

## C.  Drug Interactions

No information was found.

## D.  Toxicology

### Hepatotoxicity

The hepatotoxic effects of comfrey are due to the presence of the pyrrolizidine alkaloids (PA) with a 1,2-unsubstituted necine ring, including symphytine, echimidine, intermidine and lasiocarpine.[22] These alkaloids are generally not hepatotoxic themselves, but after ingestion undergo microsomal enzyme oxidation to the corresponding pyrrole and subsequent hydrolysis to the hepatotoxic metabolites. Pyrrolic esters are the primary metabolites and undergo either hydrolysis to the secondary metabolites, pyrrolic acids or reactions with cell constituents.[22] Acute and chronic hepatotoxicity results from the antimitotic effects of both, together with a stimulus for cell division. The N-oxides of PA are more water-soluble and less lipophilic, and considered to be less toxic. Although the N-oxides are not directly converted into the toxic pyrroles by microsomal enzymes, the reduction of these molecules to the free base by rumen and gut microflora would however subject them to subsequent microsomal oxidation to the toxic pyrrolic derivative in the liver.[23] This appears to occur in humans, based on evidence in which the intravenous administration of indicine N-oxide to humans for the treatment of refractory adult acute leukemia, resulted in the urinary excretion of the free-base indicine, and a high incidents of veno-occlusive disease.[22]

### Mutagenic/Carcinogenic effects

An acetone extract of comfrey tested positive in the Ames test, which was abolished in the presence of microsomal enzymes.[21] However, pyrrolizidine alkaloids have been shown to induce mutagenicity in assays *involving Drosophila melanogaster* and *Salmonella typhimurium*, as well as in mammalian cell lines.[22] Pyrrolizidine alkaloids induce point mutations and chromosomal aberrations, as well as sister chromatid exchange and "unscheduled" DNA sysnthesis in mammalian cells.[22]

Oral administration of comfrey leaves or roots (0.5 to 33% of diet) to rats for 480 to 600 days lead to the development of hepatocellular adenomas, with the group receiving comfrey root having the highest incidence of liver tumor.[25] Individual pyrrolizidine alkaloids have been linked to liver and lung cancers in rodents. Administration of lasiocarpine to rat (50 ppm per day in rations) for 55 weeks resulted in the development of malignant tumors in 17 of the 20 animals by weeks 48 and 59.[26] The malignancies presented were angiosarcomas of the liver (9 animals), hepatocellular carcinomas (7 animals), adnexal skin tumor (1 animal) and lymphoma (1 animal).[26]

Intraperitoneal administration of lasiocarpine (7.8 mg twice weekly) resulted in the development of 11 cases of hepatocellular carcinoma and 6 cases of squamous cell carcinoma of the skin.[27] Other malignancies observed during the investigation were pulmonary adenoma, adenocarcinoma of the small intestine, and cholangio-carcinoma and adenocarcinoma of the ileum. Veno-occlusive disease of the liver was not observed in any of the animals.[27]

Administration of symphytine, extracted from the roots of comfrey has been shown to liver cancer in rats.[28] Twenty rats were administered a dose of 10% of the $LD_{50}$ by intraperitoneal injection twice weekly for 4 weeks, and then once a week for 52 weeks. Four animals developed liver tumors, three developed hemangioen-dothelial sarcomas, and one developed liver cell adenoma.[28] Dried and milled comfrey leaves and roots were fed to rats for up to 20 months at a dose of 1 to 16% of the total diet. All animals in the treated groups had an increased incidence of tumors, specifically liver adenomas and urinary bladder papillomas and carcinomas, and hemangioendothelial sarcomas.[25]

## E. Dose and Dosage Forms

Although the German Commission E[8] has approved the use of comfrey containing preparations for topical use on intact skin, the risks of such therapy appear to out-weigh the benefits. Based on the review of the scientific and clinical literature, admin-istration of comfrey-containing preparations either externally or internally cannot be recommended due to concerns about the potential for serious hepatotoxicity and a lack of scientific data in regard to efficacy.

## References

1   Stearn WT. The Greek species of *Symphytum* (Boraginaceae*). Ann Musei Goulandris.* 1986;7:175–220.
2   Leung AY. Encyclopedia of common natural ingredients used in food, drugs, and cos-metics. John Wiley and Sons, New York, 1996.
3   Foster S, Tyler VE. *Tyler's Honest Herbal*, 4th Edition, The Haworth Herbal Press, New York, 1999.
4   Couet CE, Crews C, Hanley AB. Analysis, separation and bioassay of pyrrolizidine alkaloids from comfrey (*Symphytum officinale*). *Nat Toxins.* 1996;4:163–167.
5   Roitman JN. Comfrey and liver damage. *Lancet.* 1981;944.
6   Mattocks AR. Toxicity of pyrrolizidine alkaloids. *Nature.* 1968;217:723–728.
7   USP DI Monographs. Comfrey. U.S. Pharmacopoeia, February 1998.
8   German Commission E Monographs. Comfrey-Symphyti herba/folium, 07.27.1990.
9   Manual AM, Mariano HG. Effects of wound healing of *S. officinale* (comfrey) applied topically in cats. *MJEAC.* 1985;1:146–149.
10   Goldman RS et al. Wound healing and analgesic effects of crude drug extracts of *Sym-phytum officinale* in rats. *Fitoterapia.* 1985;56, 323–329.
11   MacAliser CJ. The medicinal uses of comfrey. In: *Comfrey. Past, present and future.* LD Hills (ed). Faber and Faber Ltd., London 1976.
12   Andres R. Relating antiphlogistic efficacy of dermatics containing extracts of *Symphy-tum officinale* to chemical profiles. *Planta Med.* 1989;55:643–644.
13   Andres R. The antiphlogistic efficacy of dermatics containing pyrrolizidine alkaloid-free extracts of *Symphytum officinale* to chemical profiles. *Planta Med.* 1990;56:664.

14    Stamford IF, Tavares IA. The effect of an aqueous extract of comfrey on prostaglandin synthesis by isolated rat stomach. *J Phar Pharmacol.* 1983;35:816–817.

15    Gracza L. Prufung der membranabdichtenden Wirkung eines Phytopharmakons und dessen Wirkstoffe. *Ztscht Phytother.* 1987;8:78–81.

16    Ridker PM. Hepatoxicity due to comfrey herb tea. *Amer J Med.* 1989;87:701.

17    Yeong ML et al. Hepatic veno-occlusive disease associated with comfrey ingestion. *J Gastroenterol Hepatol.* 1990;5:211–214.

18    Weston CF et al. Veno-occlusive disease of the liver secondary to the ingestion of comfrey. *Brit Med J.* 1987;295:183.

19    Bach N et al. Comfrey herb tea-induced hepatic veno-occlusive disease. *Amer J Med.* 1989;87:97–99.

20    McDermott WV, Ridker PM. The Budd-Chiari syndrome and hepatic veno-occlusive disease. *Arch Surg.* 1990;125:525–527.

21    Ridker PM et al. Hepatic veno-occlusive disease associated with the consumption of pyrrolizidine-containing dietary supplements. *Gastroenterology.* 1985;88:1050–1054.

22    Abbott PJ. Comfrey: assessing the low-dose health risk. *Med J Australia.* 1988;149:678–682.

23    Mattocks AR. Toxic pyrrolizidine alkaoids in comfrey. *Lancet.* 1980;2:1136–1137.

24    White RD et al. An evaluation of acetone extracts from six plants in the Ames mutagenicity test. *Toxicol Lett.* 1983;15:25–31.

25    Hirono I, Mori H. Haga M. Carcinogenic activity of *Symphytum officinale. J Natl Cancer Inst.* 1978;61:865–869.

26    Rao MS et al. Malignant neoplasms in rats fed lasiocarpine. *Brit J Cancer.* 1978;37:289–293.

27    Svoboda DJ, Reddy JK. Malignant tumors in rats given lasiocarpine. *Cancer Res.* 1972;32:908–912.

28    Hirono I et al. Induction of hepatic tumors in rats by senkirkine and symphytine. *J Natl Cancer Inst.* 1978;63:469–471.

# 7

# Cranberry

## Synopsis

Traditionally, cranberry (*Vaccinium macrocarpon*) has been used for the treatment and prevention of mild urinary tract infections (UTI). Weak to moderate evidence from the published clinical and pharmacological studies support these claims. However, considering the serious nature of UTI's, and the potential for long-term consequences if not properly treated, patients should contact their physician for an initial evaluation prior to self-medicating with cranberry. Since there are no drug interactions reported, cranberry may be used in combination with other medical interventions for the treatment of UTI's. The recommended dose is 3 fluid ounces or 90 ml of a 30% juice or equivalent product for prevention of UTI's, and 12 to 32 fluid ounces (360 to 960 ml) daily for treatment. No contraindications, precautions, or side effects have been reported in the clinical trials.

## Introduction

Cranberry, known scientifically as *Vaccinium macrocarpon*, is a member of the Heath family.[1-3] The plant is native to North America, but grows wild in Britain, Germany, the Netherlands and Switzerland, and is cultivated in north and central Europe.[1-3] Historically, the small, edible red-black berries were used by the Native American Indians as a food, and also as a clothing dye. Medicinally, the whole dried fruit was used to prepare a dressing for the treatment of wounds.[1] During the 17th century, the medicinal applications for cranberry included the relief of loss of appetite, blood and liver disorders, cancer, scurvy, stomach ailments, and dysurea.[1-3] In the 18th century, cranberry juice was commonly employed in Europe for the treatment of urinary tract infections in women. Mid-18th century scientists believed that the therapeutic effects of cranberry were due to a lowering of the urinary pH via the excretion of hippuric acid, and its metabolism to benzoic acid. However, by 1914, it was hypothesized that the antimicrobial effects of cranberry were not related to the excretion of hippuric acid or its effect on urinary pH. More recent evidence has shown that that cranberry contains chemical constituents that prevent the adhesion of specific urinary tract bacteria to bladder epithelial cells, thereby reducing the incidence of urinary tract infection.[1-3]

Currently, cranberry products in the United States are regulated as foods and beverages, and dietary supplements. Over 52 million American households consume cranberry products, in the form of beverages, foods, teas, and capsules. It is currently estimated that 400 million pounds of cranberry are used annually in the U.S., with a market value of $1.25 billion. Ocean Spray, a cooperative of growers markets approximately 85–90% of the cranberries grown in North America.[4]

## Quality Information

- The correct Latin name for cranberry is *Vaccinium macrocarpon* Ait (Ericaceae).[2] Botanical synonyms that may appear in the scientific literature include *Oxycoccus macrocarpus* (Ait.) Pers., and *O. macrocarpus* Pers.[2,5] Common names for *V. macrocarpon* include American cranberry, bear berry, black cranberry, cranberry, large cranberry and low cranberry.[2]
- The juice, standardized extracts and other commercial products of cranberry are prepared primarily from the ripe berries (fruit: fresh, frozen or dried) of *Vaccinium macrocarpon* Ait.[1,4] The bulk materials are then used to produce whole berry powders, juice, and dried juice concentrate, fluid extracts, capsules and other food products.
- The major and characteristic chemical constituents of the berries include the organic acids such as quinic, malic and citric acid; flavonoids and flavonoid glycosides such as quercetin, quercitrin, cyanidin-3,5-diglucoside, cyanidin-3-galactoside, and hyperoside.[2,6] The berries and juice also contain high levels of vitamins A and C.[2]
- For the prevention of UTI's the recommended daily dose of cranberry juice is 3 fluid ounces (90 ml of a 30% pure juice product); for the treatment of UTI's the daily dosage range is 12 to 32 fluid ounces or equivalent preparations for treatment.[6] Capsules containing a concentrated cranberry extract: 6 capsules daily equivalent to 3 fluid ounces (90 ml) cranberry juice cocktail. Patients with diabetes mellitus or on a low sugar diet, should be cautioned concerning the high concentration of sugar in cranberry juice, and be advised to look for artificially sweetened products.

## Medical Uses

Cranberry juice and extracts are used for the prevention and symptomatic treatment of chronic urinary tract infections.[8-21] Fresh cranberry has also been used to treat asthma, fevers, as well as gall bladder and liver disease, however there are no scientific data to support these claims.

## Summary of Clinical Evidence

Although 25 published studies have assessed the effects of cranberry juice in humans, only 15 clinical trials have assessed the effects of cranberry juice on urinary pH and urinary tract infections.[8–23] Of these 13 trials assessing the safety and efficacy of cranberry for the prevention and treatment of UTI's, only four were controlled and of sufficient scientific quality.[8,12,14,19]

A randomized, double-blind, placebo-controlled trial assessed the efficacy of concentrated cranberry extract in capsules (400 mg/day) for the urinary tract infections in 19 women.[19] The patients were treated for 6 months, however 9 subjects dropped out due to pregnancy, unrelated infections and loss to follow-up. Treatment with cranberry capsules significantly ($p < 0.005$) reduced the occurrence of urinary tract infections. On average, 2.4 UTI's per year were observed in the treated group, and 6.0 UTI's per year in the placebo group. Treatment was well tolerated and no side effects were reported.

A randomized, double-blind, placebo-controlled parallel trial involving 153 elderly female volunteers (mean age 78.5 years) assessed the effect of cranberry juice on bacteriuria and pyuria.[8] The subjects were randomly assigned to receive 300 ml/day of cranberry juice or a cranberry flavored placebo containing vitamin C for six months. At the end of the trial, subjects receiving the cranberry juice had a reduced frequency of bacteriuria with pyuria as compared with control (odds 42% of the control group, $P = 0.004$). These effects appeared only after 4 to 8 weeks of cranberry use. There was no evidence of urinary acidification, and the median urine pH in the cranberry treated group was 6.0. The results of this study did not support the use of cranberry for the prevention of urinary tract infections but showed that cranberry juice may be of more benefit for treatment of UTI's.[8]

A randomized single-blind, crossover study assessed the efficacy of 15 ml/kg/day of cranberry cocktail juice (30% concentrate) as prophylaxis for bacterial UTI's in 40 children with neuropathic bladder, managed by intermittent catherterization.[12] The subjects were treated for 6 months with either cranberry juice or water as control. Outcomes measured were a positive or negative urine culture with symptomatic UTI. The results of this study did not support the use of cranberry juice as prophylaxis or UTI's in children with neuropathic bladder. However, 19 subjects dropped out of the trial, and the diagnostic criteria for UTI was much lower than any other trial ($10^3$ CFU/L of a pathogenic organism).[12] The results of this trial were confirmed in a double-blind, placebo-controlled, crossover study involving 15 pediatric patients with neurogenic bladder receiving clean intermittent catheterization.[22] Two ounces of cranberry concentrate (equal to 300 ml of cranberry juice cocktail) or placebo were administered daily for 3 months, followed by a 3 month crossover (no washout period was described). Cranberry ingestion did not reduce bacteriuria or symptomatic urinary tract infections in this population.[22]

A randomized, placebo-controlled trial assessed the efficacy of 30 ml of pure cranberry juice to reduce the bacterial concentrations in the urine of elderly subjects with a mean age of 81 years.[14] Thirty-eight volunteers were treated with 30 ml of cranberry juice mixed with water or water for 4 weeks, then crossover for a further 4 weeks. Significant results were reported, with cranberry treatment decreasing the frequency of bacteriuria (p = 0.004). However, 21 patients dropped out prior to the completion of the trial. No side effects were reported.[14]

In an uncontrolled study involving 60 subjects with symptoms of acute urinary tract infection such as frequency, dysuria, urgency, and nocturia, the effects of cranberry on the bacteria levels in the urine was assessed.[17] Patients were treated with 16 ounces of cranberry juice daily for 21 days. After three weeks, a positive clinical response, no urogential complaints and fewer than 100,000 bacteria per ml of urine was noted in 32 patients (53%). Another 12 patients (20%) were "moderately improved" and 16 patients (27%) showed no bacteriological improvement or symptomatic relief.[17]

In an uncontrolled study, 28 nursing home patients were treated with 4 to 6 ounces of cranberry juice daily for seven weeks.[13] Twice weekly urine samples were examined for leukocytes and/or nitrates as a measure of urinary tract infection. At the end of seven weeks 10 patients had no leukocytes or nitrates in the urine; nine patients had trace to 2+ leukocytes and negative nitrates; nine had trace or greater leukocytes. However, this study did not include non-exposure cohort (controls).[13]

Two uncontrolled trials assessed the effects of cranberry juice on urinary pH.[20,21] In one study, 59 patients were treated with 450–720 ml of a preparation containing 80% cranberry juice per day for 12 days and the urine pH was measured. A decrease in urine pH was observed, however it was not dose-related.[20] The second study involved four healthy volunteers who were administered 1500–4000 ml/day of a 33% cranberry juice product.[21] Three of the four subjects showed transient changes in urine pH and titratable acidity. However, no controls were used in this study.

Two observational studies were reported in the literature.[11,16] One was a case report of a female patient with chronic pyelonephritis for which no drug therapy was effective.[16] She had persistent 4+ albuminuria and 4+ pyuria. The patient took 12 ounces of cranberry juice per day for two and after nine months of treatment no albumin was present in the urine. The second study involved six healthy subjects to whom cranberry sauce (22–54 grams) was administered and the composition of the urine was assessed, along with blood alkali reserve.[11] An increase in titratable and organic acids, hippuric acid, hydrogen ion concentration, and ammonia was observed. Uric acid and nitrogen levels decreased, but no effect was observed on urine pH.[11]

In an uncontrolled trial involving 13 urostomy patients, the effects of cranberry juice on skin complications were assessed.[24] Treatment of 160 to 320 grams of cranberry juice daily for 6 months did not significantly acidify the urine, but improved the skin conditions in four urostomy patients with peristomal skin disorders. A decrease in erythema, maceration and pseudoepithelial hyperplasia was observed.[24]

## Mechanism of Action

Early investigations of the mechanism of cranberry's bateriostatic activity suggested that acidification of the urine may be responsible for its activity.[16] However, more recent studies have demonstrated that urinary acidification is not the mechanism by which cranberry exerts its effects, but through a mechanism that involves the inhibition of bacterial adhesins.[25]

Cranberry juice has been shown to be a potent inhibitor of bacterial adherence.[26–28] *In vitro* studies have shown that cranberry juice decreased bacterial adherence of *E. coli* to uroepithelial cells by 60% of the 77 clinical isolates tested as compared with saline solution.[25] Furthermore, cranberry juice inhibited the adherence of *E. coli* to human urinary epithelial cells three times more than *E. coli* isolated from other clinical sources.[25,26] One study demonstrated that cranberry juice inhibited the hemagglutination activity of *E. coli* urinary isolates expressing Type I and P adhesin.[6] The inhibitory effect on hemagglutination of P fimbriated *E. coli* was dependent on the concentration of the juice.[6] Two compounds in cranberry juice inhibited lectin-mediated adherence of *E. coli* to mucosal cells.[6,27] One of the compounds was fructose and the other was a nondialyzable polymeric compound. Further investigations have found that exposure of pathogenic bacteria to the nondialyzable polymeric compound in either the gut or bladder produces a bacteriostatic effect by inhibiting specific adhesins present on the pili on the bacterial surface.[28]

Cranberry juice appears to irreversibly inhibit the expression of P-fimbriae of *E. coli in vitro* at a concentration of 25% in the medium.[29] Electron micrographic data indicates that cranberry juice acts on the cell wall and either prevents proper attachment of the fimbrial subunits or prevents the expression of normal fimbrial subunits.[29] Proanthocyanidin extracts from cranberries inhibit the adherence of P-fimbriated *E. coli* to uroepithelial cell surfaces.[30]

## Pharmacokinetics

No pharmacokinetic data were found in the scientific or medical literature.

## Safety Information

*A.  Adverse Reactions*
Ingestion of more than 4 liters of cranberry juice per day may result in gastrointestinal irritation and diarrhea.[2]

*B.  Contraindications*
There are no contraindications reported in the scientific literature.

*C.  Drug Interactions*
No drug interactions have been reported.

## D. Toxicology

No toxicological data have been reported.

## E. Dose and Dosage Forms

For the prevention of UTI's the recommended daily dose of cranberry juice is 3 fluid ounces (90 ml of a 30% pure juice product); for the treatment of UTI's the daily dosage range is 12 to 32 fluid ounces or equivalent preparations for treatment.[6] Capsules containing a concentrated cranberry extract: 6 capsules daily equivalent to 3 fluid ounces (90 ml) cranberry juice cocktail.

## References

1    Siciliano AA. Cranberry. *Herbal Gram.* 1996;38:51–54.
2    Small E, Catling PM. *Vaccinium macrocarpon* Ait. Cranberry. Canadian medicinal crops. NRC-CNRC, Ottawa, Canada, NRC Research Press, 1999; 160–168.
3    Harkins KJ. What's the use of cranberry juice. *Age and Ageing.* 2000;29:9–12.
4    USP-National Formulary. Cranberry liquid preparation, USP-24 NF-19;1999;2440.
5    Farnsworth NR, ed. *Vaccinium macrocarpon* (Cranberry). The NAPRALERT database. Copyright The Board of Trustees at the University of Illinois at Chicago, 2000.
6    Zafriri D, Ofek I, Adar R et al. Inhibitory activity of cranberry juice on adherence of type 1 and type P fimbriated *Escherichia coli* to eucaryotic cells. *Antimicrob Agents Chemother.* 1989;33:92–98.
7    Foster S, Tyler VE. *Herbs of Choice, The Therapeutic Use of Phytomedicinals.* Hawthorne Herbal Press 1999.
8    Avorn J, Monane M, Gurwitz JH et al. Reduction of bacteriuria and pyuria after ingestion of cranberry juice. *JAMA.* 1994;271:751–754.
9    Bodel PT, Cotran R, Kass E. Cranberry juice and the antibacterial action of hippuric acid. *J Lab Clin Med.* 1959;54:881–888.
10   Dungan C, Cardaciotto PS. Reduction of ammoniacal urinary odors by the sustained feeding of cranberry juice. *J Psych Nurs.* 1966;8:467.
11   Fellers CR, Redmon BC, Parrott EM. Effects of cranberries on urinary acidity and blood alkali reserve. *J Nutr.* 1933;6:455–463.
12   Foda M, Middlebrook P, Gatfield CT et al. Efficacy of cranberry in prevention of urinary tract infection in a susceptible pediatric population. *Canadian J Urol.* 1995;2:98–102.
13   Gibson L, Pike L et al. Effectiveness of cranberry juice in preventing urinary tract infections in long term care facility patients. *J Natur Med.* 1991;2:45–47.
14   Haverkorn MJ, Mandigers J. Reduction of bacteriuria and pyuria using cranberry juice. *JAMA.* 1994;272:590.
15   Jackson B et al. Effect of cranberry juice on urinary pH in older adults. *Home Health Nurse.* 1997;15:198–202.
16   Moen DV. Observations on the effectiveness of cranberry juice in urinary infections. *Wisconsin Med J.* 1962;61:282–283.
17   Papas PN, Brusch CA, Ceresia GC. Cranberry juice in the treatment of urinary tract infections. *Southwestern Med.* 1996;47:17–20.
18   Schultz A. Efficacy of cranberry juice and ascorbic acid in acidifying the urine in multiple sclerosis subjects. *Community Health Nursing.* 1984;13:159–169.
19   Walker EB et al. Cranberry concentrate: UTI prophylaxis. *J Fam Prac.* 1997;45:167–168.
20   Kinney AB, Blount M. Effect of cranberry juice on urinary pH. *Nursing Res.* 1979;28:287.

21  Kahn HD, Panariello VA. Effect of cranberry juice on urine. *J Am Diet Assoc.* 1967;51:251.
22  Schlager T et al. Effect of cranberry juice on bacteriuria in children with neurogenic bladder receiving intermittent catheterization. *J Ped.* 1999;135:698–702.
23  Prodromos PN, Brusch CA, Ceresia GC. Cranberry juice in the treatment of urinary tract infections. *Southwest Med.* 1968;47:17.
24  Tsukada K, Tokunaga K et al. Cranberry juice and its impact on peri-stomatal skin conditions of urostomy patients. *Ostomy Wound Manage.* 1994;49:60–62;64;66–68.
25  Sobota AE. Inhibition of bacterial adherence by cranberry juice: potential use for the treatment of urinary tract infections. *J Urol.* 1984;131:1013.
26  Schmidt DR, Sobota AE. An examination of the anti-adherence activity of cranberry on urinary and nonurinary bacterial isolates. *Microbios.* 1988;55:173–181.
27  Ofek I, Adar R et al. Inhibitory activity of cranberry juice on adherence of type 1 and P-fimbriated *Escherichia coli* to eukaryotic cells. *Antimicrob Agents Chemother.* 1989;33:92–98.
28  Ofek I, Goldhar J, Zafriri D et al. Anti-Escherichia adhesin activity of cranberry and blueberry juices. *N Engl J Med.* 1991;324:1599
29  Ahuja S, Kaack B, Roberts J. Loss of fimbrial adhesion with the addition of *Vaccinium macrocarpon* to the growth medium of P-fimbriated *Escherichia coli*. *J Urol.* 1998;159:559–562.
30  Howell AB, Vorsa N et al. Inhibition of the adherence of P-fimbriated *Escherichia coli* to uroepithelial cell surfaces by proanthocyanidin extracts from cranberries. *N Engl J Med.* 1998;339:1085–1086.

# 8

# *Echinacea*

## Synopsis

*Echinacea* is currently one of the best selling botanical supplements on the American and worldwide herbal market. Commercial *Echinacea* products are manufactured from three *Echinacea* species, *Echinacea purpurea* (root or herb), *E. angustifolia* (roots), and *E. pallida* (roots). *Echinacea* preparations are administered orally as supportive therapy for the prophylaxis and treatment of the common cold, bronchitis, influenza, bacterial and viral infections of the respiratory tract, and urinary tract infections. However, many of the clinical trials suffer from poor methodology, and recent controlled clinical trials have indicated that *Echinacea* products are not effective for prophylaxis of upper respiratory tract infections. The *in vitro* and *in vivo* studies indicate that the therapeutic effects of *Echinacea* are due to the stimulation of cellular immune response.

*Echinacea* may also be used externally to promote wound healing, and to soothe minor inflammatory skin conditions, such as poison ivy. The major adverse events reported are allergic reactions, ranging from contact dermatitis to anaphylaxis. Patients with an allergy to plants in the daisy family (Asteraceae) should be instructed not to use products containing *Echinacea*. There are no drug interactions reported, and *Echinacea* products may be used safely in conjunction with antibiotics or sulfa drugs. However, recent *in vitro* studies have shown that *Echinacea* extracts inhibit cytochrome P450 3A4, and thus should be used cautiously in patients taking medication metabolized via this route. According to the German Commission E, patients with autoimmune disorders, AIDS, HIV infection or tuberculosis should not use *Echinacea*, however there is little data to substantiate these contraindications. Recommendations for oral dosage and dosage forms are difficult due to the wide variety of preparations and doses used in the clinical trials (see dosage section). For external applications, a cream or ointment containing 15% expressed juice, is recommended.

## Introduction

*Echinacea* is a perennial herbaceous flowering plant, native to the prairie regions of North America.[1] The Native American Indians used *Echinacea* to treat a wide range of ailments including the common cold, sore throat, stomachache, wounds, as an antidote for rattlesnake bite, and to treat saddle sores on horses.[1,2] *Echinacea angustifolia* was also used extensively by the white settlers of North America in the 18th century. Around the middle of the 19th century, the use of *Echinacea* became very prominent among American medical practitioners. The Eclectic physicians, who relied heavily on natural remedies, used *Echinacea* to treat snakebite, typhus, diphtheria, and septicemia.[2,3] In fact, monographs for the dried roots and rhizome of *E. angustifolia* and *E. pallida* were official in the United States National Formulary from 1916 until 1950, when they were removed due to a lack of clinical evidence supporting the therapeutic claims. In Europe, *Echinacea* did not receive any attention until the early 20th century, when clinical successes were reported and *Echinacea angustifolia* was designated as an official drug.[1,4] Since then, the results from over 350 scientific investigations and clinical trials have been published, primarily in the German scientific literature. While the genus *Echinacea* is comprised of 9 species and 2 varieties,[1] its medicinal use is limited to three species *Echinacea purpurea*, *E. pallida* and *E. angustifolia*. Botanical supplements containing these *Echinacea* species now enjoy immense popularity with the American consumer, and top the list of the best-selling botanicals in the United States.

## Quality Information

- Entering into commerce are three *Echinacea* species, *Echinacea purpurea* (L.) Moench., *E. angustifolia* D.C. var. *angustifolia* or its variety *strigosa* McGregor, and *E. pallida* (Nutt.) Nutt.[5] These plants are in the daisy family (Asteraceae). The botanical synonyms for *Echinacea purpurea* include *Brauneria purpurea* (L.) Britt., *Echinacea intermedia* Lindl., *E. purpurea* (L.) Moench f., *E. purpurea* (L.) Moench var. *arkansana* Steyerm., *E. speciosa* Paxt., *Rudbeckia purpurea* L., *R. hispida* Hoffmgg., *R. serotina* Sweet.[5] The common names for *Echinacea purpurea* include coneflower, purple coneflower and red sunflower.[5]
- The botanical synonyms for *E. angustifolia* include *Brauneria angustifolia* Heller, *Echinacea pallida* var. *angustifolia* (D.C.) Cronq.[5] The common names for *Echinacea angustifolia* D.C. var. *angustifolia* include American coneflower, black sampson, cock up head, coneflower, Indian head, Kansas snakeroot, narrow-leaved purple coneflower, and purple coneflower.[5]
- The botanical synonyms for *E. pallida* include *Echinacea angustifolia* Hook, *Rudbeckia pallida* Nutt., *Brauneria pallida* Britt., and *Echinacea pallida* f. *albida* Steyerm.[5] The common names for *Echinacea pallida* (Nutt.) Nutt., include pale coneflower and pale purple cone.[5]
- Standardized extracts and other commercial preparations of *Echinacea* consist of the fresh or dried aerial parts of *Echinacea purpurea* harvested in full bloom, or the fresh or dried roots of *Echinacea angustifolia*, *E. purpurea* or *E. pallida*.[5]
- The active chemical constituents of *Echinacea angustifolia*, *E. purpurea*, and *E. pallida* fall into five groups, the alkamides, the polyalkenes, the polyalkynes,

caffeic acid derivatives, and polysaccharides.[1,5] In addition, the essential oil contains, among other compounds, borneol, bornyl acetate, pentadeca-8-en-2-one, germacrene D, caryophyllene and caryophyllene epoxide, vanillin and palmitic acid.[1] The pharmacologically active constituents of *Echinacea purpurea* include cichoric acid (2,3-O-dicaffeoyl-tartaric acid), a caffeic acid derivative, present in concentrations of 1.2–3.1%.[6] Also present are a series of alkamides with the isomeric dodeca-2,4,8,10-tetraenoic acid and isobutylamides, as the main compounds.[6] The isolated caffeic acid ester derivatives include echinacoside, cynarin, and cichoric acid methyl ester. Cynarin is present only in *E. angustifolia*, thus can be used for qualitative analysis to distinguish it from the closely related *E. pallida*.[5] The polysaccharide constituents isolated from *Echinacea purpurea* are of two types: a heteroxylan of average mw ca. 35 kD (e.g., PS-I), and an arabinorhamnogalactan of average MW ca. 45 kD (e.g., PS-II). Other constituents include trace amounts of pyrrolizidine alkaloids (tussilagine [0.006%] and isotussilagine). While other pyrrolizidine alkaloids are known to cause liver damage, these two alkaloids are not hepatotoxic, as they lack the 1,2-unsaturated necine ring in their chemical structure.[7]

## Medical Uses

Used internally as supportive therapy for colds and chronic infections of the upper respiratory tract and lower urinary tract.[5,8–13] *Echinacea* treatments may reduce the duration of illness and symptoms of these infections by enhancing the body's immune response.[5,8–13]

*Echinacea* preparations are used in topical applications for the treatment of superficial wounds and minor skin inflammations.[5,8,14]

## Summary of Clinical Evidence

A systematic review of the controlled clinical trials published between 1961 and 1993 has been published.[9] Of the 26 trials assessed, 19 investigated the efficacy of various *Echinacea* preparations for the prophylaxis and treatment of upper respiratory and urinary tract infections; four trials investigated the reduction of the adverse reactions to chemotherapy; and three trials investigated the effects on immune parameters in patients with recurrent infections.[9] Of the 26 trials reviewed, 18 were randomized, but only 11 were double-blinded. While the methodological quality of most studies was fair, the details of the randomization process were not adequately described in any trial, and the statistical evaluation was insufficient in 18 studies, and a description of the diagnostic criteria was unsatisfactorily described in 16 studies. Data from the 26 trials showed that of a total of 34 treated groups, 30 showed positive results. There was good evidence of the efficacy of a high dose of a tincture of *Echinacea purpurea* (180 drops per day) for reducing the symptoms of upper respiratory tract infections, while a dose of 90 drops per day was not statistically superior to placebo.[15] A placebo-controlled clinical trial assessed the efficacy of an ethanol *Echinacea pallida* root extract in the treatment of 160 patients with infections of the upper respiratory tract.[11] A significant decrease in the duration of illness from 13 to 9.8 days for bacterial infections, and 12.9 to 9.1 days for viral infections was observed after treatment of the patients with 90 drops/day (900 mg roots) of an aqueous-alcoholic tincture (1:5).[11] A randomized, double-blind placebo-controlled trial assessed the prophylactic effects of the pressed juice (8 ml/day) of *E. purpurea* herb in an 8-week trial in 108 subjects.[10] There was a modest reduction in the frequency and intensity of illness in the treated group, however it was not statistically significant. Sixteen controlled trials using a combination product containing the extracts of *Echinacea*, *Baptisia* and *Thuja* also had positive data.[9] While this review concluded that the existing controlled clinical trials indicate that *Echinacea* products can be efficacious immune stimulants, the authors could not provide clear therapeutic recommendations as to which preparation to use or what dose to employ for a specific therapeutic indication.[9]

Results from recent randomized controlled clinical trials are conflicting, and the results appear to be both product and dose dependent. Results have been published from five small, randomized placebo-controlled trials assessing the immunomodulatory effects of preparations containing *Echinacea* extracts in healthy volunteers.[16] Two trials tested an intravenous homeopathic preparation containing *Echinacea angustifolia* (studies 1 and 5). Two of the trials tested ethanol extracts of *E. purpurea* roots (study 2 and 3a), one trial used an extract of *E. pallida* roots (study 3b), and one trial used an extract of *E. purpurea* herb (study 4). The extracts and placebo were administered for four (study 5) or five (studies 1–4) consecutive days. A total of 134 healthy volunteers between the 18 and 40 years of age were included in the trials. The primary outcome measured was the relative phagocytic activity of polymorphonuclear neutrophil granulocytes (PNG), and the secondary outcome measured was the number of leukocytes in the peripheral venous blood. In trials 1 and 2 the phagocytic activity of PNG was significantly enhance as compared with placebo (22.7% versus 54%, respectively), while no significant effects were observed in trials 3–5. Leukocyte number was not significantly influenced in any trial.[16]

A three-armed, randomized double-blind placebo-controlled trial assessed the effect of *Echinacea* preparations on the prophylaxis of upper respiratory infections (URI) in 302 healthy volunteers recruited from German army centers.[17] The subjects were treated with an ethanol extract of *E. purpurea* root, *E. pallida* root, or placebo for 12-weeks, and the time until the first respiratory tract infection was measured. The results of this trial showed that the prophylactic use of *Echinacea* did not significantly reduce the incidence of URI as compared with placebo.[17] A randomized, double-blind, placebo-controlled clinical trial assessed the effect of a fluid extract of *Echinacea purpurea* on the incidence and severity of colds and respiratory infections[18]. A total of 109 patients with a history of more than 3 colds or respiratory infections in the preceding year received 4 ml of the fluid extract or 4 ml of matching placebo twice daily for 8 weeks. The incidence and severity of colds and respiratory infections were determined during follow-up, and the investigators graded the severity of each infection. During the test period, 65% of the patients in the *Echinacea* group and 74% of the patients in the placebo group had at least one cold or respiratory infection. The average number of cold in the *Echinacea* group was 0.78 and 0.93 in the placebo group. The median duration of colds and respiratory tract infections was 4.5 days in the *Echinacea* group and 6.5 in the placebo group. The investigators concluded that prophylactic treatment with a fluid extract of *Echinacea purpurea* did not significantly reduce the incidence, duration or severity of colds and respiratory tract infections.[19]

A double-blind, placebo-controlled clinical trial compared the therapeutic efficacy of *Echinacea pallida* root extract with placebo in the treatment of upper respiratory tract infections.[19] The subjects were 160 adults who had not been sick for longer than three days and had a symptom score of greater than or equal to 15 in the investigator's rating scale. The subjects were treated with either 90 drops of an *Echinacea pallida* root hydroalcoholic extract (equivalent to 900 mg of root) or matching placebo per day on a randomized basis. The outcome criteria measured were length of illness and resolution of symptoms. Treatment with *Echinacea* reduced the average length of infection from 13 to 9.8 days in cases of bacterial infections and to 9.1 days in cases of viral infections.[19] Another placebo-controlled, double-blind trial assessed the efficacy of an ethanol extract of *E. purpurea* aerial parts (95%) and root (5%) on treatment of the early symptoms of the common cold.[12] The subjects (n = 199) were instructed to take two tablets twice a day (1500 mg) of *Echinacea* extract or matching placebo at the first sign of a cold. The results showed that treatment with the extract produced a statistically significant reduction in clinical cold symptoms, as compared with placebo.[12] These data support the findings of a previously published randomized, double-blind, placebo-controlled trial that assessed the effects of *Echinacea* in 120 patients with the initial symptoms of acute, uncomplicated upper airway infections.[13] The subjects took 20 drops of an *E. purpurea* expressed juice product or placebo every two hours for the first day, and then three times daily thereafter for up to 10 days, at the first sign of a cold. Patients treated with the extract did not develop a full cold (40% versus 60% in the placebo group), had fewer symptoms and recovered faster (6 days versus 10 days) than those patients treated with placebo.[13]

The safety of an *Echinacea angustifolia* product (encapsulated whole dried plant) was tested in a phase I trial in HIV positive patients.[14] Fourteen patients with CD4 counts ranging from 6 to 600/mm$^3$ (mean 269) and viral loads ($log_{10}$) ranging from < 2.3 to 5.4 (mean 4.68) took part in the trial. Each patient received a 12-week course of the product at 1000 mg three times daily. At 12-weeks there was no significant difference in mean CD4 counts as compared with baseline. However, there was an overall 0.32 $log_{10}$ reduction in the viral load (mean 4.36, $p < 0.05$). No effects were observed on natural killer cell activity or direct anti-HIV killing activity. The authors concluded that the *Echinacea angustifolia* product was safe and associated with a significant reduction in viral load.[14]

A large scale longitudinal trial (4598 patients) of the effects of an ointment containing a lyophylisate of the expressed juice of *Echinacea purpurea* on inflammatory skin conditions, wounds, eczema, burns, herpes simplex and varicose ulcerations of the legs was undertaken.[14] Therapeutic benefit from the ointment was observed in 85.5% of the cases. The treatment periods ranges from 7.1 to 15.5 days.[14]

**Mechanism of Action**

Numerous investigations have demonstrated that extracts of *Echinacea purpurea, E. angustifolia* and *E. pallida* induce the activation of the cellular and humoral immune response both *in vitro* and *in vivo*.[1,5,6] *Echinacea* extracts appear to enhance immune function through multiple mechanisms including activation of phagocytosis and stimulation of fibroblast; increasing cellular respiration; and increasing the mobility of leukocytes.[1,5] *Echinacea* extracts enhanced natural killer cell activity, and antibody-dependent cellular cytotoxicity of human peripheral blood mononuclear cells, obtained from healthy subjects or patients with acquired immunodeficiency syndrome or chronic fatigue syndrome.[14,21] The addition of a lyophylisate of the expressed juice of *E. purpurea* (aerial parts harvested at full bloom) to the culture medium, significantly increased the number of phagocytizing human granulocytes, and enhanced the phagocytosis *of Candida albicans in vitro*.[22,23] Isolated chemical constituents such as the lipophilic amides, alkamides and caffeic acid derivatives appear to contribute to the immunostimulant activity of *Echinacea* extracts by stimulating phagocytosis of polymorphnuclear neutrophil granulocytes *in vitro*.[1,22] Inhibition of hyaluronidase activity, stimulation of the activity of the adrenal cortex, stimulation of the production of properdin (a serum protein which can neutralize bacteria and viruses), and stimulation of interferon production have also been reported after *Echinacea* treatments.[24]

Polysaccharides, isolated from *E. purpurea* enhance cytokine production in murine and human macrophages *in vitro*.[25] Heteroxylan, a high molecular weight polysaccharide has been shown to activate phagocytosis, and arabinogalactan, promote the release of tumor necrosis factor and increase the production of interleukin-1 and $\beta$-interferon.[22,23] Pre-treatment of mouse macrophages with arabinogalactan (AR), isolated from the supernatant of *Echinacea purpurea* cell cultures, stimulated the production of tumor necrosis factor alpha (TNF$\alpha$), interleukin-1 and interferon-$\beta_2$.[22,23] Arabinogalactan also enhanced the cytotoxicity of macrophages against tumor cells and microorganisms (*Leishmania enreittii*).[25–27] The polysaccharide induced a slight increase in the proliferation of T-cells, but did not enhance the production of interleukin-2, interferon-$\gamma$ or interferon-$\beta_2$.[25–27] Purified polysacchrides isolated from large scale cell cultures of *E. purpurea* enhanced the spontaneous motility of human polymorphnuclear neutrophil granulocytes under soft agar, and increased the ability of these cells to kill *Staphylococcus aureus*.[27] Both crude extracts of *E. purpurea* (0.025 µg/ml) and purified polysaccharides stimulated the production of TNF$\alpha$, interleukin-1 and interleukin-6 in both mouse peritoneal macrophages and human peripheral blood monocytes.[27–29] The immune stimulatory activity of the crude extract was similar to that of the purified polysaccharides. Thus, the overall immune stimulating activity of the alcoholic and aqueous *Echinacea* extracts appears to be dependent on the combined effects of several constituents.[1]

Root extracts of *E. angustifolia* and *E. purpurea* significantly enhance the phagocytosis of erythrocytes in isolated, perfused rat liver.[1] Intragastric administration of an ethanol extract of *Echinacea purpurea*, *E. angustifolia* or *E. pallida* roots significantly enhanced phagocytosis in mice as measured by the carbon clearing test.[6]

External application of an *Echinacea* extract were used traditionally to promote wound healing and inhibit the activity of hyaluronidase.[24] The effect of *Echinacea* extracts on streptococcal and tissue hyaluronidase has been demonstrated *in vivo*. Inhibition of tissue and bacterial hyaluronidase is thought to localize the infection and prevent the spread of causative agents to other parts of the body. In addition to the direct antihyaluronidase activity, an indirect effect on the hyaluronic acid-hyaluronidase system has been shown.[30] Stimulation of new tissue production by increasing fibroblasts, and stimulation of both blood and tissue-produced phagocytosis, appear to be involved in this mechanism.[30] A polysaccharide fraction, called echinacin B, appears to promote wound healing through the formation of a hyaluronic acid-polysaccharide complex that indirectly leads to the inhibition of hyaluronidase activity. *In vitro* experiments have shown that a hydroalcoholic extract (65% ethanol v/v) of *Echinacea purpurea* inhibits the contraction of collagen in mouse fibroblasts, as assessed by the collagen lattice diameter.[31]

*Echinacea purpurea* extracts have also been shown to have antiviral activity *in vitro*. Extracts (concentration range 10–100 µg/ml) protected mouse L cells (clone 929) against the cytotoxic effect of encephalomyocarditis virus, vesicular stomatitis virus (VSV) in the presence of diethylamino elkyl dextron (DEAE).[1] The antiviral activity was similar to interferon, since DEAE alone is inactive. The expressed juice of *E. purpurea* has also shown interferon-like effects and demonstrated activity against VSV, influenza and herpes viruses'.[32] Purified caffeic acid derivatives, isolated from *E. purpurea* have also shown antiviral activity *in vitro*.[1] Incubation of VSV with 125 µg/ml of cichoric acid for four hours reduced VSV in mouse L-929 cells by more than 50%.[33]

*Echinacea* extracts have also been shown to have antiinflammatory activity.[1] An alkylamide fraction from *Echinacea* roots markedly inhibited activity *in vitro* in the 5-lipoxygenase model (porcine leukocytes).[34] A crude polysaccharide extract from *E. angustifolia* has been reported to have anti-inflammatory activity in the rat hind paw edema model, after topical application.[35]

Recently, extracts of *Echinacea* have been shown to inhibit the activity of cytochrome P450 isozyme 3A4 *in vitro*. Ethanol extracts of *E. angustifolia* root, *E. pupurea* root or *E. pupurea* aerial parts had a median inhibitory concentration of 6.73, 1.05 and 8.56% respectively.[36]

## Safety Information

*A. Adverse Reactions*
A few cases of *Echinacea*-associated anaphylaxis have been reported.[37,38] One case

of anaphylaxis occurred after oral ingestion of a commercial extract containing *Echinacea angustifolia* and *E. purpurea* root in combination with numerous other dietary supplements.[37] However, the association between the allergic reaction and *Echinacea* has been disputed.[39] Other allergic reactions such as contact dermatitis, urticaria, dizziness and tongue swelling have been reported.[37,40]

## B. Contraindications

*Echinacea* products are contraindicated in patients with an allergy to plants in the daisy family (Asteraceae), as well as in those patients with progressive conditions such as tuberculosis, leukosis, collagenosis, multiple sclerosis, AIDS, HIV infection and auto-immune disorders.[5,8] Most of these contraindications are speculative and controversial. The contraindication for autoimmune disorders may be based on a non-specific stimulation of the immune response.[41] The contraindication in AIDS or HIV infection is based on the fact that *Echinacea* induces the secretion of tumor necrosis factor from macrophages, which is elevated in the serum of AIDS patients with cachexia and associated with an increased viral load.[41] However, a recent phase I trial has demonstrated that *Echinacea* treatment may actually decrease the viral load in AIDS patients.[20] Furthermore, oral administration of *Echinacea* extracts to humans did not increase serum tumor necrosis factor levels.[42] Due to a lack of safety data the use of oral *Echinacea* products during pregnancy or nursing cannot be recommended.[5]

## C. Drug Interactions

None reported. However, *Echinacea* extracts have now been shown to effect the activity of cytochrome P450 isozyme 3A4 *in vitro*. Thus, these extracts should be used cautiously in patients taking prescription medications that are known to be metabolized via cytochrome P450.

## D. Toxicology

Contrary to previously published reports,[43] *Echinacea* products are not hepatotoxic.[5] Although trace amounts of pyrrolizidine alkaloids (tussilagine [0.006%] and isotussilagine) have been found in *E. angustifolia* and *E. purpurea*, these two alkaloids do not cause liver damage, as they lack the 1,2-unsaturated necine ring in their chemical structure.[1,5]

Intragastric (15–30 g/kg) or intravenous (5–10 g/kg) administration of the expressed juice of *Echinacea purpurea* to rats and mice resulted in no abnormalities.[44] The $LD_{50}$ values for the expressed juice of *E. purpurea* are:

Rats: i.g. >15 g/kg body weight; i.v. > 5 g/kg body weight
Mice: i.g. >30 g/kg body weight; i.v. 10 g/kg body weight.[44]

No mutagenicity was observed *in vitro* in bacterial or mammalian cells.[45] Carcinogenicity study results were also negative.[45] Doses up to a polysaccharide concentration of 500 mg ml$^{-1}$ did not increase in sister chromatid exchange or structural chromosome aberrations.[45]

## E. Dose and Dosage Forms

*Echinacea* products are available in crude drug, or processed drug in liquid or solid dosage forms. Daily oral dose for *Echinacea purpurea* (herb) is 6 to 9 ml of expressed juice. For external use: semisolid preparations containing a minimum of 15% expressed juice.[5,8] Daily dose of *Echinacea pallida/angustifolia* (roots): as a tincture (1:5) prepared with 50% ethanol (v/v) from a native dry extract (50% ethanol (7–11:1), dose equivalent to 900 mg of roots.[5] Administration for longer that eight weeks is not recommended.[8]

## References

1    Bauer R, Wagner H. *Echinacea* species as potential immunostimulatory drugs. In: Wagner H, Farnsworth NR, eds. *Economic and Medicinal Plants Research.* Vol. 5, London, Academic Press, 1991:253–321.

2    Hobbs C. *Echinacea*, a literature review. *Herbal Gram.* 30:33–46, 1994.

3    Murray MT. *The Healing Power of Herbs*, 2nd edition, Rocklin: Prima Publishing, 1995:410 pp.

4    Clarke JH. *A Dictionary of Practical Materia Medica.* New Delhi: Jain Publication, 1900. (London, reprinted 1984).

5    Anon. *WHO Monographs on Selected Medicinal Plants*, Volume I, Geneva, World Health Organization, 1999.

6    Bauer R, Jurcic K, Puhlmann J, Wagner H. Immunological *in vivo* examinations of *Echinacea* extracts. *Arzneimittelforschung.* 1988;38:276–281.

7    Röder E, Wiendenfeld H, Hille T, Britz-Kirstten R. Pyrollizidine in *Echinacea angustifolia* DC and *Echinacea purpurea.* Isolation and analysis. *Deutsche Apotheker Zeitung.* 1984;124:2316–2317.

8    German Commission E Monographs, *Echinacea purpurea* herba, *Echinacea pallida* radix, *Bundesanzeiger*, 2.3.1989, Volume 42; 8.29.1992; Volume 162.

9    Melchart D et al. Immunomodulation with *Echinacea* – a systematic review of controlled clinical trials. *Phytomedicine.* 1994;1:245–254.

10   Schöneberger D. The influence of immune-stimulating effects of pressed juice from *Echinacea purpurea* on the course and severity of colds. *Forum Immunologie.* 1992;8:2–12.

11   Bräunig B, Knick E. Therapeutische Erfahrungen mit Echinaceae pallidae bei grippalen Infekten. Ergebnisse einer plazebokontrollierten Doppelblindstudie. *Naturheilpraxis.* 1993;46:72–75.

12   Brinkeborn R, Shah D, Geissbühler S, Degenring FH. Echinaforce in the treatment of acute colds. *Schweiz Zeitschrift Gansheits Medizin.* 1998;10:26–29.

13   Hoheisel O, Sandberg M, Bertram S, Bulitta M, Schäfer M. Echinagard treatment shortens the course of the common cold: a double-blind, placebo-controlled clinical trial. *Eur J Clin Res.* 1997;9:261–268.

14   Viehmann P. Results of treatment with an *Echinacea*-based ointment. *Erfahrungsheilkunde.* 1978;27:353–358.

15   Bräunig B, Dorn M, Knick E. *Echinacea purpurea* radix: zur Stärkung der körpereigenen Abwehr bei grippalen Infekten. *Zeitschrift für Phytotherapie.* 1992;13:7–13.

16   Melchart D, Linde K, Worku F, Sarkady L, Holzmann M, Jurcic K, Wagner H. Results of five randomized studies on the immunomodulatory activity of preparations of *Echinacea. Journal of Alternative and Complementary Medicine.* 1995;1:145–160.

17    Melchart D, Walther E, Linde K, Brandmaier R, Lersch C. *Echinacea* root extracts for the prevention of upper respiratory tract infections: a double-blind, placebo-controlled randomized trial. *Arch Fam Med.* 1998;7:541–545.

18    Grimm W, Muller HH. A randomized controlled trial for the effect of fluid extract of *Echinacea purpurea* on the incidence and severity of colds and respiratory infections. *Amer J Med.* 1999;106:138–143.

19    Dorn M, Knick E, Lewith G. Placebo-controlled, double-blind study of *Echinacea pallidae* radix in upper respiratory infections. *Complementary Therapies in Medicine.* 1997;3:40–42.

20    See D. Berman S, Justus J, Broumand N, Chou S, Chang J, Tilles J. A phase I study on the safety of *Echinacea angustifolia* and its effect on viral load in HIV infected individuals. *Journal of the American Nutraceutical Association.* 1998;1:14–17.

21    See D, Broumand N, Sahl I, Tilles J. *In vitro* effects of *Echinacea* and ginseng on natural killer and antibody-dependent cytotoxicity in healthy subjects and chronic fatigue syndrome or acquired immunodeficiency syndrome patients. *Immunopharmacology.* 1996;961:229–235.

22    Stotzem CD, Hungerland U, Mengs U. Influence of *Echinacea purpurea* on the phagocytosis of human granulocytes. *Med Sci Res.* 1992;20:719–720.

23    Bittner E. Die Wirkung von Echinacin auf die Funktion des Retikuloendothelialen Systems, Ph. D. Dissertation, Universität, Freiburg, 1969.

24    Büsing KH. Hyaluronidase inhibition by Echinacin. *Arzneimittelforschung.* 1952;2:467–469.

25    Stimpel M, Proksch A, Wagner H, Lohman-Matthes ML. Macrophage activation and induction of macrophage cytotoxicity by purified polysacchride fractions from the plant *Echinacea purpurea. Infect Immun.* 1984:845–849.

26    Steinmüller C, Roesler J, Grottrup E, Franke G, Wagner H, Lohman-Matthes ML. Polysaccharides isolated from plant cell cultures of *Echinacea purpurea* enhance the resistance of immunosuppressed mice against systemic infections with *Candida albicans* and *Listeria monocytogenes. Int J Immunopharmacol.* 1993;15:605–614.

27    Roesler J, Emmendorffer A, Steinmüller C, Luettig B, Wagner H, Lohman-Matthes ML. Application of purified polysaccharides from cell cultures of the plant *Echinacea purpurea* to test subjects mediates activation of the phagocyte system. *Int J Immunopharmacol.* 1991;13:931–941.

28    Luettig B, Steinmüller C, Gifford GE, Wagner H, Lohman-Matthes ML Macrophage activation by polysaccharide arabinogalactan isolated from plant cell cultures of *Echinacea purpurea. Journal of the National Cancer Institute.* 1989;81:669–675.

29    Burger RA, Torres AR, Warren RP, Caldwell VD, Hughes BG. *Echinacea*-induced cytokine production by human macrophages. *Int J Immunopharmacol.* 1997;19:371–379.

30    Koch FE, Haase H. A modification of the spreading test in animal assays. *Arzneimittelforschung.* 1952;2:464–467.

31    Zoutewelle G, van Wijk R. Effects of *Echinacea purpurea* extracts on fibroblast populated collagen lattice contraction. *Phytother Res.* 1990;4:77–81.

32    Wacker A, Hilbig W. Virus-inhibition by *Echinacea purpurea. Planta Med.* 1978;33:89–102.

33    Cheminat A et al. Caffeoyl conjugates from *Echinacea* species: structures and biological activity. *Phytochemistry.* 1988;27:2787–2794.

34    Wagner H et al. *In vitro* inhibition of arachidonate metabolism by some alkamides and prenylated phenols. *Planta Med.* 1988;55:566–567.

35    Tubaro A et al. Anti-inflammatory activity of a polysaccharidic fraction of *Echinacea angustifolia. J Pharm Pharmacol.* 1987;39:567–569.

36    Budzinski JW et al. An *in vitro* evaluation of human cytochrome P450 3A4 inhibition by selected herbal extracts and tinctures. *Phytomedicine*. 2000;7:273–282.

37    Mullins RJ. *Echinacea*-associated anaphylaxis. *Med J Australia*. 1998;168:170–171.

38    Immunallergische Reaktionen nach *Echinacea*-Extrakten (Echinacin, Esberitox N u.a.) *Arznei-Telegramm*. 1991;4:39.

39    Myers S, Wohlmuth H. *Echinacea*-associated anaphylaxis. *Med J Australia*. 1998;168:583.

40    Bruynzeel DP, Van Ketel WG, Young E. et al. Contact sensitisation by alternative topical medicaments containing plant extracts. *Contact Dermatitis*. 1992;27:278–279.

41    Bodinet C, Willigmann I, Beuscher N. Host-resistance increasing activity of root extracts from *Echinacea* species. *Planta Med*. 1993;59 (Supp):A672.

42    Elassier-Beile U et al. Cytokine production in leukocyte cultures during therapy with *Echinacea* extract. *J Clin Lab Anal*. 1996;10:441–445.

43    Miller, L. Herbal medicinals: selected clinical considerations focusing on known or potential drug-herb interactions. *Arch Intern Med*. 1998;158:2200–2211.

44    Mengs U, Clare CB, Poiley JA. Toxicity of *Echinacea purpurea*. Acute, subacute and genotoxicity studies. *Arzneimittelforschung*. 1991;41:1076–1088.

45    Kraus C, Abel G, Schimmer O. Untersuchung einiger Pyrrolizidinalkaloide auf chromosomenschädigende Wirkung in menschlichen Lymphocyten *in vitro*. *Planta Med*. 1985;51:89–91.

# 9

# *Ephedra*

## Synopsis

Various ephedrine-containing *Ephedra* species have been used since antiquity for the treatment of the common cold, asthma, bronchitis and nasal congestion. While there are no data from clinical trials to substantiate these claims, two of the primary active constituents of *Ephedra*, ephedrine, and pseudoephedrine, have been approved by the FDA as ingredients in over-the-counter drugs for bronchodilation, decongestant and anti-allergic activity. Depending on the dose administered both ephedrine and pseudoephedrine can have strong stimulating activities on the central nervous system. Therefore, preparations containing *Ephedra* must be recognized as a drug with potentially serious side effects when administered in higher than the recommended dose, or if abused. The dosage range of *Ephedra* containing products is dependent upon the ephedrine concentration of the product. The single dose for adults should correspond to 15–30 mg total alkaloid, calculated as ephedrine, with a maximum of 300 milligrams total alkaloid per day, calculated as ephedrine. For children, the single dose is 0.5 milligrams total alkaloid per kg bogy weight, to a maximum of 2 mg total alkaloid per kg of body weight. The adverse reactions associated with the ingestion of *Ephedra* and ephedrine-containing products include nervousness, headaches, insomnia, dizziness, palpitations, skin flushing and tingling, and vomiting. The principal adverse effects of ephedrine are CNS stimulation, nausea, tremors, tachycardia and urinary retention. As with all other products containing ephedrine, *Ephedra* preparations should not be administered to patients with coronary thrombosis, diabetes, glaucoma, heart disease, hypertension, thyroid disease, impaired circulation of the cerebrum, pheochromocytoma or enlarged prostate. *Ephedra* preparations should not be administered during pregnancy or nursing. Adverse drug interactions have been reported when *Ephedra* is used in combination with cardiac glycosides, halothane, guanethidine, monoamine oxidase inhibitors, ergot alkaloid derivatives or oxytocin.

## Introduction

The genus *Ephedra* is composed of 40 different plant species all belonging to the same plant family, Ephedraceae.[1] The plant is a small, evergreen, almost leafless shrub native to many parts of the world. *Ephedra* herb has a pine-like odor and an astringent taste, often having a numbing action on the tongue.[2] The common name "ma huang" is thought to refer to the astringent action "ma" and the term huang refers to the yellow color of the twigs.[3] In traditional Chinese medicine, *Ephedra* containing preparations have been used for 5,000 years for the treatment of colds, flus, fever, headache, bronchial asthma, nasal congestion coughs and wheezing.[1,4–6] Some of the *Ephedra* alkaloids, such as ephedrine and pseudoephedrine, are employed in modern medicine as drug treatment for bronchial asthma, nasal congestion, acute bronchospasm and idiopathic orthrostatic hypertension.[4,7] In the United States, the FDA has approved several *Ephedra* alkaloids as ingredients in over-the-counter nasal decongestants and bronchodilator drugs. Pseudoephedrine is approved as an oral decongestant for the symptomatic treatment of the common cold, hayfever, allergic rhinitis, upper respiratory allergies and sinusitis. Ephedrine has been approved as topical therapy for the treatment of nasal congestion, and asthma.[8] Although herbal products containing *Ephedra* are regulated as dietary supplements by the FDA,[9] it is important to recognize that *Ephedra* is a drug, that may have potentially serious side effects when doses are exceeded or if abused (see "Safety Assessment Section").

## Quality Information

- The correct Latin name for *Ephedra* is *Ephedra sinica* Stapf, although other ephedrine containing species of the same genus are also similarly used (Ephedraceae).[1] There are no other botanical synonyms used for this plant.[1] However, there are numerous vernacular (common) names used worldwide for *Ephedra*, including amsania, budshur, chewa, Chinese ephedra, ephédra, horsetail, hum, huma, joint fir, khama, ma hoàng, ma huang, máhuáng, mao, maoh, mao-kon, môc tac ma hoàng, mu-tsei-ma-huang, phok, san-ma-huang, shrubby, soma, song tuê ma hoàng, trung aa hoàng, tsao-ma-huang, and tutgantha.[1]
- Standardized extracts and other commercial products of *Ephedra* are prepared from the dried stem or aerial part of *Ephedra sinica* Stapf or other ephedrine containing species of the same genus.[1]
- *Ephedra* species are found in China, India, Mongolia, Afghanistan, as well as regions of the Mediterranean, North and Central America.[1]
- The chemical constituents of *Ephedra* include (−)-ephedrine in concentrations of 40–90% of the total alkaloid fraction, accompanied by (+)-pseudoephedrine. Other compounds in the alkaloid complex include trace amounts of (−)-norephedrine, (+)-norpseudoephedrine, (−)-methylephedrine and (+)-methylpseudoephedrine. The total alkaloid content can exceed 2% depending on the species; however, not all *Ephedra* species contain ephedrine or alkaloids.
- The daily dose of *Ephedra* containing products varies depending on the concentration of ephedrine in the preparation. For crude plant material: 1–6 g per daily generally given as a decoction. Liquid extract (1:1 in 45% alcohol): 1–3 ml. Tincture (1:4 in 45% alcohol): 6–8 ml.[1]

## Medical Uses

*Ephedra* containing preparations are used therapeutically for the symptomatic treatment of nasal congestion due to hay fever, allergic rhinitis, acute coryza, common cold and sinusitis. The drug is further used as a bronchodilator in the treatment of bronchial asthma.[1,2,10–12] Clinical evidence supporting its use for the treatment of obesity is weak and requires further clinical evaluation.[13–15]

In addition, *Ephedra* preparations have been used for the treatment of urticaria, enuresis (incontinence), narcolepsy, myasthenia gravis and chronic postural hypotension, however there are no clinical data to support these claims.

## Summary of the Clinical Evidence

There are no controlled clinical trials evaluating the effects of *Ephedra* preparations for the treatment of respiratory diseases. All clinical data is associated with one of its two main alkaloids, ephedrine and pseudoephedrine; thus the experimental pharmacology is based on these two alkaloids. Both ephedrine and pseudoephedrine are potent sympathomimetic drugs with $\alpha_1$, $\beta_1$ and $\beta_2$ receptors-agonist properties.[11,12] Ephedrine causes vasoconstriction and cardiac stimulation, which in turn may produce acute increases in blood pressure and heart rate, as well as mydriasis, insomnia, vertigo, headache and nervousness. Part of ephedrine's peripheral action is due to the release of norepinephrine, however the drug also has direct effects on receptors. Tachyphylaxis develops to its peripheral actions, and rapidly repeated doses become less effective due to the depletion of norepinephrine stores.[11] Pseudoephedrine's action is similar to ephedrine, acting via $\beta_2$-agonist activity; it enhances bronchodilation while its alpha-agonist properties relieve the symptoms of nasal congestion.

### Effects on the cardiovascular system

Ephedrine is similar to epinephrine (adrenaline) in that excites the sympathetic nervous system, causing vasoconstriction and cardiac stimulation. Ephedrine differs from epinephrine in that it is orally active, has a much longer duration of action, has more pronounced CNS activity, but is much lower in potency.[11,12] The drug stimulates the heart rate, as well as cardiac output, and increases peripheral resistance, thereby producing a lasting rise in blood pressure. The cardiovascular effects of ephedrine persist up to ten times longer than epinephrine.[11] Ephedrine elevates both the systolic and diastolic pressures, and pulse pressure. Renal and splanchnic blood flows are decreased, while coronary, cerebral, and muscle blood flows are increased.[11,12]

### Effects on the central nervous system

Ephedrine is a potent stimulator of the central nervous system, and its effects last for several hours after oral administration.[11,12] For this reason, preparations containing *Ephedra* have been promoted for their use in weight reduction and thermogenesis (fat burning). However, conflicting results have been reported in a number of controlled clinical trials assessing the effects of a combination of ephedrine and caffeine combinations for the treatment of obesity.[13–15] In a randomized double blind, placebo-controlled clinical trial assessed the effects of a proprietary formula containing ephedrine, caffeine plus aspirin on weight loss was assessed.[13] In this investigation, 24 obese subjects were treated with a mixture (EAC) of ephedrine (75–150 mg), caffeine (150 mg) and aspirin (330 mg) daily in divided doses without caloric restriction. After 8 weeks the mean weight loss was 3.2 kg in the EAC group as compared with 1.3 kg for the placebo group.[13] No significant changes in heart rate, blood pressure or other side effects were reported during this trial, suggesting that therapeutic doses of ephedrine and caffeine were not achieved. Treatment of obese patients with ephedrine alone was no better than placebo for inducing weight loss.[14] In addition, due to the potential serious side effects associated with *Ephedra* use, the safety and effectiveness of these preparations is currently an issue of debate, and requires further investigation.[12]

Ephedrine has been used for the treatment of the pain associated with dysmenorrhea, in part because the activity of the smooth muscles of the uterus is usually reduced by ephedrine.[11,12] Ephedrine also stimulates the α-adrenergic receptors of the smooth muscle cells of the bladder base, thereby increasing the resistance to the outflow of urine.[12] Thus, *Ephedra* containing preparations have been used for the treatment of urinary incontinence and nocturnal enuresis.

*Effects on the respiratory system*
Ephedrine, like epinephrine, relaxes bronchial muscles and is a potent bronchodilator due to its activation of the β-adrenergic receptors in the lungs.[11,12] Bronchial muscle relaxation is less pronounced, but more sustained with ephedrine than with epinephrine. As a consequence, ephedrine should only be used in patients with mild cases of acute asthma and in chronic cases that require maintenance medication. Like other sympathomimetic drugs with α-receptor activity, ephedrine causes vasoconstriction and blanching when applied topically to nasal and pharyngeal mucosal surfaces.[11,12] Continued, prolonged use of these preparations (> 3 days) may cause rebound congestion and chronic rhinitis.[1] Both ephedrine and pseudoephedrine are useful orally as nasal decongestants in cases of allergic rhinitis, but may not be very effective for the treatment of nasal congestion due to colds.

Topical application of ephedrine (3 to 5%) to the eye induces mydrasis, which lasts for only a few hours.[16] However, in the presence of inflammation, ephedrine is of little value as a topical mydriatic.

## Pharmacokinetics

The pharmacokinetics and cardiovascular effects of *Ephedra sinica* were investigated in 12 healthy normotensive subjects.[6] The subjects were administered four 375 mg capsules containing *Ephedra* and serial plasma samples were analyzed. The pharmacokinetic parameters of ephedrine were determined from plasma concentration profiles. Six subjects experienced a significant increase in heart rate, but the effects on blood pressure were variable. The mean ephedrine concentration-time data were best fit to a one-compartment, open model with first-order elimination and first order absorption and lag time of 0.25 hours. The area under the concentration-time curve was 798 ng per hr/ml, with a t½ of 5.20 hours. The apparent volume of distribution at steady state was 182.3 L with an apparent clearance of 24.3 L per hour. The time to reach maximum concentration was 3.9 hours and the maximum plasma concentration was 81.0 ng/ml. The pharmacokinetic parameters for ephedrine are similar to that previously reported.[6]

## Safety Information

### A. Adverse Reactions
*Ephedra* and ephedrine containing products can cause nervousness, headaches, insomnia, dizziness, palpitations, skin flushing and tingling, and vomiting.[10] The principal adverse effects of ephedrine are CNS stimulation, nausea, tremors, tachycardia and urinary retention.[10] One case of hypersensitivity myocarditis has been associated with the use of *Ephedra* for three months in combination with unspecified vitamins, pravastatin and furosemide.[17] One case report of acute hepatitis occurred after prolonged use (> 3 days) of topical preparations containing *Ephedra* or ephedrine, for the treatment of nasal congestion, may cause rebound congestion and chronic rhinitis.[16] Continued prolonged use of oral preparations may cause development of dependency.[16]

Since 1994, the FDA has received an increasing number of reports of adverse reactions associated with the use of *Ephedra* containing products. The adverse reactions were linked to ephedrine toxicity and 17 deaths.[8] These reports varied in severity from mild reactions such as gastrointestinal irritations, dizziness, headache, and nervousness to severe life threatening reactions such as myocardial infarction, stroke and death. Adverse reactions have been reported in healthy individuals as well as patients with chronic conditions such as hypertension and cardiovascular disease. Chronic administration of *Ephedra* or ephedrine-containing products has also been associated with the development of nephrolithiasis.[18] Analysis of kidney stones from eight patients positively identified the substrate of renal calculi as combinations of ephedrine, norephedrine and pseudoephedrine.[18] Insomnia may occur with continued use of *Ephedra* or ephedrine containing preparations.[18]

### B. Contraindications and Warnings
As with all other products containing ephedrine, *Ephedra* preparations should not be administered to patients with coronary thrombosis, diabetes, glaucoma, heart disease, hypertension, thyroid disease, impaired circulation of the cerebrum, pheochromocytoma or enlarged prostate.[1,10,12] Reduce the dose or discontinue use if nervousness, tremor, sleeplessness, loss of appetite or nausea occurs. *Ephedra* containing preparations should not be administered to children under 6 years old. Keep out of the reach of children. Continued, prolonged use may cause dependency.[10]

While extracts of *Ephedra sinica* have not been shown to be abortifacient in rats,[19] there is no clinical data available. Therefore, *Ephedra* containing preparations should not be administered during pregnancy or nursing without consulting a physician.[1]

### C. Drug and Laboratory Test Interactions

When administered in combination with:

- Cardiac glycosides or halothane, *Ephedra* may cause heart rhythm disturbances.[10]
- Guanethidine, *Ephedra* may enhance the sympathomimetic effect.[21]
- Monoamine oxidase inhibitors, *Ephedra* may cause severe, possibly fatal hypertension.[16]
- Ergot alkaloid derivatives or oxytocin, *Ephedra* may increase risk for high blood pressure.[10]
- Ephedrine may cause a false-positive amphetamine EMIT assay.[7]

### D. Toxicology

Acute oral toxicity studies of *Ephedra* extracts and ephedrine have been performed in mice.[20] The LD50 values of the extract and ephedrine were 5.3 g/kg and 698 mg/kg, respectively. The extract used in this study contained 2.27, 2.14 and 0.057% of ephedrine, pseudoephedrine and norephedrine, respectively. At the lethal dose, ephedrine caused kidney damage due to vasoconstriction, however these injuries were not observed with the extract.[20] Extracts of *Ephedra sinica* are not mutagenic in the Salmonella/microsome reversion assay.[21] *Ephedra sinica* did not have any teratogenic effects in rats at a dose of 500 mg/kg beginning on the 13th day of pregnancy.[19]

### E. Dose and Dosage Forms

*Ephedra* preparations include powdered herb (crude drug) and extracts. The dosage range for adults is preparations corresponding to 15–30 mg total alkaloid, calculated as ephedrine for a single dose to a maximum of 300 milligrams total alkaloid per day, calculated as ephedrine. For children, the dose is 0.5 milligrams total alkaloid per kg bogy weight, to a maximum of 2 mg total alkaloid per kg of body weight.[10]

### References

1    Anon. Herba Ephedrae. *WHO Monographs on Selected Medicinal Plants*. Volume I, WHO, Geneva, Switzerland 1999.
2    Blumenthal M, King P. A review of the botany, chemistry, medicinal uses safety concerns and legal status of *Ephedra* and its alkaloids. *Herbal Gram* 1995;34:22–57.
3    Tyler VE, Brady LR, Robbers JE, eds. *Pharmacognosy*, 9th ed., Philadelphia, Lea and Febiger, 1988.
4    Kasahara Y et al. Antiinflammatory actions of ephedrines in acute inflammations. *Planta Med.* 1985;51:325–330.
5    Kalix P. The pharmacology of psychoactive alkaloids from *Ephedra* and Catha. *J Ethnopharmacol.* 1991;32:201–208.
6    White LM et al. Pharmcokinetics and cardiovascular effects of Ma-huang (*Ephedra sinica*) in normotensive adults. *J Clin Pharmacol.* 1997;37:116–122.

7     Lacy C et al. *Drug Information Handbook*, 5th ed. Lexi-Comp, Inc., Hudson, OH 1998.

8     Adverse events with *Ephedra* and other botanical dietary supplements. *FDA Medical Bulletin*, 1994;24:3.

9     Hutchinson K, Andrews KM. The use and availability of *Ephedra* products in the United States. Drug Enforcement Administration internal memo. February 1995.

10    German Commission E monograph. *Ephedra*, (Ephedrae herba) *Bundesanzeiger*. 17.01.1991.

11    Goodman LS, Gilman A. *Goodman and Gilman's, The Pharmacological Basis of Therapeutics*, 6th ed., New York, Mc Millian Publishing Co., 1985:169–170.

12    Goodman LS et al. *Goodman and Gilman's, The Pharmacological Basis of Therapeutics*, 8th ed., New York, Mc Millian Publishing Co., 1993:213–214.

13    Toubro S. et al. Safety and efficacy of long-term treatment with ephedrine, caffeine and ephedrine/caffeine mixture. *Int J Obesity*. 1993;17:S69–S72.

14    Daley PA et al. Ephedrine, caffeine and aspirin: safety and efficacy for the treatment of human obesity. *Int J Obesity*. 1993;17(Suppl. 1):S73-S78.

15    Pardoe AU, Gorecki DKJ, Jones D. Ephedrine alkaloid patterns in herbal products based on Ma Huang (*Ephedra sinica*). *Int J Obesity*. 1993;17 (Suppl. 1):S82.

16    *Handbook of Non-Prescription Drugs*, 10th ed., Washington DC: American Pharmaceutical Association, 1993.

17    Zaacks SM et al. Hypersensitivity myocarditis associated with *Ephedra* use. *Clin Toxicol*. 1999;37:485–489.

18    Powell T, Hsu FF, Turk J et al. Ma-huang strikes again: Ephedrine nephrolithiasis. *Am J Kid Dis*. 1998;32:153–159.

19    Lee EB. Teratogenicity of the extracts of crude drugs. *Korean Journal of Pharmacognosy*. 1982;13:116–121.

20    Minematsu S et al. Acute Ephedra Herba and ephedrine poisoning in mice. *Japanese J Toxicol*. 1990;4:143–149.

21    Yin Xue-jun et al. A study on the mutagenicity of 102 raw pharmaceuticals used in Chinese traditional medicine. *Mut Res*. 1991;260:73–82.

# 10

# Evening Primrose Oil

## Synopsis

Evidence from controlled clinical trials support the use of evening primrose oil (EPO) for the symptomatic treatment of atopic eczema, diabetic neuopathy and cyclical mastalgia. However, its therapeutic effects on rheumatoid arthritis, and premenstrual and menopausal symptoms have not been adequately demonstrated in clinical trials. Treatment with EPO should last at least 12 weeks before evaluating the therapeutic response. The recommended daily dose is 320–480 mg (expressed as γ-linolenic acid) in divided doses for atopic eczema, and 240–320 mg (expressed as γ-linolenic acid) in divided doses for the treatment of mastalgia. Adverse reactions associated with the ingestion of evening primrose oil include headaches, nausea, soft stools and diarrhea. Oral administration of the seed oil may precipitate symptoms of undiagnosed temporal lobe epilepsy in schizophrenic patients taking epileptogenic drugs, such as the phenothiazines. Evening primrose oil has been shown to inhibit platelet aggregation in animal studies, and inhibits platelet-activating factor in human studies. While no drug interactions have been reported, patients taking other anticoagulant or antiplatelet drugs in conjunction with the seed oil should be closely monitored for bleeding times. In traditional medicine, evening primrose oil has been used as an emmenogogue and a method of inducing labor. While one clinical trial has indicated that EPO ingestion does not shorten gestation or decrease the overall length of labor, its administration was associated with an increase in the incidence of prolonged rupture of membranes, oxytocin augmentation, arrest of decent and vacuum extraction. Therefore, evening primrose oil should be administered during pregnancy only on the advice of a physician.

## Introduction

The evening primrose (*Oenothera biennis*) is a native American wildflower, which was introduced into Europe in the early 17th century.[1] Its large, delicate yellow blooms usually last for only one evening. The seeds of the flower are rich in essential fatty acids including γ-linolenic acid, an intermediate in the synthesis of prostaglandins in humans. Evening primrose, also known as the "King's cureall" was widely utilized throughout history as an herbal medicine. Traditional preparations used the whole plant, while more contemporary products contain the oil obtained from the seeds of the plant. The Native American Indians used an infusion of the whole plant as an astringent and a sedative, for the treatment of coughs, gastrointestinal disorders, pain and wounds. Modern applications of the seed oil include its use for the treatment of atopic eczema, and cyclical and non-cyclical mastalgia, premenstrual syndrome, psoriasis, rheumatoid arthritis, chronic fatigue syndrome and diabetic neuropathy. Evening primrose oil is currently regulated as a dietary supplement in the United States.[1]

## Quality Information

- The correct Latin name for the plant is *Oenothera biennis* L. (Onagraceae).[2] Botanical synonyms that may appear in the scientific literature include *Oenothera communis* Léveillé, *O. graveolens* Gilib., *Onagra biennis* Scop, *Onagra vulgaris* Spach.[2] The vernacular (common) names for the plant include enotera, evening primrose oil, la belle de nuit, ligetszépeolaj, king's cureall, mematsuyoigusa, Nachtkerzenöl, onagre, and Teunisbloem.[2]
- Commercial products of evening primrose consist of the fixed oil obtained from the seeds of *Oenothera biennis* L. (Onagraceae).[2]
- Evening primrose is native to North America and naturalized in Western Europe and many other areas.[1,2]
- The major chemical constituents include *cis*-linolenic acid (65–80%), *cis*-γ-linolenic acid (GLA, 8–14%), oleic acid (6–11%), palmitic acid (7–10%), and stearic acid (1.5–3.5%). Other constituents of the oil include sterols and triterpene alcohols.[2]
- Daily dose: Expressed as γ-linolenic acid: 320–480 mg in divided doses for atopic eczema and 240–320 mg in divided doses for mastalgia.[2]

## Medical Uses

For the symptomatic treatment of atopic eczema and dermatitis,[4–16] diabetic neuropathy,[17,18] and cyclical mastalgia.[19–21]

The clinical evidence supporting the use of the seed oil for the treatment of rheumatoid arthritis is conflicting,[22–25] as are the data for its use in the treatment of premenstrual syndrome.[26–30] Further, well-designed clinical trials are necessary to clarify data. Results from controlled clinical trials do not substantiate the use of the seed oil for the treatment of menopausal flushing or psoriasis.[13,31]

## Summary of Clinical Evidence

### Effect on Skin Disorders

Seventeen controlled clinical trials have assessed the efficacy of evening primrose oil for the treatment of atopic eczema and dermatitis.[4-16] A meta-analysis of nine of the controlled clinical trials, four with parallel and five with crossover designs, involving 311 subjects with atopic eczema was performed.[11] Both physicians and patients assessed the severity of eczema by scoring measures of inflammation, dryness, scaliness, pruritus and overall skin involvement. The individual symptom scores were combined to give a single global score at each assessment point. Analysis of the parallel trials showed a highly significant improvement over baseline ($p < 0.0001$), and concluded that the seed oil was more effective than placebo.[11] Similar results in the crossover trials was observed but failed to reach statistical significance except for the symptom of itching where there was a highly significant response to the seed oil ($p < 0.0001$) as compared with placebo.[11] In general, there was a positive correlation between an improvement in the clinical scores and an increase in the plasma concentrations of the free fatty acids, dihomo-γ-linolenic and arachidonic acids. Unfortunately, crossover designs for evening primrose oil are problematic since the therapeutic effects of the oil are not observed until several months of treatment and last for several months after cessation of therapy.[32] Thus, short washout times of a couple of weeks before crossover is not enough time to allow for the therapeutic effects to fade and hence would increase the placebo effects. Moreover, two double-blind, placebo-controlled studies that were not included in the meta-analysis, one crossover trial in 123 patients, and one more recent parallel trial in 102 patients, both of which reported negative results.[4,5]

In a double blind, placebo-controlled parallel group trial, 58 children with atopic dermatitis were treated with either placebo or evening primrose oil (2 to 3 grams per day) for 16 weeks.[9] The plasma concentrations of essential fatty acids increased significantly in the seed oil treated group. Symptomatic improvements occurred in both groups, however no significant difference was found between the two treatments.[9] A significant flaw of this particular clinical trial was the use of a "placebo" containing sunflower oil, which has a similar spectrum of essential fatty acids as evening primrose oil. A double-blind placebo-controlled clinical trial assessed the effects of two doses of the evening primrose oil in children with atopic dermatitis.[7] Fifty-one children were treated with placebo or 0.5 g/kg/day of the seed oil or a combination of 50% placebo and 50% seed oil (0.5 g/kg/day) for 16 weeks. A significant improvement in the overall severity of the clinical symptoms was observed in children treated with the seed oil (0.5 g/kg/day). The treatment also increased the concentration of n-6 fatty acids in erythrocyte cell membranes.[7] In a double blind parallel trial, oral administration of the seed oil to 37 patients with psoriasis (dose 430 mg/day) showed no significant improvement in symptoms.[13]

A double blind, placebo-controlled clinical trial involving 39 adult subjects assessed the efficacy of the seed oil for the treatment of chronic hand dermatitis.[14]

Each subject received 6 g of seed oil or placebo daily for 16 weeks. An improvement was observed in both groups, however no statistical difference was observed between the two groups.[14] The effects of oral supplementation with evening primrose oil (EPO) on plasma fatty acid concentrations and symptoms (dryness, pruritus, erythema) of uremic skin disorders was evaluated in a double-blind study.[16] Patients treated with EPO for six weeks exhibited a significant ($p < 0.05$) increase in plasma dihomo-$\gamma$-linolenic acid and a decrease in uremic pruritus.[16]

*Rheumatoid Arthritis*

Four randomized parallel clinical trials assessed the efficacy of evening primrose oil for the treatment of rheumatoid arthritis in small numbers of patients.[22–25] Two of the four studies were unable to establish any statistically significant benefit of evening primrose oil for the treatment of rheumatoid arthritis. However, two of the clinical trials reported a significant improvement in symptoms, although only one of the trials concluded that the evening primrose oil offered some therapeutic benefit, as it reduced the dosage of non-steroidal anti-inflammatory drugs, when given concomitantly.[24] Oral administration of the seed oil (20 ml/day) to 10 patients with rheumatoid arthritis decreased prostaglandin $E_2$ levels in four patients and increased 6-keto-$PG_{F1\alpha}$ levels in 3 patients after 12 weeks of treatment.[33]

A double blind, placebo-controlled study assessed the effects of a combination of evening primrose oil and/or fish oil for the treatment of rheumatoid arthritis in 34 patients taking NSAIDS.[22] Following 12 months of treatment, a significant subjective improvement was observed in those patients receiving the seed oil and/or fish oil as compared with placebo. In addition, after 12 months, the patients treated with $\gamma$-linolenic acid (540 mg daily, administered as seed oil) with or without fish oil, had markedly reduced their intake of NSAIDS.[22]

*Effects on PMS and Menopausal Symptoms*

A review of four clinical published clinical trials (3 with crossover design) has reported positive assessments for use of the seed oil in the treatment of (PMS).[30] A double-blind placebo-controlled crossover study assessed the efficacy of the seed oil for the treatment of premenstrual syndrome PMS.[28] All the major clinical features of PMS were scored. After eight weeks, symptoms in both groups improved, patients treated with the seed oil had a 60% improvement, and a 40% improvement was observed in the placebo group. The symptoms of irritability and depression were notably improved in the seed oil group.[28] In an open unpublished study of 196 women with PMS, the subjects were asked to score their symptoms one cycle prior to treatment and two cycles after treatment. The patients received two capsules of the seed oil (500 mg) twice daily during the luteal phase of the cycle. The symptoms of irritability, depression, headache, breast pain and tenderness, and ankle swelling showed highly significant improvements ($p < 0.001$) from the initial cycle to the last one. Irritability improved by 77%, depression in 74%, breast tenderness and pain in 76%, headache in 71% and ankle swelling in 63%.[28] An uncontrolled clinical trial assessed the efficacy of the seed oil for symptomatic treatment in 68 women with severe PMS, who had failed to respond to at least one other therapeutic regime.[26] The patients were treated with a graduated dosage of the seed oil, starting with two 0.5 g capsules twice daily in the luteal phase only and working up to four capsules twice daily during the whole cycle if there was no response. The results showed that 61% of patients had a total remission of symptoms and a further 23% had a partial remission. Of the thirty-six of the women who also experienced breast pain as part of the PMS syndrome, 26 had total relief of this symptom, 5 had partial relief, and 5 showed no improvement.[26]

The efficacy of evening primrose oil for the treatment of premenstrual syndrome has been reviewed.[32,34] Seven placebo-controlled trials were identified, but only five of the trials were randomized. Five of the seven trials reported positive improvements in the symptoms of PMS. However, two trials failed to show any beneficial effects of evening primrose oil on PMS.[27,29] A randomized, double blind, placebo-controlled crossover study assessed the efficacy of the seed oil in the treatment of 27 women with PMS and 22 symptom free controls.[27] The first menstrual cycle was used as a diagnostic assessment. For women with PMS, placebos were administered during the second cycle. This was followed by randomization to four cycles of treatment with the oil and four cycles of placebo, with a crossover after completion of the sixth cycle. However, no wash out period was used between treatments. Each subject received 12 capsules of the seed oil daily (500 mg each) or placebo. Treatment with the seed oil did not reduce PMS symptoms or symptom cyclicity.[27] The second trial found no difference between the seed oil and placebo in a crossover study (after 3 cycles) involving 38 women.[29]

The clinical effects of the seed oil were investigated in an open study involving women with PMS.[35] The patients were treated with four capsules (500 mg each) of the seed oil twice daily for five cycles. A reduction in the scores of individual symptoms (i.e., irritability, swollen abdomen, breast discomfort, depression, anxiety, fatigue and edema), and total PMS scores was observed after one cycle, and improvements continued over all five cycles.[35] The clinical and biochemical effect of the seed oil was investigated in 30 women with severe, incapacitating PMS.[36] The women were treated with 3 g of the seed oil daily or placebo, beginning on day 15 of their cycle until the next menstrual period. Treatment with the seed oil alleviated the premenstrual symptoms in general as compared with placebo. No changes were found in the plasma levels of 6-keto-prostaglandin $F_{1\alpha}$ levels or follicle stimulating hormone, LH, prolactin, progesterone, estradiol or testosterone.[36]

The effects of the seed oil on mastalgia, one of the symptoms of premenstrual syndrome, was assessed in a randomized, double blind, crossover study.[20] Seventy-three women were treated with the seed oil or placebo for 3 months. In the seed oil treated group, both pain and tenderness were significantly reduced in both the cyclical and non-cyclical groups (p < 0.02–0.05). A placebo-controlled, double blind clinical trial assessed the efficacy of the seed oil for the treatment of cyclic breast symptoms and pain in 42 women.[37] Patients were treated with eight 500 mg capsules per day for 12 weeks. The seed oil was more effective than placebo for treatment of the following symptoms: nodularity, breast tenderness, feeling of well being, and irritability (p < 0.05).[37]

A review of the randomized trials and open studies in 291 women with severe persistent mastalgia was performed.[21] Patients were treated with bromocriptine (5 mg/day), danazol (200 mg/day) or the seed oil (6 × 500 mg/day) for 3–6 months. In those patients with cyclical mastalgia, good responses were obtained in 70% of patients treated with danazol, 47% of patients treated with bromocriptine and 45% of patients treated with the seed oil. The response rate in patients with non-cyclical mastalgia was 31%, 20% and 27%, respectively. Thirty-three percent of patients treated with bromocriptine reported adverse reactions, 22% with danazol and 2% with the seed oil.[21]

A review of 17 years of drug treatment for mastalgia in the Cardiff Mastalgia Clinic, Cardiff, UK described the efficacy of the danazol (200 mg daily), bromocriptine (5 mg daily) and the seed oil (6 × 500 mg daily) in a total of 490 patients.[19] Of the 490 patients 324 with cyclical mastalgia and 90 women with non-cyclical mastalgia completed the study. Oral administration of danazol (79%) was more effective than either bromocriptine (54%) or the seed oil (58%), with bromocriptine and the seed oil being equally effective. However, 30% of patients treated with danazol reported significant adverse reactions; 35% of the patients treated with bromocriptine reported adverse reactions, and only 4% of the patients treated with the seed oil reported adverse reactions.[19]

*Miscellaneous Activity*

Oral administrations of evening primrose oil (20 g/person, enriched with Vitamin E) to 10 healthy or 9 diabetic male subjects for one week enhanced erythropoeisis in both groups.[38] Administration of the seed oil also changed the fatty acid profiles in diabetic and normal male subjects after one week of 20 grams per day.[38] Inhibition of platelet activating factor 4 and plasma β-thromboglobulin was also observed in these subjects.[38]

Dietary supplementation with the seed oil was associated with a clinical, neurophysiological, and quantitative sensory improvement in 22 patients with diabetic polyneuropathy.[17] In a preliminary trial involving twenty-two patients with diabetes, positive effects were reported on many neurological and neurophysiological end points in a well performed parallel double-blind study in 111 patients with mild diabetic neuropathy.[18]

The effect of evening primrose oil on menopausal flushing was evaluated in a randomized, double blind, placebo controlled study.[31] Fifty-six women were treated with four capsules twice daily of evening primrose oil (500 mg, containing 10 mg natural vitamin E) or placebo for six months. Of the 35 women completing the study, no significant improvement in menopausal flushing was observed as compared with placebo.[31]

## Pharmacokinetics

The time courses of 8 fatty acids concentrations were measured in the serum after oral administration of evening primrose oil to 6 healthy volunteers.[39] Six capsules of the seed oil were administered in the morning and evening and the serum concentrations of fatty acids were profiled for 24 hours. The fatty acid concentrations in the serum were determined as their methyl esters by gas chromatography mass spectrometry. After administration of the seed oil, γ-linolenic acid showed an absorption-elimination pattern, and its area under the curve at 24 hours and $C_{max}$ were significantly increased over baseline values. After the evening dose, the half-life was shorter (2.7 h) than after the morning administration (4.4 h). The serum levels of dihomo-γ-linolenic acid and arachidonic acid did not increase after administration of the seed oil.[39]

## Mechanism of Action

The efficacy of evening primrose oil is attributed to the essential fatty acid content of the oil and the involvement of these fatty acids in the biosynthesis of prostaglandins.[28] Both γ-linolenic acid and dihomo-γ-linolenic acid are precursors to prostaglandins $PGE_1$ and $PGE_2$. Prostaglandin $PGE_1$ has anti-inflammatory, vasodilatory, and platelet aggregation inhibitory properties. Subcutaneous administration of the seed oil to rats (4 mg/kg) suppressed adjuvant-induced arthritis when administered on days 1–15 after adjuvant injection.[40] The effect of enhanced prostaglandin biosynthesis by dietary supplementation of evening primrose oil on vascular reactivity to renin and angiotensin II was determined *in vivo*.[41] Rats were treated with the seed oil (1 ml/day for 3 months), while the control group received olive oil. Treatment with the seed oil diminished vascular reactivity to renin and angiotensin II and increased the formation of vascular prostacyclin-like activity (p < 0.05).[41]

External application of a cream containing 10% of evening primrose oil to the skin of female pigs enhanced cell proliferation when used at a concentration of 10%.[42] There was an increase in the size of the rete pegs in the epidermis after 6 weeks of treatment, however this did not translate into an increase in the total thickness of the viable epidermis.[42]

Administration of the seed oil to rats (5 ml/kg) inhibited adenosine diphosphate-induced platelet aggregation *ex vivo*.[41] However, no effect on adenosine diphosphate-induced platelet aggregation was observed in rats fed a diet containing the seed oil (100 g/kg diet).[43] Administration of evening primrose oil to rats, as 10% w/w of diet, corrected a decrease in conduction velocity, but did not prolong hypoxic time to conduction failure, in animals with streptozotocin-induced diabetes after one month of treatment.[44] Capillary density of the endoneurium increased in the treatment group as well. Administration of flurbiprofen, a cyclooxygenase inhibitor, reduced the effect of evening primrose oil.[44] Dietary supplementation of the seed oil (5% w/w of diet) to streptozotocin-induced diabetic rats prevented the development of motor nerve conduction velocity deficit without affecting the levels of nerve sorbitol, fructose and myo-inositol or the deficit in axonal transport of substance P.[45] Administration of the seed oil (1 g/kg, gavage) to male rats improved conduction velocity in the sciatic nerve and increased sciatic endoneural blood flow.[46] Treatment of pregnant rats with ethanol and the seed oil (0.6 ml, gavage) on days 4 through 8 of gestation led to a significant reduction in the embryotoxic activity of ethanol.[47]

## Safety Information

### A.  Adverse Reactions
Headaches, nausea, production of a soft stool and diarrhea have been reported.[1] Oral administration of the seed oil precipitated symptoms of undiagnosed temporal lobe epilepsy in schizophrenic patients taking epileptogenic drugs, in particular phenothiazines.[48,49]

### B.  Contraindications and Warnings
Traditional applications of the seed oil include its use as an emmenogogue and as a means of inducing labor.[50] A retrospective study of 108 pregnant woman investigated the effect of oral administration of evening primrose oil on the length of pregnancy and selected intrapartum outcomes in low-risk nulliparous women.[51] The results showed that evening primrose ingestion did not shorten gestation or decrease the overall length of labor. However, evening primrose use was associated with an increase in the incidence of prolonged rupture of membranes, oxytocin augmentation, arrest of decent and vacuum extraction.[51] Therefore, evening primrose oil should be administered during pregnancy only on the advice of a physician.

Oral administration of evening primrose oil may precipitate symptoms of undiagnosed temporal lobe epilepsy particularly in schizophrenic patients and/or in subjects receiving epileptogenic drugs such as phenothiazines.[48,49]

### C.  Drug Interactions
Evening primrose oil has been shown to inhibit platelet aggregation in vivo,[41,52] and inhibits platelet-activating factor in human studies.[38] Therefore, patients taking anticoagulant drugs in conjunction with the seed oil should be closely monitored for bleeding times.

### D.  Toxicology
In clinical trials evening primrose oil is well tolerated, with few side effects. Animal studies have indicated that the seed oil is not teratogenic.[2]

### E.  Dose and Dosage Forms
Daily dose: Expressed as gamma-linolenic acid: 320–480 mg in divided doses for atopic eczema and 240–320 mg in divided doses for mastalgia.[3]

## References

1   Briggs CJ. Evening primrose. *Canadian Pharmaceutical Journal*, 1986;119:249–254.
2   Anon. *WHO Monographs on Selected Medicinal Plants*, Volume II, Oleum Oenotherae Biennis, World Health Organization, Geneva, Switzerland, in press.
3   *Martindale, The Extra Pharmacopoeia*, 30th ed., JEF Reynolds, ed., London, The Pharmaceutical Press, 1996.
4   Bamford JTM et al. Atopic eczema unresponsive to evening primrose oil (linoleic and gamma linolenic acids). *J Amer Acad Dermatol*. 1985;13:959–965.
5   Berth-Jones J, Graham-Brown RAC. Placebo-controlled trial of essential fatty acid supplementation in atopic dermatitis. *Lancet*. 1993;341:1557–1560.
6   Biagi PL et al. A long-term study on the use of evening primrose oil (Efamol) in atopic children. *Drugs Exper Clin Res*. 1988;14:285–290.
7   Biagi PL et al. The effect of gamma-linolenic acid on clinical status, red cell fatty acid composition and membrane microviscosity in infants with atopic dermatitis. *Drugs Exper Clin Res*. 1994;20:77–84.
8   Bordoni A et al. Evening primrose oil (Efamol) in the treatment of children with atopic eczema. *Drugs Exper Clin Res*. 1987;14:291–297.
9   Hederos CA, Berg A. Epogam evening primrose oil treatment in atopic dermatitis and asthma. *Archives of Disease in Childhood*. 1996;75:494–497.
10  Gehring W, Bopp R, Rippke F, Gloor M. Effect of topically applied evening primrose oil on epidermal barrier function in atopic dermatitis as a function of vehicle. *Arzneimittelforschung*. 1999;49:635–642.
11  Morse PF et al. Meta-analysis of placebo-controlled studies of the efficacy of Epogam in the treatment of atopic eczema. Relationship between plasma essential fatty acid changes and clinical response. *Brit J Dermatol*. 1989;121:75–90.
12  Stewart JCM et al. Treatment of severe and moderately severe atopic dermatitis with evening primrose oil (Epogam), a multicenter study. *Journal of Nutritional Medicine*. 1991;2:9–15.
13  Oliwiecki S, Burton JL. Evening primrose oil and marine oil in the treatment of psoriasis. *Clin Exp Dermatol*. 1994;19:127–129.
14  Whitaker DK et al. Evening primrose oil (Epogam) in the treatment of chronic hand dermatitis: disappointing therapeutic results. *Dermatology*. 1996;193:115–120.
15  Wright S, Burton JL. Oral evening primrose seed oil improves atopic eczema. *Lancet*. 1982;1120–1122.
16  Yoshimoto-Furuie K et al. Effects of oral supplementation with evening primrose oil for six weeks on plasma fatty acids and uremic skin symptoms in hemodialysis patients. *Nephron*. 1999;81:151–159.
17  Jamal GA et al. The effect of gamma linolenic acid on human diabetic peripheral neuropathy: a double-blind placebo-controlled trial. *Diabetic Medicine*. 1990;7:319–323.
18  Gamma-Linolenic acid multicenter trial group. Treatment of diabetic neuropathy with gamma-linolenic acid. *Diabetes Care*. 1993;16:8–15.
19  Gateley CA et al. Drug treatments for mastalgia: 17 years experience in the Cardiff mastalgia clinic. *J Royal Soc Med*. 1992;85:12–15.
20  Pashby NL et al. A clinical trial of evening primrose oil in mastalgia. *Brit J Surg*. 1981;68:801–824.
21  Pye JK et al. Clinical experience of drug treatments for mastaglia. *Lancet*. 1985;373–376.
22  Belch JJF et al. Effects of altering dietary essential fatty acids on requirements for non-steroidal anti-inflammatory drugs in patients with rheumatoid arthritis: a double-blind placebo controlled study. *Annals of Rheumatic Diseases*. 1988;47:96–104.

23   Brzeski M et al. Evening primrose oil in patients with rheumatoid arthritis and side effects of non-steroidal anti-inflammatory drugs. *Brit J Rheumatol.* 1991;30:370–372.

24   Joe LA, Hart LL. Evening primrose oil in rheumatoid arthritis. *Annals of Pharmacotherapy.* 1993;27:1475–1477.

25   Leventhal LJ et al. Treatment of rheumatoid arthritis with gamma-linolenic acid. *Annals Int Med.* 1993;119:119:867–873.

26   Brush MG. Evening primrose oil in the treatment of premenstrual syndrome. In: DF Horrobin, ed. *Clinical uses of essential fatty acids.* Montreal, Eden Press, 1983.

27   Collins A et al. Essential fatty acids in the treatment of premenstrual syndrome. *Obstetrics and Gynecology.* 1993;81:93–98.

28   Horrobin DF. The role of essential fatty acids and prostaglandins in the premenstrual syndrome. *J Reprod Med.* 1983;28:465–468.

29   Khoo SK et al. Evening primrose oil and treatment of premenstrual syndrome. *Med J Australia.* 1990;153:189–192.

30   O'Brian PMS et al. Premenstrual syndrome: clinical studies on essential fatty acids. In: Horrobin DF, ed. *Omega-6-essential fatty acids. Pathophysiology and Roles in Clinical Medicine.* New York, Wiley-Liss 1990:523–545.

31   Chenoy R et al. Effect of oral gamolenic acid from evening primrose oil on menopausal flushing. *Brit Med J.* 1994;308:501–503.

32   Kleijnen J. Evening primrose oil. *Brit Med J.* 1994;309:824–825.

33   Jantti J et al. Evening primrose oil and olive oil in the treatment of rheumatoid arthritis. *Clinical Rheumatol.* 1989;8:238–244.

34   Budeiri D et al. Is evening primrose oil of value in the treatment of premenstrual syndrome? *Controlled Clinical Trials.* 1996;17:60–68.

35   Larsson B et al. Evening primrose oil in the treatment of premenstrual syndrome. *Curr Ther Res.* 1989;46:58–63.

36   Poulakka J et al. Biochemical and clinical effects of treating the premenstrual syndrome with prostaglandin synthesis precursors. *J Reprod Med.* 1985;30:149–153.

37   Mansel RE et al. The use of evening primrose in mastalgia. In: DF Horrobin, ed. *Clinical uses of essential fatty acids.* Montreal, Eden Press, 1983.

38   Van Doormaal JJ et al. Effects of short-term high dose intake of evening primrose oil on plasma and cellular fatty acid composition, alpha-tocopherol levels, and erythropoiesis in normal and type 1 (insulin-dependent) diabetic men. *Diabetologia.* 1988;31:576–584.

39   Martens-Lobenhoffer J, Meyer FP. Pharmacokinetic data of gamma-linolenic acid in healthy volunteers after the administration of evening primrose oil (Epogam). *International Journal of Clinical Pharmacology and Therapeutics.* 1998;36:363–366.

40   Delbarre F, De Gery A. Immunomoderated effect of lipids from *Oenothera* seed extracts on adjuvant polyarthritis in the rat. *Rhumatologie.* 1980;10:361–363.

41   Schölkens BA et al. Evening primrose oil, a dietary prostaglandin precursor, diminishes vascular reactivity to renin and angiotensin II in rats. *Prostaglandins, Leukotrienes and Medicine.* 1982;8:273–285.

42   Morris GM et al. Modulation of the cell kinetics of pig skin by the topical application of evening primrose oil or lioxasol. *Cell Proliferation.* 1997;30:311–323.

43   Sugano M et al. Influence of Korean pine (*Pinus koraiensis*) seed oil containing *cis*-5, *cis*-9, *cis*-12-octadecatrienoic acid on polyunsaturated fatty acid metabolism, eicosanoid production and blood pressure of rats. *Brit J Nutr.* 1994;72:775–783.

44   Cameron NE et al. The effects of evening primrose oil on nerve function and capillarization in streptozotocin-diabetic rats: modulation by the cyclo-oxygenase inhibitor flurbiprofen. *Brit J Pharmacol.* 1993;109:972–979.

45    Tomlinson DR et al. Essential fatty acid treatment-effects on nerve conduction, polyol pathway and axonal transport in streptozotocin diabetic rats. *Diabetologia.* 1989;32:655–659.

46    Dines KC et al. Comparison of the effects of evening primrose oil and triglycerides containing gamma-linolenic acid on nerve conduction and blood flow in diabetic rats. *J Pharmacol Exp Ther.* 1995;273:49–55.

47    Varma PK et al. Protection against ethanol-induced embryonic damage by administering gamma-linolenic and linoleic acids. *Prostaglandins, Leukotrienes and Medicine.* 1982;8:641–645.

48    Holman CP et al. A trial of evening primrose oil in the treatment of chronic schizophrenia. *J Orthomol Psych.* 1983;12:302–304.

49    Vaddadi KS. The use of gamma-linolenic acid and linoleic acid to differentiate between temporal lobe epilepsy and schizophrenia. *Prostaglandins and Medicine.* 1981;6:375–379.

50    Farnsworth NR, ed., *Oenothera biennis.* Napralert database, University of Illinois at Chicago (an on-line database available directly through the University of Illinois at Chicago or through the Scientific and Technical Network of Chemical Abstracts) 1998.

51    Dove D, Johnson P. Oral evening primrose oil: its effect on length of pregnancy and selected intrapartum outcomes in low-risk nulliparous women. *J Nurse Midwifery.* 1999;44:320–324.

52    De La Cruz JP. Effect of evening primrose oil on platelet aggregation in rabbits fed an atherogenic diet. *Thrombosis Res.* 1997;87:141–149.

# 11

# Feverfew

## Synopsis

Results from clinical trials indicate that dried feverfew (*Tanacetum parthenium*) leaves (not extracts) may be used for the prophylaxis of migraine headaches. Placebo-controlled clinical trials have demonstrated that oral administration of feverfew reduces the frequency and severity of migraine attacks, as well as the degree of nausea and vomiting. Patients should be directed to use dried encapsulated feverfew leaf products and not feverfew leaf extracts, as the extracts (even standardized extracts) are not effective, as shown in clinical trials. The recommended dosage form is an encapsulated fresh or dried leaf preparation, with a daily dose equivalent to 0.2–0.6 mg of parthenolide. Treatment for 1–3 months may be required before the therapeutic results are apparent. Administration of feverfew is contraindicated during pregnancy or nursing, in children under the age of 12 years, and in those patients with an allergy to plants in the daisy family. No drug interactions have been reported in the medical literature. Adverse reactions include dizziness, heartburn, indigestion, inflammation of the mouth and tongue, accompanied by swelling of the lips; loss of taste, mouth ulceration; and weight gain. Mouth ulceration, while rare, appears to be due to a systemic reaction and requires discontinuation of feverfew administration.

## Introduction

Feverfew (*Tanacetum parthenium*) is a perennial flowering plant with numerous, small daisy-like heads of yellow flowers with outer white petals. The plant is a member of the Asteraceae (Compositae), which is also commonly referred to as the daisy family.[1] Feverfew leaves have a very characteristic pungent camphorous smell and bitter taste.[1] Since the time of Dioscorides (50 AD), feverfew preparations have been used as a self-medication for the prevention of migraine, as well as for the treatment of other types of headaches. In fact, in keeping with its medical use, feverfew has been called "the aspirin of the 18th century".[2]

Traditionally, feverfew leaves were also administered for the treatment of fevers, stomach and toothaches, insect bites, arthritis and asthma, as well as to regulate menstruation.[1] However, there is little scientific evidence to support any of these latter traditional uses.[1] Since the 1980s the focus of scientific research has been directed toward the use of feverfew for the prevention and treatment of migraine headaches, and for the treatment of rheumatoid arthritis. Over the past ten years there has been resurgence in the use of feverfew products for the treatment of headache, particularly in the United States, Canada and the United Kingdom.[1] Currently, feverfew is regulated as a dietary supplement in the United States, and has over-the-counter drug status in Canada.

## Quality Information

- The correct Latin name for the plant is *Tanacetum parthenium* L. Schultz Bip. (Asteraceae).[1] Botanical synonyms found in the scientific literature include *Chrysanthemum parthenium* (L.) Bernh., *Leucanthemum parthenium* (L.) Gren & Gordon, *Matricaria eximia* Hort., *M. parthenium* L., *Pyrethrum parthenium* (L.) Sm.[1] Common names include: acetilla, altamisa mexicana, altamza, amargosa, bachelor's buttons, boulet, bouton d'argent, camamieri, camomilla, camoumida, featherfew, featherfoil, feather-fully, febrifuge plant, feverfew, feverfew tansy, flirtwort, hierba Santa Maria, grande camomille, manzanilla, matricaria, matricaria comum, midsummer daisy, Moederkruid, Mutterkraut, natsushirogiku, Santa Maria, varadika, and vettervoo.[1]
- Commercial products of feverfew are prepared from dried leaves,[1] or aerial parts of *Tanacetum parthenium* (L.) Schultz Bip. (Asteraceae).[1,5]
- Feverfew is indigenous to Southeast Europe to Caucasius, but also commonly found in Europe and the USA.[1]
- The major chemical constituents in feverfew include the germacranolide sesquiterpene lactone, parthenolide, which occurs at a concentration of up to 0.9%.[1] Parthenolide, and other characteristic sesquiterpene lactones, including members of the guaianolides (e.g canin, artecanin), contain an $\alpha$-methylenebutyrolactone structure. To date, more than 45 sesquiterpenes have been reported from the leaves and stems of feverfew, and monoterpenes, flavonoids and polyacetylenes have also been isolated.[1]
- Daily dosage: encapsulated fresh or dried leaf preparations, equivalent to 0.2–0.6 mg of parthenolide (as a chemical marker).[3,6–9]

## Medical Uses

Placebo-controlled clinical trials support the use of dried feverfew leaf products for the prophylaxis of migraine headache.[4,8–11] [Although the herb has also been recommended for treatment of rheumatoid arthritis, a clinical study failed to prove any beneficial effects of the herb in the treatment of this disease].[12]

## Summary of Clinical Evidence

To date, at least five randomized, double-blind, placebo-controlled clinical trials have assessed the efficacy of various feverfew products for the prophylaxis and treatment of migraine headaches.[7-9,13,14] Three of the clinical trials were performed with an encapsulated dried or freeze-dried leaf product.[7-9] The fourth clinical trial assessed the effects of a 90% ethanol extract of a commercial feverfew sample bound to microcrystalline cellulose.[13] The fifth study was published only as an abstract, and the feverfew preparation used was not defined.[14]

In the trial by Johnson and co-workers,[7] 17 subjects suffering from migraines who had been treating themselves with feverfew for at least 3 months were recruited for a 6-month trial. Patients were administered an oral dose of 50 mg/day (concentration of parthenolide was not stated) of the freeze-dried leaf preparation or placebo. The average frequency of migraine headaches in the treatment group was 1.69 over the 6-month treatment period and 1.5 during the final 3 months of the study, as compared with 3.13 and 3.43, respectively in the placebo group. Nausea and vomiting were reported on 39 occasions in the treatment group as compared with 116 occasions in the placebo groups. The results were statistically significant ($p < 0.05$).[7]

A randomized, double-blind, placebo-controlled crossover trial assessed the effects of a feverfew leaf product for migraine prevention.[8] Fifty-nine patients suffering from migraine headaches were treated with an encapsulated feverfew product containing 70–114 mg of dried leaf per day (equivalent to 0.545 mg of parthenolide) or placebo after a 1-month placebo run-in. Patients received the dried leaf for 4 months and then placebo for 4 months in a crossover design. During the treatment phase of the study, a 24% decrease in the number of migraine attacks was observed in the treatment group as compared with placebo. No change in the duration of the migraine attacks or the number of attacks associated with an aura was observed in the treatment group, however a significant reduction in the nausea and vomiting associated with the attacks was reported ($p < 0.02$). Global assessments of efficacy also demonstrated that the dried leaf product was significantly superior to placebo in prevention of migraines ($p < 0.0001$).[8]

A controlled clinical trial assessed the efficacy of a dried feverfew leaf product in the prophylaxis of migraine headache.[9] Migraine severity was determined using a numerical self-assessment pain scale and linked symptoms were recorded on a numerical analogue scale. Fifty-seven patients (47 women and 10 men) were divided into two groups for phase one of the trial, which was an open label study. Patients were treated with 100 mg of the encapsulated dried leaf preparations (standardized to 0.2% parthenolide) for 60 days. After phase one was completed, phases 2 and 3 were designed as a randomized, double-blind, placebo-controlled, crossover study. Patients were randomized into two groups, group A (n = 30) continued to receive 100 mg of the leaf preparation and group B (n = 27) received a placebo. After 30 days of treatment, the group A received placebo and group B received the leaf preparation for a further 30 days. No washout period was allowed between experimental phases. Results of the open-labeled phase showed a statistically significant reduction in pain intensity of the migraine attacks and symptoms such as vomiting or sensitivity to light or noise was noted ($P < 0.001$). In the double-blind phase of the study, patients in the treatment group reported a decrease in the pain intensity of migraines, while patients in the placebo group noted an increase in pain intensity ($p < 0.01$). Similar results were reported after the crossover procedure occurred. In the double-blind phase of the study, a statistically significant decrease in vomiting and sensitivity was observed in the treatment group as compared with placebo ($p < 0.001$ and $p < 0.017$).[9]

A randomized, double-blind, placebo-controlled crossover study assessed the efficacy of a 90% dried ethanol extract of the leaves and stems of feverfew bound to microcrystalline cellulose (143 mg standardized to contain 0.5 mg parthenolide) with placebo for the prophylaxis of migraine headache in 44 patients.[13] Diagnosis was performed in accordance to the International Headache Society diagnostic criteria. After a one-month washout period, patients were treated with the extract or placebo for 4 months, then crossed over to placebo (or vice versa) for 4 months. The average response to the two treatments was the same and the extract did not have any effect on migraine prophylaxis. Statistical significance was not reported.[13]

A systematic review assessed the five randomized placebo-controlled clinical trails to determine the clinical effectiveness of feverfew in the prevention of migraine.[11] Two independent reviewers evaluated each of the five trials, and the data was extracted in a predefined, standardized fashion. Each of the clinical trials was scored from using the Jadad scoring system.[15] The study concluded that the data favored feverfew over placebo for the prevention of migraine, however the shortcomings of the clinical studies assessed, such as small sample size, no washout period and poor definition of inclusion criteria, prevented a firm conclusion.[11]

An open study demonstrated that platelet aggregation in 10 patients taking preparations of the herb for 3.5 to 8 years were the same as a control group of 4 patients who had stopped taking the herb for at least 6 months prior to being tested.[16] A double-blind placebo-controlled clinical trial assessed the efficacy of the herb for the treatment of arthritis. Forty women with rheumatoid arthritis were treated with either 70–86 mg of dried leaf or placebo for six weeks. No beneficial effects were observed.[12]

## Mechanism of Action

The mechanism of action of feverfew in the prophylaxis of migraine is currently a matter of debate.[3,4] However, based on pharmacological investigations of the extracts, and one of its chemical constituent's parthenolide, the mechanism appears to involve anti-inflammatory effects, inhibition of platelet aggregation, and the inhibition of serotonin reuptake.

Various extracts of feverfew, or pure parthenolide, inhibit the biosynthesis of prostaglandins, leukotrienes and thromboxanes, which are potent mediators of inflammation. An aqueous extract of feverfew (50 μg/ml) inhibited the activity of lipoxygenase *in vitro*, thereby reducing the biosynthesis of prostaglandins and thromboxane B-2 in rat leukocytes.[17] A chloroform extract of the leaves inhibits N-formyl-methionyl-leucyl-phenylalanine-, or calcium ionophore A23187-induced leukotriene B-4 and thromboxane B-2 synthesis in human and rat leukocytes *in vitro* ($IC_{50} < 50$ μg/ml).[18] An extract of the dried leaves prepared with a Tris-HCl buffer (pH 7.4) inhibited the activity of phospholipase $A_2$ (30 μl, 1:20 extract), the enzyme responsible for facilitating the release of arachidonic acid from the phospholipid cell membrane.[19] The extract further prevented both the release of arachidonic acid and the formation of thromboxane in human blood platelets *in vitro*.[19,20]

Chloroform extracts of the dried leaves inhibit anti-IgE- or calcium ionophore A23187-induced histamine release in rat peritoneal mast cells *in vitro*.[21] Parthenolide inhibited the expression of cyclooxygenase (COX) and the proinflammatory cytokines, tumor necrosis factor-alpha and interleukin-1, in lipopolysaccharide (LPS)-stimulated murine macrophages *in vitro*.[22] Parthenolide further suppressed LPS-stimulated protein tyrosine phosphorylation in the macrophages that correlated with its inhibitory effect on the expression of COX and the cytokines.[22] A chloroform-methanol extract (1:3) of the dried leaves inhibited the release of vitamin $B_{12}$-binding protein from polymorphonuclear leukocytes induced by N-formyl-methionyl-leucyl-phenylalanine or sodium arachidonate.[23] An acetone, chloroform or saline extract of the leaves inhibited phorbol 12-myristate 13-acetate-induced oxidative burst in human polymorphonuclear leukocytes *in vitro* ($IC_{50}$ 0.79 mg/ml).[24,25]

Feverfew extracts and the constituent sesquiterpene lactones, have been shown to inhibit platelet aggregation and serotonin release from blood platelets in response to various chemical stimuli.[3,26–28] Acetone, chloroform or chloroform-methanol (1:3) extracts of the leaves inhibited arachidonic acid-, adenosine diphosphate, collagen-, and adrenalin-induced serotonin release from human blood platelets and polymorphonuclear leucocytes *in vitro*.[23,29,30] The chloroform-methanol extract however, did not inhibit serotonin release from platelets or polymorphonucleocytes induced by the calcium ionophore A23184.[23] A 95% ethanol extract of the leaves inhibited serotonin release from bovine platelets *in vitro* ($IC_{50}$ 1.3–2.9 mg/ml).[31] The ability of the freeze-dried or air-dried leaf extracts to inhibit serotonin release from human blood platelets has been correlated with the concentration of parthenolide in the extract.[3,32]

Current evidence also suggests that serotonin receptor-based mechanisms may be involved in the etiology of migraine headache.[33] *In vitro* studies have demonstrated that parthenolide displaces radioligand binding from serotonin receptors isolated from rat and rabbit brains, as well as cloned serotonin receptors, indicating that parthenolide may be a low-affinity antagonist.[33]

An aqueous, chloroform or chloroform-methanol (1:3) extract of the leaves inhibits human platelet aggregation induced by arachidonic acid-, collagen-, or adrenalin-stimulation, at a concentration up to 100 μl.[19,23,30,34] A chloroform extract of the fresh leaf completely inhibited human blood platelet function *in vitro*.[35] After fractionation of the extract, only those fractions containing constituents with an α-methylenebutyrolactone functional group were active. Parthenolide was the most active, while canin, tanaparthin-α-peroxide, and *cis*-cycloheptane lactone ester were partially active.[35] While the exact mechanism of platelet aggregation inhibition is unknown, it has been suggested that their ability to undergo Michael additions with thiol groups may influence their biological activity. The following experimental data support this hypothesis: addition of cysteine or 2-mercaptoethanol to the crude extract or to pure parthenolide completely suppressed their ability to inhibit platelet aggregation. Furthermore, the inhibitory effects are both dose and time-dependent, and treatment of the platelets with the extract or parthenolide caused a dramatic reduction in the number of thiol groups present.[27,28,35]

The antispasmodic activity of feverfew may also be partly responsible for the biological activity. A chloroform extract of the fresh leaves, containing parthenolide and other sesquiterpene lactones, inhibited the contractile response to exogenously applied agonists (angiotensin, phenylephrine, serotonin, thromboxane mimetic U48819 or thromboxane $A_2$) in isolated rings of rabbit aorta.[36,37] However, the chloroform extract of the dried leaves, which did not contain parthenolide or other sesquiterpene lactones, was not active.[37]

## Pharmacokinetics

No information was found.

## Safety Information

### A. Adverse Reactions

Dizziness, heartburn, indigestion, inflammation of the mouth and tongue, accompanied by swelling of the lips; loss of taste, mouth ulceration; and weight gain have been reported.[7,8,38] In a survey of 300 feverfew users, 11.3% of users reported mouth ulceration (aphthous ulcers) from chewing fresh leaves, which disappear after discontinuation of the drug and appear to be systemic in origin.[6,7] Mouth ulceration is a systemic reaction to the herb, and requires discontinuation of the product. Inflammation of the mouth and tongue with swelling of the lips appears to be a local reaction that may be overcome by using encapsulated herb products. Abdominal bloating, heart palpitations, constipation, diarrhea, flatulence, heavier menstrual flow, nausea and skin rashes have also been reported to a lesser degree.[7,8,13] Allergic reactions, such as contact dermatitis, have also been reported.[38] A cross-sensitivity between *Parthenium hysterophorus* (American feverfew) and ragweed pollen allergens has been reported.[39]

In one study, approximately 10% of patients who were switched to placebo after taking feverfew for several years experienced "post-feverfew syndrome" which involved a cluster of nervous system reactions such as rebound migraine headaches, anxiety, poor sleep patterns and muscle and joint stiffness.[7]

## B. Contraindications

Due to the traditional use of feverfew preparations as an emmenagoge, its should not be administered during pregnancy. Due to a lack of safety and efficacy information in children, feverfew products should not be administered during nursing or to children under the age of 2 years old.[1] Feverfew preparations should not be administered to patients with an allergy to plants of the Asteraceae.[1]

## C. Drug Interactions

No drug interactions have been reported.

## D. Toxicology

No significant differences were observed in the frequency of chromosomal aberrations and sister chromatid exchanges in lymphocytes, or in the mutagenicity of urine, in tests comparing chronic feverfew users with non-users.[40] An *in vitro* study demonstrated that an extract of the herb or parthenolide were cytotoxic to mitogen-induced human peripheral blood mononuclear cells and interleukin 1-stimulated synovial cells.[41] Parthenolide-induced cytotoxicity was due to the inhibition of thymidine incorporation into deoxyribonucleic acid.[42,43] Intragastric administration of 100 times the normal daily dose of powdered leaf to rats did not result in loss of appetite or weight.[38]

## E. Dose and Dosage Forms

Daily dosage: encapsulated fresh or dried leaf preparations, equivalent to 0.2–0.6 mg of parthenolide (as a chemical marker).[1,7–9]

## References

1   Mahady GB, Fong HHS, Farnsworth NR. Herba Tanaceti Parthenii. *WHO Monographs on Selected Medicinal Plants*, World Health Organization, Traditional Medicine Programme, Geneva, Switzerland, 1999.

2   Berry MI. Feverfew faces the future. *Pharm J*. 1984;611–614.

3   Heptinstall S, Awang D. Feverfew: A review of its history, its biology and medicinal properties, and the status of commercial preparations of the herb. In: Bauer R, Lawson L (eds.). ACS symposium series, *Phytomedicines of Europe, chemistry and biological activity*. Washington, American Chemical Society, 1998;158–175.

4   Awang D. Prescribing therapeutic feverfew (*Tanacetum parthenium* (L.) Schultz Bip. Syn. *Chrysanthemum parthenium* (L.) Bernh.). *Integrative Med*. 1998;1:11–13.

5   *United States Pharmacopoeia National Formulary*: Feverfew. Rockville, MD, The United States Pharmacopeial Convention, Inc., 1998.

6   Awang DVC. Parthenocide: The demise of a facile theory of feverfew activity. *Journal of Herbs, Spices and Medicinal Plants*. 1998;5:95–98.

7   Johnson ES et al. Efficacy of feverfew as prophylactic treatment of migraine. *Brit Med J*. 1985;291:569–573.

8   Murphy JJ et al. Randomized double-blind placebo-controlled trial of feverfew in migraine prevention. *Lancet*. 1988;189–192.

9       Palevitch D et al. Fever (*Tanacetum parthenium*) as a prophylactic treatment for migraine: a double-blind placebo-controlled study. *Phytother Res.* 1997;11:508–511.

10      Hylands DM et al. Efficacy of feverfew as prophylactic treatment of migraine, reply. *Brit Med J.* 1985;291:1128.

11      Vogler BK et al. Feverfew as a preventative treatment for migraine: a systematic review. *Cephalagia.* 1998;18:704–708.

12      Pattrick M et al. Feverfew in rheumatoid arthritis: a double-blind, placebo-controlled study. *Ann Rheumatic Dis.* 1989;48:547–549.

13      De Weerdt CJ et al. Herbal medicines in migraine prevention: randomized double-blind placebo-controlled crossover trial of a feverfew preparation. *Phytomedicine.* 1996;3: 2250–230.

14      Kuritzky A et al. Feverfew in the treatment of migraine: its effect on serotonin uptake and platelet activity. *Neurology.* 1994;44(Suppl 2):293P (Abstract).

15      Jadad AR et al. Assessing the quality of reports of randomized clinical trials: is blinding necessary? *Controlled Clinical Trials.* 1996;17:1–12.

16      Biggs MJ et al. Platelet aggregation in patients using feverfew for migraine. *Lancet.* 1982;ii:776.

17      Capasso F. The effect of an aqueous extract of *Tanacetum parthenium* L. on arachidonic acid metabolism by rat peritoneal leukocytes. *J Pharm Pharmacol.* 1986;38:71–72.

18      Summer H et al. Inhibition of 5-lipoxygenase and cyclooxygenase in leukocytes by feverfew. Involvement of sesquiterpene lactones and other components. *J Pharm Pharmacol.* 1992;44:737–740.

19      Makheja AN, Bailey JM. A platelet phophoslipase inhibitor from the medicinal herb feverfew (*Tanacetum parthenium*). *Prostaglandins and Leukotrienes in Medicine.* 1982;8:653–660.

20      Jain MK, Jahagirdar DV. Action of phospholipase A-2 on bilayers. Effects of inhibitors. *Biochimica et Biophysica Acta.* 1985;814:319–326.

21      Hayes NA, Foreman JC. The activity of compounds extracted from feverfew on histamine release from rat mast cells. *J Pharm Pharmacol.* 1987;39:466–470.

22      Hwang D et al. Inhibition of the expression of inducible cyclooxygenase and proinflammatory cytokines by sesquiterpene lactones in macrophages correlates with the inhibition of MAP kinases. *Biochemical and Biophysical Research Communications.* 1996;226:810–818.

23      Heptinstall S et al. Extracts of feverfew inhibit granule secretion in blood platelets and polymorphonuclear leukocytes. *Lancet.* 1985;I:1071–1073.

24      Brown AMG et al. Pharmacological activity of feverfew (*Tanacetum parthenium* (L.) Schultz-Bip.): Assessment by inhibition of human polymorphonuclear leukocyte chemoluminescence *in vitro. J Pharm Pharmacol.* 1997a, 49:558–561.

25      Brown AMG et al. Effects of extracts of *Tanacetum* species on human polymorphonuclear leukocyte activity *in vitro. Phytother Res.* 1997b;11:479–484.

26      Groenewegen WA et al. Compounds extracted from feverfew that have anti-secretory activity contain an alpha-methylene butyrolactone unit. *J Pharm Pharmacol.* 1986;38:709–712.

27      Heptinstall S et al. Extracts of feverfew may inhibit platelet behavior via neutralization of sulphydryl groups. *J Pharm Pharmacol.* 1987;39:459–465.

28      Heptinstall S et al. Studies on feverfew and its mode of action. In: FC Rose, ed. *Advances in Headache Research.* London, John Libbey, 1987b:129–134.

29      Heptinstall S et al. Inhibition of platelet behavior by feverfew: a mechanism of action involving sulphydryl groups. *Folia haematologia* (Leipzig). 1988;115:447–449.

30    Groenewegen WA, Heptinstall S. A comparison of the effects of an extract of feverfew and parthenolide, a component of feverfew, on human platelet activity *in vitro*. *J Pharm Pharmacol*. 1990;42:553–557.

31    Marles RJ et al. A bioassay for inhibition of serotonin release from bovine platelets. *J Nat Prod*. 1992;55:1044–1056.

32    Heptinstall S et al. Parthenolide content and bioactivity of feverfew *(Tanacetum parthenium* (L.) Schultz-Bip.). Estimation of commercial and authenticated feverfew products. *J Pharm Pharmacol*. 1992;44:391–395.

33    Weber JT et al. Activity of parthenolide at 5HT$_{2A}$ receptors. *J Nat Prod*. 1997;60:651–653.

34    Loesche W et al. Effects of an extract of feverfew (*Tanacetum parthenium*) on arachidonic acid metabolism in human blood platelets. *Biomed Biochim Acta*. 1988;47:5241–5243.

35    Hewlett MJ et al. Sesquiterpene lactones from feverfew, *Tanacetum parthenium*: isolation, structural revision, activity against human blood platelet function and implication for migraine therapy. *J Chem Soc. Perkin Tansactions I*. 1996;16:1979–1986.

36    Barsby RWJ et al. Feverfew extracts and parthenolide irreversibly inhibit vascular responses of the rabbit aorta. *J Pharm Pharmacol*. 1992;44:737–740.

37    Barsby RWJ et al. Feverfew and vascular smooth muscle: extracts from fresh and dried plants show opposing pharmacological profiles, dependent upon sesquiterpene lactone content. *Planta Med*. 1993;59:20–25.

38    Hausen BM. Sesquiterpene lactones – *Tanacetum parthenium*. In: De Smet PAGM, Keller K, Hänsel R, Chandler RF. *Adverse Effects of Herbal Drugs 1*. Berlin, Springer-Verlag, 1994.

39    Sriramaras P et al. Allergenic cross-reactivity between *Parthenium* and ragweed pollens allergens. *International Archives of Allergy and Immunology*. 1993;100:79–85.

40    Anderson D et al. Chromosomal aberrations and sister chromatid exchanges in lymphocytes and urine mutagenicity of migraine patients: a comparison of chronic feverfew users and matched non-users. *Human Toxicol*. 1988;7:145–152.

41    O'Neill LAJ et al. Extracts of feverfew inhibit mitogen-induced human peripheral blood mononuclear cell proliferation and cytokine mediated responses: a cytotoxic effect. *Brit J Clin Pharmacol*. 1987;23:81–83.

42    Woynarowski JW et al. Induction of deoxyribonucleic acid damage in HeLa S-3 cells by cytotoxic and antitumor sesquiterpene lactones. *Biochem Pharmacol*. 1981a;30:3305–3307.

43    Woynarowski JW et al. Inhibition of DNA biosynthesis in HeLa cells by cytotoxic and antitumor sesquiterpene lactones. *Molecular Pharmacol*. 1981b;19:97–102.

# 12

# Garlic

## Synopsis

Clinical evidence suggests that garlic supplementation may reduce the risk factors associated with atherosclerosis and cardiovascular disease by inhibiting platelet aggregation, increasing fibrinolysis, reducing blood pressure and reducing serum lipid levels. The overall results from clinical trials, and meta-analyses of these trials, indicate that garlic may reduce total cholesterol levels by 6–10%, after 3–6 months of supplementation. Recent clinical trials have shown that the concomitant use of garlic supplements with a low cholesterol diet will not further reduce cholesterol levels. The prevention and treatment of mild hypertension or ischemic diseases with garlic supplements requires further clinical evaluation. Two cases of increased international normalized ratio results, attributed to the ingestion of garlic products, were reported in patients previously stabilized on warfarin. Adverse reactions associated with garlic consumption are mainly limited to contact dermatitis, gastrointestinal complaints and garlic odor. However, one case of spontaneous spinal epidural hematoma, secondary to a qualitative platelet disorder due to excessive garlic ingestion, has been reported. A few cases of post-operative bleeding after chronic ingestion of garlic have also been noted. Garlic products should be used with caution in patients receiving anticoagulant or antiplatelet therapy, or having bleeding disorders. Patients should avoid garlic supplementation for 10–14 days prior to any surgical procedure. The recommended daily dosage is 900 mg/day of garlic powder, 2–5 mg/day of garlic oil, aged garlic extract (7.2 g/day) or 2 to 4 g of fresh garlic.

## Introduction

Garlic (*Allium sativum*, Liliaceae), also known as "the spice of life", was one of the earliest documented examples of a food plant also used for the prevention and treatment of disease.[1-3] The plant is a perennial, erect bulbous herb, with the bulb, giving rise to a number of narrow, keeled, grass-like leaves above ground.[1] Botanical researchers believe that garlic originated in Central Asia,[1,3] however, its botanical name, *Allium sativum*, may have been derived from the Celtic word, "áll" which means "warm or pungent".[2,4] Currently, garlic is commercially cultivated in Argentina, China, Egypt, France, Hungary, India, Italy, Japan, Mexico, Spain, USA and the former Czechoslovakia.[4,5]

The medical history of garlic dates back at least 4000 years, where its therapeutic uses were described in the ancient Chinese, Indian and Sumerian literature. In 1,550 BC the importance of garlic in Egyptian medical practice was shown by the fact that the *Papyrus Ebers*, a famous Egyptian papyri, recorded over 800 medical formulas, 22 of which contained garlic for treatment of various ailments including body weakness, headaches and throat tumors.[2,4] In addition, cloves of garlic were often found among the ruins of the tombs of the ancient Egyptian Pharaohs, including, Tutankhamen.[4] During the 1st century, Pliney the Elder, a Roman naturalist, advocated garlic for the treatment of epilepsy, hoarseness, hemorrhoids, and tuberculosis. Discorides, a Roman physician, recommended garlic to clean the arteries, and Hippocrates (460–370 BC), the father of modern medicine, was known to prescribe garlic for a wide variety of ailments.[4] The therapeutic properties of garlic are also mentioned in the ancient holy writings of the Bible and the Talmud. In medieval Europe, during the 14–17th centuries, garlic was purported to instill immunity to the bubonic plague, and individual resistance to the plague was often attributed to its consumption.[2,3]

## Quality Information

*   The correct Latin name for garlic is *Allium sativum* L. (Liliaceae).[1,5,6] Botanical synonyms that may appear in the scientific literature include *Porvium sativum* Rehb.[1] Due to its wide spread use throughout the world, there are numerous vernacular (common) names for garlic. The following represent a small portion from many countries around the world: ail, ail common, ajo, akashneem, Allium, alubosa elewe, ayo-ishi, ayu, banlasun, camphor of the poor, dai tóan, dasuan, dawang, dra thiam, foom, gartenlauch, hom khaao, hom kía, hom thiam, hua thiam, kesumphin, kitunguu-sumu, Knoblauch, kra thiam, krathiam, krathiam cheen, lahsun, lai, lashun, lasuna, lauch, lay, Layi, lobha, majo, naharu, Nectar of the Gods, ninniku, Pa-se-waa, Poor Man's Treacle, rason, rasonam, rasun, rustic treacles, seer, stinking rose, sudulunu, ta-suam, tafanuwa, tellagada, vallaippundu, and velluli.[1–6]
*   Standardized extracts and other commercial products of garlic are prepared from the fresh or dried bulbs of *Allium sativum* L. (Liliaceae).[1,7,8]
*   The important chemical constituents of garlic bulbs are the organosulfur compounds.[1,9,10] Approximately 82% of the total sulfur content of a garlic bulb are comprised of the cysteine sulfoxides (e.g., alliin) and the non-volatile γ-glutamylcysteine peptides. The thiosulfinates (e.g., allicin), ajoenes (e.g., *E*-ajoene, *Z*-ajoene), vinyldithiins (e.g., 2-vinyl-(4H)–1,3-dithiin, 3-vinyl-(4H)-1,2-dithiin), and sulfides (e.g., diallyl disulfide, diallyl trisulfide), however, are not naturally occurring compounds. These compounds are degradation products

that are produced from the naturally occurring cysteine sulfoxide, alliin. When a garlic bulb is crushed, minced, or otherwise processed, the compartmentalized alliin comes in contact with the enzyme, alliinase, from the adjacent vacuoles, resulting in hydrolysis and immediate condensation of the reactive intermediate (allylsulfenic acid) to form allicin. Allicin is an unstable compound, and will undergo additional reactions to form other derivatives, depending on environmental/processing conditions.[1,9,10] Analysis of various commercial garlic products shows the varying sulfur chemical profiles which are reflective of the processing procedure. For example, processed bulb or dried garlic bulb powder products contain mainly alliin, and allicin. While the volatile oil, contains almost entirely diallyl sulfide, diallyl disulfide, diallyl trisulfide, and diallyl tetrasulfide. Oil macerates, on the other hand, contain mainly 2-vinyl-[4H]–1,3-dithiin, 3-vinyl-[4H]–1,3-dithiin, *cis*-ajoene, and *trans*-ajoene.[9,11–15]

## Medical Uses

Over 26 clinical trials, and three meta-analyses support the use of garlic and garlic preparations as an adjunct to dietetic management of hyperlipidemia, and for the prevention of atherosclerotic (age associated) vascular changes.[16–29] However, results from recent clinical trials, that included a low cholesterol diet run-in for 2–4 weeks prior to garlic treatments, failed to show any benefits of garlic supplementation, when used in conjunction with a cholesterol lowering diet.[22–26,28]

Results from clinical trials have further indicated that garlic may be of some benefit in the dietary management of mild hypertension[16,19] and for the treatment of patients with increased risk of juvenile ischemic attack.[18–20] Other medical uses claimed for garlic include treatment of asthma, bronchitis, dyspepsia, fever, upper respiratory and lower urinary tract infections, ringworm, and rheumatism.[1,3–5] However, there are no clinical data to support these claims.

## Summary of Clinical Evidence

Early human studies (1969–1980) have suggested that the consumption of garlic or garlic products is inversely related to the incidence and mortality of coronary heart disease.[20,30–32] As determined by epidemiological studies, cardiovascular disease is correlated with atherosclerosis, of which hypercholesterolemia, hypertension, diabetes, smoking and male gender, are the major risk factors.[33] Thus, prevention and treatment of CVD has focused on reducing the risk factors that are associated with atherosclerosis. Results from scientific investigations have indicated that garlic has both antiatherosclerotic (treatment) and antiatherogenic (preventative) activities.[2,18,33]

Since 1980, over 30 clinical trials, one systematic review and three meta-analyses of the clinical trials have assessed the antihypercholesterolemic and antihyperlipidemic effects of garlic.[16–29] A meta-analysis, assessing the lipid lowering effects of garlic and its preparations, reviewed a total of twenty-five randomized, controlled trials [published and unpublished].[20] Sixteen of the clinical trials, with data from 952 subjects, met the minimum criteria and were included in the analyses. Fourteen of the trials used a parallel group design while two studies used a crossover design. Two of the studies were uncontrolled; two others were single-blind, and the remaining 12 trials were double-blinded. The total daily dose of garlic per trial ranged from 600–900 mg of dried garlic powder; 10 g of raw garlic; 18 mg of garlic oil; or aged garlic extracts (dose not stated), and the median duration of treatment was 12 weeks. The pooled mean difference in the absolute change (from baseline to the final measurement in mmol/l) of total serum cholesterol, triglycerides, and high-density lipoprotein cholesterol was compared between subjects treated with garlic supplements as compared with placebo or other agents. The mean difference in reduction of total cholesterol between garlic treatments and placebo was –0.77 mmol/l (95% CI: –0.65, –0.89 mmol/l). The conclusion reached by this meta-analysis was that garlic supplementation (all dosage forms) reduced cholesterol levels by an average 12%, which was evident one month after therapy and lasted for at least six months. However, patients treated with dried garlic powder, there was no significant difference in the size of the reduction across the dosage range of 600–900 mg/day. Dried garlic powder preparations also significantly lowered serum triglyceride levels by 0.31 mmol/l (13%) as compared with placebo (95% CI: –0.14, –0.49). High-density lipoprotein cholesterol was not significantly reduced. However, the authors further indicated that in general the overall quality of the clinical trials was poor, and more rigorously designed trials were needed before garlic could be recommended as a lipid lowering agent for routine clinical use.[20]

A second meta-analysis of 28 controlled clinical trials assessed the antihypercholesterolemic effects of garlic preparations.[21] Five trials met the inclusion criteria which included only randomized, placebo-controlled trials were 75% of patients had cholesterol levels greater than 5.17 mmol/L (200 mg/dl), and there was enough data included to compute the size effect. Analysis of the five trials showed a net decrease in cholesterol levels (0.59 mmol/L, 95% CI: 0.44 to 0.74), which was attributable to garlic consumption.[21] A systematic review of eight clinical trials that assessed the lipid lowering effects of a garlic powder product (tablet) had data on 500 subjects.[22] In seven of the eight studies reviewed, a daily dose of 600 to 900 mg of garlic powder, reduced serum cholesterol and triglyceride levels by 5–20%. The systematic review concluded that garlic powder preparations do have lipid-lowering activity.[22] A randomized, double-blind comparison trial assessed the efficacy and tolerance of a standardized garlic preparation (1.3% alliin) for the treatment of primary hypercholesterolemia with that of bezafibrate, a lipid-lowering drug.[34] Ninety-eight patients were treated with 900 mg/day of the garlic powder or 600 mg of bezafibrate for 12 weeks after a 6-week placebo run-in. All patients followed a low-cholesterol "step 1 diet" for the duration of the trial. Both treatments reduced total cholesterol, LDL and triglycerides and increased HDL cholesterol in a statistically significant manner. However, no significant differences were observed between the two treatment groups.[34]

A randomized, double-blind, placebo-controlled clinical trial assessed the effects of a dried garlic powder product (600 mg/day, 15 weeks), in 68 subjects with normal blood lipid patterns.[35] After ten weeks of treatment, total cholesterol levels decreased significantly from 223 to 214 mg/dl (p < 0.05), and triglyceride levels decreased from 124 to 118 mg/dl in the garlic treated patients as compared with placebo. Blood pressure levels remained unchanged throughout the study.[35] A randomized placebo-controlled study of thirty-five renal transplant patients with a total serum cholesterol of greater than 240 mg/dl and LDL cholesterol levels of greater than 160 mg/dl assessed the efficacy of garlic supplementation for the treatment of hypercholes-terolemia.[23] Patients were treated with 680 mg of a standardized garlic powder preparation twice daily or matching placebo for 12 weeks, along with a Step One National Cholesterol Education Program reduction diet. After six weeks of therapy, the patients treated with the garlic preparation had an average of a 14-point drop in total serum cholesterol that was still observed at 12 weeks. The patients in the placebo group had an initial five point decrease in total serum cholesterol at six weeks, which was no longer observable at 12 weeks. Serum triglycerides increased in both groups at six weeks, and decreased at 12 weeks.[23] A randomized, double-blind placebo-controlled parallel clinical trial assessed the efficacy the effect of 900 mg/day of dried garlic powder (tablets, standardized to 1.3% allicin) in reducing total cholesterol.[24] One hundred and fifteen subjects with repeat total cholesterol levels of 6.0–8.5 mmol/l and LDL cholesterol of 3.5 mmol/l or above after six weeks of dietary intervention participated in the 6-month study. Patients received 300 mg of the standardized garlic product three times daily or a matching placebo (lactose tablets coated with an outer layer impregnated with garlic powder). At the end of the trial, no statistically significant difference in the mean concentrations of serum lipids, lipoproteins or apolipoproteins A1 or B was observed in the garlic treated group as compared with placebo.[24] The results from this trial were included in a re-analysis of a previously published meta-analysis.[20] The pooled estimate for the effect of the standardized garlic product compared to placebo remained statistically significant, but was reduced in magnitude from –0.75 mmol/l (95% CI: –0.88 to –0.63) to –0.65 mmol/l (95% CI: –0.53 to –0.76).[24]

A double-blind, placebo-controlled, randomized crossover study assessed the effects of a standardized garlic product on plasma lipid and lipoprotein levels in 28 subjects with mild to moderate hypercholesterolemia (6.0–7.8 mmol/l, after 28 days on standard dietary advice).[25] The participants were the randomly assigned to receive either 300 mg of the garlic product three times daily or matching placebo for 12 weeks, followed by 28 day washout, followed by a 12 week crossover on the alter-native preparation. No significant differences in plasma cholesterol, LDL cholesterol, HDL cholesterol, plasma triglycerides, lipoprotein concentrations or blood pressure were observed between the treated group or the group treated with placebo.[25] Another double-blind, placebo-controlled, randomized parallel study of similar design found no significant effect of garlic in 50 patients with hyperlipidemia.[26] A randomized, double-blind, placebo-controlled cross-over trial assessed the effects of a steam-distilled garlic oil preparation (5 mg twice a day) versus placebo in 25 patients with moderate hypercholesterolemia.[28] After 12 weeks of treatment (with wash-out periods of 4 weeks), no significant effects on serum lipoproteins, cholesterol absorption, or cholesterol metabolism were observed in either group.[28] A double-blind placebo-controlled crossover study assessed the effect of aged garlic extract on blood lipids of 41 men with moderate hypercholesterolemia (5.7 to 7.5 mmol/l).[27] After a 4-week dietary run-in, patients received 7.2 g daily (in divided doses of aged garlic extract or placebo for 6 month, followed by a 4-month crossover period. A reduction in total serum cholesterol of 6.1% (4 months) or 7.0% (6 months) was observed in the garlic-treated patients as compared with placebo. A 5.5% reduction in systolic blood pressure was also observed in the garlic treated subjects.[27] A randomized double-blind, placebo-controlled clinical trial assessed the efficacy and safety of a standardized garlic powder product in 30 pediatric patients with familial hypercholesterolemia.[36] The children were treated for eight weeks with 900 mg/day of the garlic product or matching placebo. No effect of garlic extract on fasting total cholesterol levels or the levels of high-density lipoprotein and triglycerides was observed.[36]

An epidemiological cross-sectional observational study assessed the protective effect of chronic garlic intake on aortic elasticity in healthy elderly patients (50 to 80 years).[30] One hundred and one patients who were taking ⩾ 300 mg/d of a standardized garlic preparation for ⩾ 2 years were matched with 101 control subjects. Pulse wave velocity (PWV) and pressure-standardized elastic vascular resistance (EVR) was used to measure the elastic properties of the aorta. Both PWV and EVR were lower in the garlic-treated group as compared with controls. These data suggest that chronic garlic intake attenuates age-related increases in aortic stiffness.[30]

It has been suggested that regular dietary consumption of garlic inhibits platelet aggregation, increases fibrinolytic activity and inhibits atherosclerotic plaque formation.[4,9,37] An increase in fibrinolytic activity in the serum of patients suffering from atherosclerosis was observed after oral administration of an aqueous garlic extract, the essential oil, or garlic powder products.[38,39] Results from clinical studies have demonstrated that oral administration of garlic activates endogenous fibrinolysis, which is detectable for several hours after administration of garlic, and the effect increases as garlic is taken regularly for several months.[40,41] Acute hemorheological effects, including a decrease in plasma viscosity, tissue plasminogen activator activity, and the hematocrit level, were observed after oral administration of 600–1200 mg of a standardized garlic powder product.[22] Results from a clinical study of patients with hypercholesterolemia demonstrated that treatment with a garlic-oil-macerate for three months significantly decreased both platelet adhesion and aggregation.[42] Treatment of 41 hypercholesterolemic men with 7.2 grams/day of aged garlic extract (AGE) for 10 months reduced epinephrine- and collagen-induced platelet aggregation *ex vivo*.[43] Platelet adhesion to fibrinogen was also reduced by approximately 30% in subjects treated with AGE compared with placebo. A decrease in the susceptibility of lipoproteins to oxidation was also noted during this study.[43]

A number of clinical trials have assessed the effects of acute and chronic administration of garlic supplements on fibrinolysis and platelet-aggregation.[44–47] A randomized, double-blind, placebo-controlled trial assessed the effects of garlic administration on platelet aggregation, serum thromboxane levels and lyso-platelet activating factor (lysoPAF) in 14 healthy men.[44] The subjects were treated with a garlic oil extract or matching placebo for 6 days after a familiarization period of two weeks. No significant differences in platelet aggregation, or serum thromboxane and lysoPAF levels were observed in the garlic treated group as compared with placebo.[44] Three randomized, double-blind, placebo-controlled cross-over studies investigated effects of garlic on fibrinolysis and platelet aggregation in healthy subjects.[45,46,47] Oral administration of 900 mg/day of a standardized garlic powder product for 14 days to 12 healthy subjects significantly increased total euglobulin fibrinolytic and tissue plasminogen activator activity as compared with placebo.[47] In addition, *ex vivo* platelet aggregation induced by adenosine diphosphate or collagen was significantly inhibited at 2 and 4 hours after garlic ingestion, and remained lower for 7 to 14 days post treatment.[47]

A randomized, double-blind, placebo-controlled parallel study assessed the effect of garlic on platelet aggregation in 60 subjects with increased risk of juvenile ischemic attack.[48] Oral administration of 800 mg of a standardized garlic powder product for 4-weeks significantly decreased the increased ratio of circulating platelet aggregates by 10.3% (p < 0.01), and also decreased spontaneous platelet aggregation by 56.3% (p < 0.01) as compared with the placebo group.[48] In a double-blind, placebo-controlled trial, patients with stage II peripheral arterial occlusive disease treated with 800 mg/day of a standardized garlic powder product for 4 weeks, had an increase in capillary erythrocyte flow rate, and a decrease in plasma viscosity and fibrinogen levels.[49] In a randomized placebo-controlled study of patients suffering from platelet dysfunction, oral administration of 800 mg of garlic powder over 5-weeks decreased spontaneous platelet aggregation and the number of circulating aggregates as compared with placebo.[50] In a three-year intervention study, 432 patients with myocardial infarction were treated with an ether-extracted garlic oil (0.1 mg/kg/day, corresponding to 2 g fresh garlic daily) or placebo.[51] The results of this study showed that patients treated with the garlic extract had a 35% decrease in new heart attacks and a 45% reduction in total deaths as compared with the placebo group. The garlic treated group also had a reduction in serum lipids.[51]

A meta-analysis assessed the efficacy of oral garlic administration for the treatment of mild hypertension.[19] Eleven controlled clinical trials (published and unpublished) were assessed. Eight of the trials with data from 415 subjects were included in the analysis and three trials were excluded due to a lack of data. In the trials, patients were treated with dried garlic powder (tablets) at a dose of 600–900 mg daily (equivalent to 1.8–2.7 g/day fresh garlic), or placebo. The median duration of the trials was 12 weeks. Only three of the trials were conducted specifically with hypertensive subjects, and many of the studies suffered from methodological flaws. Of the seven studies that compared garlic with placebo, three reported a reduction in systolic blood pressure, and four studies reported a decrease in diastolic blood pressure.[19] Pooled results of the trials showed a decrease of 10 mm Hg in systolic blood pressure and a decrease of 7 mm Hg in diastolic blood pressure in garlic treated groups. The meta-analysis concluded that garlic might have some clinical usefulness in mild hypertension, but that there is still insufficient evidence to recommend garlic as routine clinical therapy for the treatment of hypertension.[19] A randomized, double-blind, placebo-controlled crossover trial assessed the effects of a standardized garlic powder preparation on the diameter of conjunctival vessels (arterioles, venules and capillaries) in 20 healthy volunteers.[52] Oral administration of a standardized garlic powder preparation (900 mg/day) resulted in a significant increase in the mean diameter of the arterioles (4.2%, $p < 0.002$) and venules (5.9%, $p < 0.0001$) as compared with placebo.[52] In a randomized, placebo-controlled double-blind crossover study of 10 healthy volunteers, a 55% increase in capillary skin perfusion was observed 5 hours after oral administration of a standardized garlic powder product (900 mg).[53]

A double-blind, placebo-controlled clinical trial assessed the effects of a standardized garlic powder product (800 mg/day) on blood glucose levels in 120 patients.[30] Patients were treated with 800 mg/day of the garlic preparation or placebo for 4 weeks. The average blood glucose levels in the garlic treated patients decreased by 11.6% as compared with placebo.[30] One other study found no such activity in non-insulin dependent patients dosed with 700 mg/day of a sprayed dried garlic preparation (this product was not standardized).[54]

**Mechanism of Action**

As demonstrated by *in vitro* and *in vivo* studies, there are multiple mechanisms by which garlic reduces cardiovascular risk and exerts its antiatherosclerotic effects. Possible mechanisms of action include: reduction of serum lipid levels; inhibition of cholesterol biosynthesis; inhibition of vessel wall lipid accumulation; inhibition of smooth muscle phenotypic change and proliferation; reduction in BP, inhibition of platelet aggregation; activation of fibrinolysis; reduction of fibrinogen levels; reduction of plasma and blood viscosity; and reduction of lipoprotein oxidation.[4,33,37,55–76]

A reduction of cholesterol and lipid levels, and an inhibition of cholesterol biosynthesis by fresh garlic, garlic juice, aged garlic extracts or volatile oil preparations have been reported in both *in vitro* and *in vivo* studies.[55–73] Aqueous garlic extracts inhibit cholesterol biosynthesis in a dose-dependent manner in isolated primary rat hepatocytes, human HepG2 cells and in cultured human aortic cells *in vitro*.[55–57,72] Various garlic extracts inhibit the activities of hepatic hydroxymethylglutaryl-CoA reductase and acyl-CoA: cholesterol acyltransferase, as well as induce a remodeling of plasma lipoproteins and cell membranes *in vitro*.[72,73] The addition of a garlic powder extract to cultured human aortic cells, lowered lipid levels, and inhibited lipid biosynthesis in both normal and atherosclerotic cells.[72] A standardized garlic extract, in low concentrations [< 0.5 mg/ml], inhibited the activity of hepatic hydroxymethylglutaryl-CoA reductase (HMG-CoA reductase).[73] Furthermore, the same extract, at higher concentrations [> 0.5 mg/ml], also inhibited enzymes in the latter stages of cholesterol biosynthesis.[73] Alliin was not active in this assay, however both allicin and ajoene inhibited the activity of HMG-CoA reductase *in vitro* ($IC_{50}$ 7 and 9 mM, respectively).[56] In addition, both allyl mercaptan (50 mM) and diallyl disulfide (5 mM) enhanced palmitate-induced inhibition of cholesterol biosynthesis *in vitro*.[57] Aqueous garlic extracts do not contain allicin, ajoene or allyl mercaptan, therefore other constituents of garlic, such as nicotinic acid and adenosine, which also inhibit HMG-CoA reductase activity and cholesterol biosynthesis, may be involved.[74]

Intragastric administration of minced garlic bulbs, aqueous, ethanol, petroleum ether, or methanol garlic extracts, the essential oil, aged garlic extracts, or the fixed oil to various animal models (rats, rabbits, chickens, pigs) fed a high-cholesterol diet reduced serum cholesterol and lipid levels.[58–69] Intragastric administration of allicin to rats, for two months, lowered both serum and liver levels of total cholesterol, lipids, phospholipids, and triglycerides.[70] Intraperitoneal administration of a combination of diallyl disulfide and diallyl trisulfide to rats reduced the plasma concentration of total cholesterol and lipids.[71] Treatment of rabbits fed a 1% cholesterol enriched diet for 6 weeks with 800 µL/kg/day of an aged garlic extract (AGE) reduced fatty streak development, vessel wall cholesterol accumulation and the development of fibro fatty plaques in neointimas.[75] In addition, AGE treatment prevented vascular smooth muscle phenotypic changes from the contractile, high volume fraction of myofilament (Vvmyo) state and inhibited smooth muscle cell proliferation *in vitro*.[75]

The antihypertensive effects of garlic have been demonstrated in a variety of animal models. Intragastric administration of minced garlic bulbs, or aqueous or alcoholic extracts of the bulbs, to dogs, guinea pigs, rabbits or rats significantly lowered blood pressure.[59,77–79] The mechanism of action appears to involve a decrease in peripheral vascular resistance, due to a direct relaxant effect of garlic on smooth muscle.[79] Both aqueous garlic and ajoene induced membrane hyperpolarization in the cells of isolated vessel strips.[80,81] The potassium channels opened frequently causing hyperpolarization, which resulted in vasodilatation due to a blockade of the calcium channels.[80,81] Garlic may also exert its antihypertensive activity via an increased production of nitric oxide (also known as endothelium-derived relaxant factor). Nitric oxide regulates vascular homeostasis by controlling vascular resistance, blood pressure, cell-to-cell contact and cellular proliferation. Aqueous and alcoholic extracts of garlic activate nitric oxide synthase (eNOS) in human tissues,[82] and inhibit L-N-nitro-L-arginine-methyl-ester-(L-NAME) induced arterial hypertension in rats.[77,83] Treatment of isolated rat intralobar pulmonary arteries with aqueous garlic extracts induced vasorelaxation *in vitro*.[84] Mechanical disruption of the endothelium or pretreatment with L-NAME decreased garlic-induced relation by 30–40%.[84] Both allicin and ajoene inhibit the expression of inducible nitric oxide synthase (iNOS) in lipopolysaccharide-stimulated macrophages.[85] The activity of this enzyme is induced under inflammatory conditions in human atherosclerotic lesions and results in the formation of peroxynitrite, which is a potent oxidant that initiates LDL oxidation as well as platelet aggregation.[85]

Inhibition of platelet aggregation by garlic extracts and individual chemical constituents has been demonstrated in both *in vitro* and *in vivo* studies. An aqueous, chloroform or methanol extract of garlic inhibited collagen-, ADP-, arachidonic acid, epinephrine-, thrombin-induced platelet aggregation *in vitro*.[86–90] Chronic intragastric administration (3 months) of the essential oil or a chloroform extract of garlic inhibited platelet aggregation in rabbits.[86,91–92] Adenosine, alliin, allicin, and the transformation products of allicin, the ajoenes, the vinyldithiins and the dialkyloligosulfides appear to be the chemical constituents responsible for inhibition of platelet adhesion and aggregation.[4,40,93–95] Furthermore, methyl allyl trisulfide, a minor constituent of garlic oil, inhibited platelet aggregation *in vitro*.[96]

The mechanism of platelet aggregation inhibition of garlic and its constituents involves the inhibition of the arachidonic acid cascade and of the activity of platelet cyclic AMP phosphodiesterase.[87,89,97] Ajoene, inhibits platelet aggregation induced by adenosine diphosphate, arachidonic acid, A23187, collagen, epinephrine, platelet activating factor, and thrombin,[97,98] and its activity is potentiated by prostacyclin, forskolin, indomethacin and dipyridamole.[97] The antiplatelet activity of ajoene appears to involve an inhibition of the activity of both cyclooxygenase and lipoxygenase, preventing the formation of thromboxane $A_2$ and 12-hydroxyeicosatetraenoic acid in the arachidonic acid cascade.[97] The specific mechanism of ajoene's antiplatelet activity may involve an interaction with primary agonist-receptor complex of the fibrinogen receptors, through specific G-proteins involved in the signal transduction system on the platelet membrane.[94] The interaction of ajoene with a hemoprotein involved in platelet activation and modification of the hemoprotein binding to its ligands has also been proposed as a mechanism.[98]

Aqueous garlic extracts and garlic oil have been shown to alter the plasma fibrinogen level, coagulation time, and fibrinolytic activity in animal models.[1,99–100] Serum fibrinolytic activity increased after intragastric administration of a dry garlic or garlic extracts to animals that were artificially rendered arteriosclerotic.[99–100] Although adenosine was thought to be the active constituent, it does not exert an effect in whole blood.[1]

The antihyperglycemic activities of various garlic extracts and isolated constituents have been assessed *in vivo*.[9,101–107] Intragastric administration of an aqueous, ethanol, petroleum ether or chloroform extract, or the essential oil of garlic lowered blood glucose levels in both rabbits and rats.[9,101–104] However, three of the *in vivo* studies found no effect of garlic supplementation on blood glucose levels.[105–107]

Garlic extracts and its organosulphur constituents have antioxidant and free radical scavenging activities.[108–110] Diallyl disulfide, isolated from garlic, inhibited liver microsomal lipid peroxidation induced by NADPH, ascorbate and doxorubicin *in vitro*.[109] S-allyl cysteine, the major chemical constituent of aged garlic extract, inhibited free radical production, lipid peroxidation and neuronal damage in rat brain ischemia models.[110]

Numerous *in vitro* studies have shown that garlic extracts have antibacterial and anti-fungal activities.[111–116] Aqueous and ethanol extracts, the essential oil and garlic juice inhibit the growth of *Aspergillus niger*, several *Bacillus* species, *Candida* species, *Cryptococcus* species, *Erwinia carotovora*, *Escherichia coli*, *Helicobacter pylori*, *Mycobacterium tuberculosis*, *Pasteurella multocida*, *Proteus* species, *Pseudomonas aeruginosa*, *Rhodotorula rubra*, *Shigella sonnei*, *Staphylococcus aureus*, *Streptococcus faecalis*, *Toruloposis* species, *Trichosporon pullulans in vitro*.[111–117] Allicin, ajoene and diallyl trisulfide, appear to be the antibacterial and antifungal constituents of garlic.[4]

## Safety Information

### A. Adverse Reactions

Allergic reactions such as contact dermatitis, IgE-mediated urticaria and occupational asthmatic [after inhalation of powdered garlic] have been reported.[118–121] Garlic-sensitive patients may also have a cross-reaction with onion or tulip.[121] Ingestion of fresh garlic bulbs, extracts or oil on an empty stomach may occasionally cause heartburn, nausea, vomiting and diarrhea.[121] Garlic odor from breath and skin may be perceptible.[3] One case of spontaneous spinal epidural hematoma associated with chronic excessive ingestion garlic cloves has been reported.[122] Postoperative bleeding has been reported after chronic garlic consumption.[123–125]

### B. Contraindications

Patients with an allergy to plants of the Lilliaceae (Tulip family).[1]

### C. Drugs Interactions

Prolongation of blood clotting times (approximately double) has been reported in patients using garlic supplements in conjunction with warfarin.[126]

### D. Toxicology

Acute toxicity: allicin, median lethal dose ($LD_{50}$) 120 mg/kg s.c. or 60 mg/kg i.v. (mice). Garlic extract, $LD_{50}$ 0.5 ml/kg to 30 ml/kg p.o., s.c., and i.v. (mice and rats).[121]

Chronic toxicity studies with garlic oil or garlic extract have shown growth inhibitory effects on young rats, and induction of anemia in rats, cats, and dogs.[121]

Garlic extracts were not mutagenic in the Ames test (*Salmonella* microsome reversion) or in *Escherichia coli in vitro*.[127,128]

### E. Dose and Dosage Forms

Fresh bulbs, dried powder, volatile oil, oil macerates, juice, aqueous or alcoholic extracts, aged garlic extracts [minced garlic that is incubated in aqueous alcohol (15–20%) for 20 months, then concentrated], and odorless garlic products [garlic products in which the allinase has been inactivated by cooking; or in which chlorophyll has been added as a deodorant; or aged garlic preparations that have low concentrations of water soluble sulfur compounds].[1,4]

The recommended average daily dose:[1]

Fresh garlic: 2–5 g (1–4 cloves)

Dried powder: 0.4–1.2 g

Oil: 2–5 mg

Extract: 300–1000 mg (as solid material).

Other preparations: Corresponding to 4–12 mg of alliin (= ca. 2–5 mg of allicin).

## References

1    Anon. Bulbus Allii Sativi, *WHO Monographs on Selected Medicinal Plants*, Volume I, World Health Organization, Geneva, Switzerland, 1999.

2    Srivastava KC, Bordia A, Verma SK. Garlic (*Allium sativum*) for disease prevention. *South African Journal of Science.* 1995;91:68–77.

3    Bradley PR, ed. *British Herbal Compendium.* Vol 1, Guildford, UK, British Herbal Medicine Association, 1992.

4    Koch HP, Lawson LD. Garlic, The Science and Therapeutic Application *of Allium sativum* L. and Related Species, 2nd ed., Williams & Wilkins, Baltimore, 1996.

5    *African Pharmacopoeia*, Vol. 1, 1st ed., Lagos, Nigeria, Organization of African Unity, Scientific Technical & Research Commission, 1985.

6    *European Pharmacopoeia*, 3rd ed., Strasbourg, Council of Europe, 1996.

7    *Indian Pharmaceutical Codex, Vol. I – Indigenous Drugs*, New Delhi, Council of Scientific & Industrial Research, 1953:8–10.

8    Youngken HW. *Textbook of Pharmacognosy.* Philadelphia, The Blakiston Company, 1950:182–183.

9    Reuter HD, Sendl A. *Allium sativum* and *Allium ursinum*: Chemistry, Pharmacology, and Medicinal Applications, In: Wagner H, Farnsworth NR, (eds.). *Economic and Medicinal Plants Research*, Vol. 6, London, Academic Press, 1994:55–113.

10   Sendl A. *Allium sativum* and *Allium ursinum*: Part 1. Chemistry, analysis, history, botany. *Phytomedicine.* 1995;4:323–339.

11   Iberl B. Quantitative determination of allicin and alliin from garlic by HPLC, *Planta Med.* 1990;56:320–326.

12   Ziegler SJ, Sticher O. HPLC of S-alk(en)yl-L-cysteine derivatives in garlic including quantitative determination of (+)-S-allyl-L-cysteine sulfoxide (alliin), *Planta Med.* 1989;55:372–378.

13   Mochizuki E et al. Liquid chromatographic determination of alliin in garlic and garlic products. *J Chromatogr.* 1988;455:271–277.

14   Freeman F, Kodera Y. Garlic chemistry: Stability of S-(2-propenyl-2-propene-1-sulfinothioate (allicin) in blood, solvents and simulated physiological fluids. *J Agr Food Chem.* 1995;43:2332–2338.

15   Lawson L. HPLC analysis of allicin and other thiosulfinates in garlic clove homogenates. *Planta Med.* 1991;57:263–270.

16   Auer W, Eiber A, Hertkorn E, Hoehfeld E, Koehrle U, Lorenz A, Mader F, Merx W, Otto G, Schmid-Otto B, Taubenheim H. Hypertension and hyperlipidaemia: garlic helps in mild cases. *Brit J Clin Prac.* 1990;69:3–6.

17   Bordia A, Verma SK, Srivastava KC. Effect of garlic (*Allium sativum*) on blood lipids, blood sugar, fibrinogen and fibrinolytic activity in patients with coronary artery disease. *Prostaglandins, Leukotrienes and Essential Fatty Acids.* 1998;58:257–263.

18    Neil HA, Silagy CA. Garlic: its cardioprotectant properties. *Current Opinions in Lipidology.* 1994;5:6–10.

19    Silagy CA, Neil HA. A meta-analysis of the effect of garlic on blood pressure. *J Hypertension.* 1994;12:463–468.

20    Silgay C, Neil A. Garlic as a lipid lowering agent – a meta-analysis. *Journal of the Royal College of Physicians of London.* 1994;28:39–45.

21    Warshafsky S, Kamer RS, Sivak SL. Effect of garlic on total serum cholesterol. A meta-analysis. *Ann Int Med.* 1993;119:599–605.

22    Brosche T, Platt D. Garlic as a phytogenic lipid lowering drug – a review of clinical trials with standardized garlic powder preparation. *Fortschritte der Medizin.* 1990;108:703–706.

23    Lash JP, Cardoso LR, Mesler PM, Walczak DA, Pollak R. The effect of garlic on hypercholesterolemia in renal transplant patients. *Transplantation Proceed.* 1998;30: 189–191.

24    Neil HA, Silagy CA, Lancaster T, Hodgeman J, Vos K, Moore JW, Jones L, Cahill J, Fowler GH. Garlic powder in the treatment of moderate hyperlipidaemia: a controlled trial and meta-analysis. *Journal of the Royal College of Physicians* (London). 1996;30:329–334.

25    Simons LA, Balasubramaniam S, von Konigsmark M, Parfitt A, Simons J, Peters W. On the effect of garlic on plasma lipids and lipoproteins in mild hypercholesterolaemia. *Atherosclerosis.* 1995;113:219–225.

26    Isaacsohn JL, Moser M, Stein EA, Dudley K, Davey JA, Liskov E, Black HR. Garlic powder and plasma lipids and lipoproteins: a multicenter, randomized, placebo-controlled trial. *Arch Int Med.* 1998;158:1189–1194.

27    Steiner M, Khan AH, Holbert D, Lin, RI. A double-blind crossover study in moderately hypercholesterolemic men that compared the effect of aged garlic extract and placebo administration on blood lipids. *Amer J Clin Nutr.* 1996;64:866–870.

28    Berthold HK, Sudhop T, von Bergmann K. Effect of garlic oil preparation on serum lipoproteins and cholesterol metabolism: a randomized controlled trial. *JAMA.* 1998;279:1900–1902.

29    Breithaupt-Grögler K, Ling M, Boudoulas H, Belz GG. Protective effect of chronic garlic intake on elastic properties of aorta in the elderly. *Circulation.* 1997;96:2649–2655.

30    Lau BH, Moses AA, Sanchez A. *Allium sativum* (garlic) and atherosclerosis: a review. *Nutr Res.* 1983;3:119–128.

31    Keys A. Wine, garlic and CHD in seven countries. *Lancet.* 1980;1:145–146.

32    Magyrar E. Incidence of coronary sclerosis and mycocardial infarction in Hungary in the light of statistical data derived from autopsy material. *Acta Med Hung.* 1969;25:263–269).

33    Orekhov AN, Grünwald J. Effects of garlic on atherosclerosis. *Nutrition.* 1997;13:656–663.

34    Holzgartner H, Schmidt U, Kuhn U. Comparison of the efficacy and tolerance of a garlic preparation vs. benzafibrate. *Arzneimittelforschung.* 1992;42:1473–1477.

35    Saradeth T, Seidl S, Resch KL, Ernst E. Does garlic alter the lipid pattern in normal volunteers? *Phytomedicine.* 1994;1:183–185.

36    McCrindle BW, Helden E, Conner WT. Garlic extract therapy in children with hypercholesterolemia. *Arch Pediatr Adolesc Med.* 1998;152:1089–1094.

37    Orekhov AN, Tertov VV, Sobenin IA, Pivovarova EM. Direct anti-atherosclerosis-related effects of garlic. *Ann Med.* 1995;27:63–65.

38    Harenberg J, Giese C, Zimmermann R. Effects of dried garlic on blood coagulation, fibrinolysis, platelet aggregation, and serum cholesterol levels in patients with hyperlipoproteinemia. *Atherosclerosis.* 1988;74:247–249.

39   Bordia A et al. Effect of essential oil of garlic on serum fibrinolytic activity in patients with coronary artery disease. *Atherosclerosis*. 1977;26:379–386.

40   Lawson LD, Hughes BG. Inhibition of whole blood platelet-aggregation by compounds in garlic clove extracts and commercial garlic products. *Thrombosis Res*. 1992;65:141–156.

41   Chutani SK, Bordia A. The effect of fried versus raw garlic on fibrinolytic activity in man. *Atherosclerosis*. 1981;38:417–421.

42   Bordia A. Klinische Untersuchung zur Wirksamkeit von Knoblauch. *Apotheken-Magazin*. 1986;6:128–131.

43   Steiner M, Lin RS. Changes in platelet function and susceptibility of lipoproteins to oxidation associated with administration of aged garlic extract. *J Cardiovasc Pharmacol*. 1998;31:904–908.

44   Morris J, Burke V, Mori TA, Vandongen, Beilin LJ. Effects of garlic extract on platelet aggregation: a randomized placebo-controlled double-blind study. *Clin Exp Pharmacol Physiol*. 1995;22:414–417.

45   Kiesewetter H. Effect of garlic on thrombocyte aggregation, microcirculation, and other risk factors. *Int J Clin Pharmacol Ther Toxicol*. 1991;29:151–155.

46   Kiesewetter H, Jung F, Mrowietz C, Pindur G, Heiden M, Wenzel E. Effects of garlic on blood fluidity and fibrinolytic activity: a randomized, placebo-controlled, double-blind study. *Brit J Clin Prac*. 1990;69:24–29.

47   Legnani C, Frascaro M, Guazzaloca G, Ludovici S, Cesarano G, Coccheri S. Effects of dried garlic preparation on fibrinolysis and platelet aggregation in healthy subjects. *Arzneimittelforschung*. 1993;43:119–121.

48   Kiesewetter H et al. Effect of garlic on platelet aggregation in patients with increased risk of juvenile ischaemic attack. *Eur J Clin Pharmacol*. 1993;45:333–336.

49   Kiesewetter H, Jung F. Beeinflusst Knoblauch die Atherosklerose? *Medizinische Welt*. 1991;42:21–23.

50   Jung H, Kiesewetter H. Einfluss einer Fettbelastung auf Plasmalipide und kapillare Hautdurchblutung unter Knoblauch. *Die Medizinische Welt*. 1991;42:14–17.

51   Bordia A. Knoblauch und koronare Herzkrankheit: Wirkungen einer dreijährigen Behandlung mit Knoblauchextrakt auf die Reinfarkt-und Mortalitätsrate. *Deutsche Apotheker Zeitung*. 1989;129:16–17.

52   Wolf S, Reim M. Effect of garlic on conjunctival vessels: a randomised, placebo-controlled, double-blind trial. *Brit J Clin Prac*. 1990;44:36–39.

53   Jung EM, Jung F, Mrowietz C, Kiesewetter H, Pindur G, Wenzel E. Influence of garlic powder on cutaneous microcirculation. *Arzneimittelforschung*. 1991;41:626–630.

54   Sitprija S et al. Garlic and diabetes mellitus phase II clinical trial. *J Med Assoc Thailand*. 1987;70:223–227.

55   Gebhardt R, Beck H. Differential inhibitory effects of garlic-derived organosulfur compounds on cholesterol biosynthesis in primary rat hepatocyte cultures. *Lipids*. 1996;31:1269–1276.

56   Gebhardt R, Beck H, Wagner KG. Inhibition of cholesterol biosynthesis by allicin and ajoene in rat hepatocytes and HepG2 cells. *Biochim Biophys Acta*. 1994;1213:57–62.

57   Gebhardt R. Amplification of palmitate-induced inhibition of cholesterol biosynthesis in cultured rat hepatocytes by garlic-derived organosulfur compounds. *Phytomedicine*. 1995;2:29–34.

58   Yeh YY, Yeh SM. Garlic reduces plasma lipids by inhibiting hepatic cholesterol and triacylglycerol synthesis. *Lipids*. 1994;29:189–193.

59   Chi MS, Koh ET, Stewart TJ. Effects of garlic on lipid metabolism in rats fed cholesterol or lard. *J Nutr*. 1982;112:241–248.

60 Qureshi AA et al. Inhibition of cholesterol and fatty acid biosynthesis in liver enzymes and chicken hepatocytes by polar fractions of garlic. *Lipids.* 1983;18:343–348.

61 Thiersch H. The effect of garlic on experimental cholesterol arteriosclerosis of rabbits. *Zeitschrift für die gesamte experimentelle Medizin.* 1936;99; 473–477.

62 Zacharias NT et al. Hypoglycemic and hypolipademic effects of garlic in sucrose fed rabbits. *Indian J Physiol Pharmacol.* 1980;24:151–154.

63 Gupta PP, Khetrapal P, Ghai CL. Effect of garlic on serum cholesterol and electro-cardiogram of rabbit consuming normal diet. *Indian J Med Sci.* 1987;41:6–11.

64 Sodimu O, Joseph PK, Angusti KT. Certain biochemical effects of garlic oil on rats maintained on high fat-high cholesterol diet. *Experientia.* 1984;40:78–79.

65 Kamanna VS, Chandrasekhara N. Effect of garlic (*Allium sativum* Linn.) on serum lipoproteins and lipoprotein cholesterol levels in albino rats rendered hypercholes-teremic by feeding cholesterol. *Lipids.* 1982;17:483–488.

66 Kamanna VS, Chandrasekhara N. Hypocholesterolic activity of different fractions of garlic. *Indian J Med Res.* 1984;79:580–583.

67 Chi MS. Effects of garlic products on lipid metabolism in cholesterol-fed rats. *Proceedings of the Society of Experimental Biology and Medicine.* 1982;171:174–178.

68 Qureshi AA. Influence of minor plant constituents on porcine hepatic lipid metabolism. *Atherosclerosis.* 1987;64:687–688.

69 Lata S. Beneficial effects of *Allium sativum*, *Allium cepa*, and *Commiphora mukul* on experimental hyperlipidemia and atherosclerosis: a comparative evaluation. *Journal of Postgraduate Medicine.* 1991;37:132–135.

70 Augusti KT, Mathew PT. Lipid lowering affect of allicin (diallyl disulfide-oxide) on long-term feeding to normal rats. *Experientia.* 1974;30:468–470.

71 Pushpendran CK. Cholesterol-lowering effects of allicin in suckling rats. *Indian J Exp Biol.* 1980;18:858–861.

72 Orekhov AN, Tertov VV. *In vitro* effect of garlic powder extract on lipid content in normal and atherosclerotic human aortic cells. *Lipids.* 1997;32:1055–1060.

73 Beck H, Wagnerk G. Inhibition of cholesterol biosynthesis by allicin and ajoene in rat hepatocytes and Hep62 cells. *Biochim Biophys Acta.* 1994;1213:57–62.

74 Platt D, Brosche T, Jacob BG. Cholesterin-senkende Wirkung von Knoblauch? *Deutsche Medizinische Wochenschrift.* 1992;117:962–963.

75 Efendy JL, Simmons DL, Campbell GR, Campbell JH. The effect of aged garlic extract, 'Kyloic' on the development of experimental atherosclerosis. *Atherosclerosis.* 1997;132:37–42.

76 Ogawa H, Suezawa K, Meguro T, Sasagawa S. Effect of garlic powder on lipid metabolism in stroke-prone spontaneously hypertensive rats. *Nippon Eiyo, Shokuryo Gakkaishi.* 1993;46:417–423.

77 Pedraza-Chaverri J, Tapia E, Medina-Campos ON, de los Angeles Granados M, Franco M. Garlic prevents hypertension induced by chronic inhibition of nitric oxide synthesis. *Life Sci.* 1998;62:71–77.

78 Brandle M, Al Makdessi S, Weber RK, Dietz K, Jacob R. Prolongation of life span in hypertensive rats by dietary interventions. Effects of garlic and linseed oil. *Basic Res Cardiol.* 1997;92:223–232.

79 Ozturk Y et al. Endothelium-dependent and independent effects of garlic on rat aorta. *J Ethnopharmacol.* 1994;44:109–116.

80 Siegel G et al. Potassium channel activation, hyperpolarization, and vascular relaxation. *Zeitschrift für Kardiologie.* 1991;80:9–24.

81 Siegel G et al. Potassium channel activation in vascular smooth muscle. In: Frank GB, ed. *Excitation-contraction coupling in skeletal, cardiac, and smooth muscle*, New York, Plenum Press, 1992:53–72.

82   Das I, Khan NS, Sooranna SR. Nitric oxide synthetase activation is a unique mechanism of garlic action. *Biochemical Society Transactions.* 1995;23:S136.

83   Das I, Khan NS, Sooranna SR. Potent activation of nitric oxide synthetase by garlic: a basis for its therapeutic applications. *Curr Med Res Opin.* 1995;13:257–263.

84   Ku DD, Abdel-Razek, Tarek T, Dai J, Fallon MB, Abrams GA. Mechanisms of garlic induced pulmonary vasorelaxation: role of allicin. *Circulation.* 1997;96:6-I.

85   Dirsch VM, Kiemer AK, Wagner H, Vollmar AM. Effect of allicin and ajoene, two compounds of garlic, on inducible nitric oxide synthase. *Atherosclerosis.* 1998;139: 333–339.

86   Makheja AN, Vanderhoek JY, Bailey JM. Inhibition of platelet aggregation and thromboxane synthesis by onion and garlic. *Lancet.* 1979;1:781.

87   Castro RA. Effects of garlic extract and three pure components from it on human platelet aggregation, arachidonate metabolism, release reaction and platelet ultrastructure. *Thrombosis Res.* 1983;32:155–169.

88   Makheja AN, Bailey JM. Antiplatelet constituents of garlic and onion. *Agents Actions.* 1990;29:360–363.

89   Srivastava KC, Justesen U. Isolation and effects of some garlic components on platelet aggregation and metabolism of arachidonic acid in human blood platelets. *Wiener Klinische Wochenschrift.* 1989;101:293–299.

90   Bordia T, Mohammed N, Thomson M, Ali M. An evaluation of garlic and onion as antithrombotic agents. *Prostaglandins, Leukotrienes and Essential Fatty Acids.* 1996; 54:183–186.

91   Chauhan LS et al. Effect of onion, garlic and clofibrate on coagulation and fibrinolytic activity of blood in cholesterol fed rabbits. *Ind Med J.* 1982;76:126–127.

92   Ariga T, Oshiba S. Effects of the essential oil components of garlic cloves on rabbit platelet aggregation. *Igaku To Seibutsugaku.* 1981;102:169–174.

93   Agarwal KC. Therapeutic actions of garlic constituents. *Med Res Rev.* 1996;16: 111–124.

94   Jain MK, Apitz-Castro R. Garlic: A product of spilled ambrosia. *Current Sci.* 1993;65:148–156.

95   Mohammad SM, Woodward SC. Characterization of a potent inhibitor of platelet aggregation and release reaction isolated from *Allium sativum* (garlic). *Thrombosis Res.* 1986;44:793–806.

96   Ariga T, Oshiba S, Tamada T. Platelet aggregation inhibitor in garlic. *Lancet.* 1981;1:150–151.

97   Srivastava KC, Tyagi OD. Effects of a garlic-derived principal (ajoene) on aggregation and arachidonic acid metabolism in human blood platelets. *Prostaglandins, Leukotrienes, and Essential Fatty Acids.* 1993;49:587–595.

98   Jamaluddin MP, Krishnan LK, Thomas A. Ajoene inhibition of platelet aggregation: possible mediation by a hemoprotein. *Biochem Biophys Res Com.* 1988;153:479–486.

99   Bordia A. Effect of essential oil of onion and garlic on experimental atherosclerosis in rabbits. *Atherosclerosis.* 1977;26:379–386.

100  Bordia A, Verma SK. Effect of garlic on regression of experimental atherosclerosis in rabbits. *Artery.* 1980;7:428–437.

101  Sheela CG, Kumud K, Augusti KT. Anti-diabetic effects of onion and garlic sulfoxide amino acids in rats. *Planta Med.* 1995;61:356–357.

102  Jain RC, Vyas CR, Mahatma OP. Hypoglycemic action of onion and garlic. *Lancet.* 1973;2:1491.

103  Jain RC, Vyas CR. Garlic in alloxan-induced diabetic rabbits. *Amer J Clin Nutr.* 1975;28:684–685.

104 Zacharias NT. Hypoglycemic and hypolipidemic effects of garlic in sucrose fed rats. *Indian J Physiol Pharmacol.* 1980;24:151–154.

105 Farva D. Effects of garlic oil on streptozotocin-diabetic rats maintained on normal and high fat diets. *Indian J Biochem Biophys.* 1986;23:24–27.

106 Venmadhi S, Devaki T. Studies on some liver enzymes in rats ingesting ethanol and treated with garlic oil. *Med Sci Res.* 1992;20:729–731.

107 Kumar CA. *Allium sativum:* Effect of three weeks feedings in rats. *Indian J Pharmacol.* 1981;13:91.

108 Ide N, Nelson AB, Lau BHS. Aged garlic extract and its constituents inhibit $Cu^{2+}$-induced oxidative modification of low density lipoprotein. *Planta Med.* 1997;63: 263–264.

109 Dwivedi C, John LM, Schmidt DS, Engineer FN. Effects of oil-soluble organosulfur compounds from garlic on doxorubicin-induced lipid peroxidation. *Anticancer Drugs.* 1998;9:291–294.

110 Numagami Y, Ohnishi ST. S-allyl cysteine inhibits free radical production, lipid peroxidation and neuronal damage in rat cerebral ischemia. In: *Nutritional and Health Benefits of Garlic as a Supplement,* Abstract, Conference Proceedings, Newport Beach, CA 1998.

111 Fitzpatrick FK. Plant Substances Active against Mycobacterium tuberculosis. *Antibiotics and Chemother.* 1954;4:528–529.

112 Naganawa R, Iwata N, Ishikawa K, Fukuda H, Fujino T, Suzuki A. Inhibition of microbial growth by ajoene, a sulfur-containing compound derived from garlic. *Appl Environm Microbiol.* 1996;62:4238–4242.

113 Arunachalam K. Antimicrobial activity of garlic, onion and honey. *Geobios.* 1980;71:46–47.

114 Moore GS, Atkins RD. The antifungistatic effects of an aqueous garlic extract on medically important yeast-like fungi. *Mycologia.* 1977;69:341–345.

115 Caporaso N, Smith SM, Eng RHK. Antifungal activity in human urine and serum after ingestion of garlic (*Allium sativum*). *Antimicrobial Agents and Chemotherapy.* 1983;5:700–702.

116 Abbruzzese MR, Delaha EC, Garagusi VF. Absence of antimycobacterial synergism between garlic extract and antituberculosis drugs. *Diagnosis and Microbio Infect Dis.* 1987;8:79–85.

117 Cellini L, Di Campli E, Masulli M, Di Bartolomeo S, Allocati N. Inhibition of *Helicobacter pylori* by garlic extract (*Allium sativum*). *FEMS Immunol Med Microbiol.* 1996;13:273–277.

118 Asero R, Mistrello G, Roncarolo D, Antoniotti PL, Falagiani P. A case of garlic allergy. *J Allergy Clin Immunol.* 1998;101:427–428.

119 Anibarro B, Foneela JL, La Hoz F. Occupational asthma induced by garlic dust. *J Allergy Clin Immunol.* 1998;100:734–738.

120 Delaney TA, Donnelly AM. Garlic dermatitis. *Australas J Dermatol.* 1996;37:109–110.

121 Siegers CP. *Allium sativum.* In: De Smet PAGM, Keller K, Hänsel R, Chandler RF, (eds.). *Adverse Effects of Herbal Drugs,* Vol 1, Berlin, Springer-Verlag, 1992:73–76.

122 Rose KD et al. Spontaneous spinal epidural hematoma with associated platelet dysfunction from excessive garlic ingestion: A case report. *Neurosurgery.* 1990;26:880–882.

123 Burnham BE. Garlic as a possible risk for postoperative bleeding. *Plastic Resconstructive Surgery.* 1995;95, 213.

124 Petry JJ. Garlic and postoperative bleeding. *Plastic Reconstructive Surgery.* 1995;96: 483–484.

125 German K, Kumar U, Blackford HN. Garlic and the risk of TURP bleeding. *Brit J Urol.* 1995;76:518.

126   Sunter WH. Warfarin and garlic. *Pharmaceutical J.* 1991;246:722.
127   Schimmer O et al. An evaluation of 55 commercial plant extracts in the Ames muta-
      genicity test. *Pharmazie.* 1994;49:448–451.
128   Zhang YS, Chen XR Yu YN. Antimutagenic effect of garlic (*Allium sativum*) on 4NQO-
      induced mutagenesis in *Escherichia coli* WP2. *Mutation Research.* 1989;227:215–219.

# 13

# German Chamomile

## Synopsis

The clinical evidence supporting the topical applications of German chamomile preparations for the symptomatic treatment of inflammations and irritations of the skin and mucosa is at best, weak. In addition, the use of German chamomile for the treatment of gastrointestinal irritations, dyspepsia and for mild cases of insomnia is based primarily on traditional use, with some support from *in vitro* and *in vivo* studies. Nevertheless, German chamomile products continue to be important worldwide and are widely used, both internally and externally, for these therapeutic indications. The recommended dose of German chamomile preparations varies depending on the application. The average daily dose of chamomile varies depending on the product and the application. For internal use, adults: 2 to 8 g of the flower heads are prepared as an infusion (tea) and administered three times a day. In addition there is a fluid extract (1:1 in 45% ethanol) available, which is administered in a dose of 1 to 4 ml three times a day. For children: 2 g of the flower heads prepared as an infusion (tea) three times daily. For external use German chamomile compresses, rinses or gargles, use 3–10% infusion or 1% fluid extract or 5% tincture. For semi-solid preparations, such as creams, the preparation should contain a hydroalcoholic extract of the flower heads corresponding to at least 3–10% of the extract. Adverse reactions to German chamomile are generally limited to allergic reactions, ranging from contact dermatitis to anaphylaxis. Therefore, patients with allergies to plants in the daisy family (also referred to as Asteraceae or Compositae) should be cautioned not use preparations containing German chamomile.

## Introduction

*Chamomilla recutita*, also known as German chamomile is a perennial herbaceous flowing plant in the daisy family (Asteraceae).[1–3] In much of the scientific and medical reference literature German chamomile is often erroneously referred to as the plant species *Matricaria chamomilla*. However, according to the rules of botanical nomenclature, *Chamomilla recutita* (L.) Rauschert is the legitimate name for this species.[4] Thus, *Matricaria chamomilla* is actually a synonym for this plant. A related plant, known as Roman chamomile, *Anthemis nobilis*, is used similarly in Britain, however German chamomile is the plant that is commonly used in the United States.[2] Both German and Roman chamomile are considered safe for human consumption and are listed on the *Generally Regarded as Safe* (GRAS) list produced by the FDA.

German chamomile flowers have been used medicinally for at least a thousand years for the treatment of minor inflammations of the skin and mucus membranes, and for the treatment of gastric irritation and dyspepsia. Hippocrates, Galen and Asclepios all recommended the use of chamomile tea for the treatment of these ailments.[2] Today, chamomile still ranks as one of the most important medicinal plants worldwide and it has been monographed in many pharmacopoeias worldwide. The German Commission E issued a monograph for chamomile in 1984, and approved it for the treatment of minor gastrointestinal and skin irritations.[2,3]

## Quality Information

*   The correct Latin name for German chamomile is *Chamomilla recutita* (L.) Rauschert (Asteraceae).[1] Synonyms that may occur in the scientific literature include *Matricaria chamomilla* L., *M. recutita* L., *M. suaveolens* L.[1] The vernacular (common) names for this plant species include baboonig, babuna, babunah camomile, camamilla, chamomile, camomilla, chamomille allemande, campomilla, chamomille commune, camomille sauvage, fleurs de petite camomille, Flos Chamomillae, German chamomile, Hungarian chamomile, Kamille, Kamillen, Kamitsure, Kamiture, manzanilla, manzanilla chiquita, manzanilla comun, manzanilla dulce, matricaire, *Matricaria* flowers, pin heads, sweel false chamomille, sweet feverfew, and wild chamomile.[1,4,5]
*   Standardized extracts and other commercial products are prepared from the dried flowering heads of *Chamomilla recutita* (L.) Rauschert (Asteraceae).[1]
*   German chamomile is native to northern Europe and grows wild in central European countries, especially abundant in Hungary and Eastern Europe. The plant is also found in the Mediterranean region of North Africa, the USA and western Asia and is cultivated in many countries.[1]
*   German chamomile flowers contain an essential oil (0.4–1.5%), which has an intense blue color due to its high chamazulene content (1–15%), which arises from the decomposition of the sesquiterpene lactone matricin. Up to 50% of the essential oil of German chamomile consists of several sequiterpenes with a bisabolane skeleton including: (−)-α-bisabolol, it's A and B oxides and related sequiterpenes.[6,7] The essential oil also contains spirononenoid dicycloethers with double or triple bonds, formed by the cyclization of polyalkynes. Apigenin and related flavonoid glycosides including apigenin-7-O-β-glycoside, constitute up to 8% (dry weight) of the dry weight are also present.[1]

- The average daily dose of German chamomile is as follows:

    For internal use: Adults: crude drug (flower head), the average daily dose 2 to 8 g prepared as an infusion (tea) taken three times a day; or the fluid extract 1:1 in 45% ethanol in a dose of 1–4 ml taken three times a day. For children, flower heads (2 g), prepared as an infusion (tea) administered three times daily.

    For external use: compresses, rinses or gargles, 3–10% w/v infusion or 1% fluid extract or 5% tincture. For baths use 5 g/L of water or 0.8 g/L of alcoholic extract. For semi-solid preparations hydroalcoholic extracts, corresponding to 3–10% w/w of the final dosage form.

    For vapor inhalation: 6 g of the drug or 0.8 g of alcoholic extract per liter of hot water.[1]

## Medical Uses

External applications of preparations containing German chamomile are used for the symptomatic treatment of inflammation and irritations of the skin and mucosa (skin cracks, bruises, frostbite, and insect bites).[8,9,11] Traditional applications for chamomile include the symptomatic treatment of digestive ailments such as dyspepsia, epigastric bloating, impaired digestion, and flatulence. Infusions of chamomile flowers have also been used in the treatment of restlessness and in mild cases of insomnia due to nervous disorders.[1–3]

## Summary of Clinical Evidence

In a randomized, double blind placebo-controlled clinical trial the effects of standardized chamomile extract, containing 3 mg chamazulene and 50 mg α-bisabolol, was assessed in 14 male subjects. The outcomes measured were re-epithelialization and drying of wound weeping after dermabrasion due to tattooing. External application of the chamomile extract for 18 days produced a statistically significant decrease in the wound size and drying tendency as compared with placebo.[8]

In clinical trials, topical application of a chamomile extract in a cream base was found to be superior to hydrocortisone 0.25% for reducing skin inflammation.[9] In an international multicenter trial the efficacy of chamomile cream was compared with hydrocortisone 0.25%, fluocortin butyl ester 0.75% and bufexamac 5% for the treatment of eczema of the extremities.[9] The chamomile cream was shown to be as effective as hydrocortisone and superior to the other two treatments, however the results were not subjected to statistical analysis. The efficacy of chamomile preparations in the treatment of acute irradiation reactions of the skin and oral mucosa has been investigated.

The safety and efficacy of German chamomile cream was compared with almond oil ointment on acute radiation skin reaction in a randomized, single blind controlled clinical trial.[11] Fifty woman, undergoing radiation for breast cancer treatment, were included in the trial. The preparations were applied twice daily, once prior to treatment and once at bedtime. Neither cream prevented radiation skin reaction, and there was no statistical difference between the efficacy of both preparations.[11] In an uncontrolled trial chamomile preparations were reported to be beneficial in the symptomatic treatment of radiation mucositis due to head and neck radiation and systemic chemotherapy.[10]

A Phase III randomized double blind placebo-controlled clinical trial assessed the efficacy of a chamomile mouthwash for prevention of 5-fluorouracil-induced oral mucositis.[12] Each patient (n = 164) received a chamomile or placebo mouthwash three times daily for 14 days while undergoing 5-FU treatment. Chamomile mouthwash did not decrease 5-FU induced stomatitis as compared to placebo. In an uncontrolled clinical trial involving 12 hospitalized patients, administration of 180 ml of a chamomile tea induced a deep sleep, lasting for 90 minutes.[13]

**Mechanism of Action**

Extracts of German chamomile and its active constituent, bisabolol, have antispasmodic and antipeptic activity *in vitro*.[14,15] Myotropic and neurotropic spasms in the isolated guinea pig ileum induced by barium chloride and acetylcholine were inhibited by chamomile extract ($ED_{50}$ 1–3 mg/ml).[1] The essential oil had smooth muscle relaxant activity in guinea pig ileum and trachea ($ED_{50}$ 10.5 and 55.0 mg/L, respectively), and inhibited bradykinin-induced contractions in the guinea pig ileum ($ED_{50}$ 2.24 mg/ml).[16,17] The spasmolytic activity of chamomile has been attributed to the chemical constituents apigenin, apigenin-7-O-glucoside and $(-)$-$\alpha$-bisabolol, which have activity similar to papaverin. In rodents, bisabolol, one of the active constituents of chamomile flowers inhibited the occurrence of indomethacin-, stress- and ethanol-induced ulcers with an $ED_{50}$ range of 29.5 to 145 mg/kg.[18] Aqueous extracts of chamomile flowers blocked the aggregation of *Helicobacter pylori* but did not have any antibacterial activity against the bacterium *in vitro*.[19,20]

A hydroalcoholic extract of German chamomile flowers (concentration range 1.25 to 10 mg/ml) inhibited the growth of *Staphylococcus aureus, Streptococcus mutans*, group B *Streptococcus* and *Streptococcus salivarius*, and had a bactericidal effect on *Bacillus megatherium* and *Leptospira icterohaemorrhagiae in vitro*.[21] The volatile oil of German chamomile flowers also inhibited *Staphylococcus aureus* and *Bacillus subtilis in vitro*.[22]

German chamomile extracts have anti-inflammatory activity both *in vitro* and *in vivo*. Both the extracts and isolated constituents inhibit the activities of two enzymes involved in the arachidonic acid cascade, cyclooxygenase and lipooxygenase,[23] and thus inhibit the production of prostaglandins and leukotrienes, known inducers of inflammation. The sesquiterpenes, bisabolol and bisabolol oxide also inhibited the activity of 5-lipoxygenase, with bisabolol being the more active of the two compounds.[24] In animal models, the anti-inflammatory effects of chamomile extract, the essential oil and the isolated constituents have been evaluated against yeast-induced fever in rats and against ultraviolet erythema in guinea pig models.[25] External application of (−)-α-bisabolol to cutaneous burns in guinea pigs resulted in a significant reduction in healing time.[6] Applications of chamomile extract to the intact skin of guinea pigs increased the content of adenosine triphosphate, creatine phosphate and glucose-6-phosphate.[26]

The principal anti-inflammatory and antispasmodic constituents of chamomile appear to be the terpene compounds, matricin, chamazulene, (−)-α-bisabololoxides A and B, and (−)-α-bisabolol.[27–34] While matricin and (−)-α-bisabolol have been isolated from the plant, chamazulene is actually an artifact formed during the heating of the flowers, when an infusion or the essential oil is prepared.[1,2] The anti-inflammatory effect of these terpenes was assessed in various animal models. Both matricine and (−)-α-bisabolol inhibited carrageenan-induced rat paw edema, while chamazulene and guajazulene were much less active. However, both matricine and (−)-α-bisabolol were somewhat less than salicylamide.[34] Topical applications of either the total chamomile extract, or a flavonoid fraction of the extract, reduced inflammation in the croton oil dermatitis mouse model.[29] The flavonols, apigenin and luteolin were more active than either indomethacin or phenylbutazone. The activity of the flavonoids decreased in the following order: apigenin > luteolin > quercetin > myricetin > apigenin-7-glucoside > rutin.[31] Intradermal application of liposomal apigenin-7-glucoside inhibited skin inflammations dose-dependently, in rats induced by xanthine-oxidase and cumene hydroperoxide.[30]

Intraperitoneal administration of a lyophilized infusion of chamomile (360 mg/kg) to mice decreased basal motility, exploratory and motor activities, and potentiated hexobarbital-induced sleep.[35] These results demonstrated that chamomile has a depressant action on the central nervous system in mice.[35] One of the constituents of an aqueous extract of the flowers, apigenin, competitively inhibited the binding of flunitrazepam in the benzodiazepine-receptor binding assay.[36] Intraperitoneal administration of apigenin (10 mg/kg) had anxiolytic effects in mice as measured in the elevated plus-maze without sedative or muscle relaxing side effects. However sedative effects were observed at doses of 100 mg/kg.[36]

## Pharmacokinetics

An *in vivo* skin penetration study of the Chamomile flavones apigenin, luteolin and apigenin-7-O-β-glucoside was performed in a trial involving nine healthy female volunteers. The study concluded that the flavonoids are absorbed through the skin, and will penetrate into the deeper skin layers.[37]

## Safety Information

### A. Adverse Reactions
A number of allergic reactions to German chamomile have been reported.[38–48] These reactions range from reports of contact dermatitis,[38–43] to rhinoconjunctivitis[44] to a

few cases of anaphylactic reactions in sensitive individuals.[45–48] Respiratory reactions include bronchitis, asthma and dyspnea.[49] The presence of trace amounts of anthecotulide, a sesquiterpene lactone in chamomile-based preparations may be the cause allergic reactions.[49]

## B.  Contraindications
Chamomile is contraindicated in patients with a known sensitivity or allergy to plants of the Asteraceae (Compositae) such as ragweed, asters and chrysanthemums.[1]

## C.  Drug Interactions
None reported.

## D.  Toxicology
The acute toxicity of $(-)$-$\alpha$-bisabolol, a main active constituent of German chamomile is low in rodent models (15 ml/kg) and dogs and rhesus monkeys after intragastric administration.[50] In a 4-week subacute toxicity test, the toxic dose in rats and dogs ranged between 1.0 and 2.0 ml/kg after intragastric administration.[49,50] No effects on prenatal development were observed in rats and rabbits treated with bisabolol in doses up to 1.0 ml/kg body weight, and no teratogenic effects were observed.[50,51] The acute oral $LD_{50}$ of chamomile essential oil in rats and acute dermal $LD_{50}$ in rabbits are greater than 5 g/kg.[2] No mutagenic effects were found in *Salmonella typhimurium* strains TA 97a, TA 98, TA 100 and TA 104, with or without metabolic activation.[51] However, in one similar study the crude drug extract of German chamomile was mutagenic in *Salmonella typhimurium* strains TA 98 and TA.[49]

## E.  Dose and Dosage Forms
The average daily dose of chamomile is as follows:

*Internal use:*
Adults:
> Flower head: Average daily dose 2 to 8 g prepared as an infusion (tea) three times a day.[1]
> Fluid extract 1:1 in 45% ethanol: Dose 1–4 ml (three times a day).[1]

Children:
> Flower head: 2 g prepared as an infusion (tea) three times daily.
> Fluid extract (ethanol 45–60%): Single doses 0.6–2ml.[1] (Due to the high alcohol content, the fluid extract should not be administered to children under the age of 3 years old.)

*External use:*
> For compresses, rinses or gargles, 3–10% m/V infusion or 1% fluid extract or 5% tincture.[1]
> For baths: 5 g/L of water or 0.8 g/L of alcoholic extract.

For semi-solid preparations hydroalcoholic extracts corresponding to 3–10% w/w of the drug.

For vapor inhalation: 6 g of the drug or 0.8 g of alcoholic extract per liter of hot water.[1]

## References

1   Anon. Flos Chamomillae. *WHO Monographs on Selected Medicinal Plants.* Volume I. WHO Publications, Geneva, Switzerland.

2   Carle R, Gomaa K. Chamomile: a pharmacological and clinical profile. *Drugs of Today.* 1992;28:559–565.

3   Carle R, Isaac O. Die Kamille – Wirkung and Wirksamkeit. *Zeitschrift für Phytotherapie.* 1987;8:67–77.

4   Rauschert S. Nomenklatorische Probleme in der Gattung *Matricaria* L. *Folia Geobot Phytotax.* 1990;9:249–260.

5   Farnsworth NR, ed. *NAPRALERTsm Database,* University of Illinois at Chicago, IL, October 8, 1999 production. An on-line database available directly through the University of Illinois at Chicago or through the Scientific and Technical Network (STN) of Chemical Abstracts Services (1999).

6   Isaac O. Pharmakologische Untersuchungen von Kamillen-Inhaltsstoffen. *Planta Med.* 1979;35:118–124.

7   Hormann HP, Korting HC. Evidence for the efficacy and safety of topical herbal drugs in dermatology: Part I: anti-inflammatory agents. *Phytomedicine.* 1994;1:161–171.

8   Glowania HJ, Raulin C, Svoboda M. The effect of chamomile on wound healing – a controlled clinical-experimental double-blind study. *Zeitschrift für Hautkrankheiten.* 1986;62:1262–1271.

9   Aertgeerts P et al. Vergleichende Prüfung von Kamillosan® Creme gegenüber steroidalen (0.25% Hydrocortison, 0.75% Fluocortinbutylester) und nichtsteroidalen (5% Bufexamac) Externa in der Erhaltungstherapie von Ekzemerkrankungen. *Zeitschrift für Hautkrankheiten.* 1985;60:270–277.

10  Carl W, Emrich LS. Management of oral mucositis during local radiation and systemic chemotherapy: A study of 98 patients. *J Prost Dent.* 1991;66:361–369.

11  Maiche AG, Grohn P, Maki-Hokkonen H. Effect of chamomile cream and almond ointment on acute radiation skin reaction. *Acta Oncol.* 1991;30:395–396.

12  Fidler P et al. Prospective evaluation of a chamomile mouthwash for prevention of 5-FU induced oral mucositis. *Cancer.* 1996;77:522–525.

13  Gould L et al. Cardiac effect of chamomile tea. *J Clin Pharmacol.* 1973;13:475–479.

14  Thiemer VK, Stadler R, Isaac O. Biochemische Untersuchungen von Kamillen-Inhaltsstoffen. I. *Arzneimittelforschung.* 1972;22:1086–1087.

15  Isaac O, Thiemer K. Biochemische Untersuchungen von Kamilleninhaltsstoffen. *Arzneimittelforschung.* 1975;25:1086–1087.

16  Reiter M, Brandt W. Relaxant effects on tracheal and ileal smooth muscles of the guinea pig. *Arzneimittelforschung.* 1985;35:408–414.

17  Achterrath-Tuckermann U et al. Pharmacological investigation with compounds of chamomile. V. Investigations of the spasmolytic effects of compounds of chamomile and Kamillosan on the isolated guinea-pig ileum. *Planta Med.* 1980;39:38–50.

18  Szelenyi I et al. Pharmakologische Untersuchungen von Kamillen-Inhaltsstoffen. *Planta Med.* 1979;35:218–227.

19  Annuk H et al. Effect of cell surface hydrophobicity and susceptibility of *Helicobacter pylori* to medicinal plant extracts. *FEMS Microbiol Lett.* 1999;172:41–45.

20    Turi M et al. Influence of aqueous extracts of medicinal plants on surface hydropho-
      bicity of *Escherichia coli* strains of different origin. *APMIS.* 1997;105:956–962.
21    Cinco M et al. A microbiological survey on the activity of a hydroalcoholic extract of
      camomile. *Int J Crude Drug Res.* 1983;21:145–151.
22    Aggag ME, Yousef RT. Study of antimicrobial activity of chamomile oil. *Planta Med.*
      1972;22:140–144.
23    Wagner H, Wierer M, Bauer R. *In vitro* inhibition of prostaglandin biosynthesis by
      essential oils and phenolic compounds. *Planta Med.* 1986:184–187.
24    Ammon HPT, Kaul R. Pharmakologie der Kamille und ihrer Inhaltsstoffe. *Deutsche
      Apotheker Zeitung.* 1992;132 (Suppl. 27):3–26.
25    Jakovlev V et al. Pharmacological investigations with compounds of chamomile. II.
      New investigations on the antiphlogistic effects of (−)-[??missing]-bisabolol and bis-
      abolol oxides. *Planta Med.* 1979;35:125–240.
26    Thiemer VK, Stadler R, Isaac O. Biochemische Untersuchungen von Kamillen-
      Inhaltsstoffen. II. *Arzneimittelforschung.* 1973;23:756–759.
27    Jakovlev V, Isaac O, Flaskamp E. Pharmakologische Untersuchungen von Kamillen-
      inhaltsstoffen. VI. Untersuchungen zur antiphlogistischen Wirkung von Chamazulen
      und Matricin. *Planta Med.* 1983;49:67–73.
28    Tubaro A et al. Evaluation of anti-inflammatory activity of camomile extract after top-
      ical application. *Planta Med.* 1984;51:359.
29    Della Loggia R. Lokale antiphlogistische Wirkung der Kamillen-Flavone. *Deutsche
      Apotheker Zeitung.* 1985;125 (Suppl. 1):9–11.
30    Della Loggia R et al. Evaluation of the anti-inflammatory activity of chamomile prepa-
      rations. *Planta Med.* 1990;56:657–658.
31    Lang W, Schwandt K. Untersuchung über die glykosidischen Bestandteile der Kamille.
      *Deutsche Apotheker Zeitung.* 1957;97:149–151.
32    Mann C, Staba J. The Chemistry, Pharmacology, and Commercial Formulations of
      Chamomile. In *Herbs, Spices, and Medicinal Plants: Recent Advances in Botany, Hor-
      ticulture and Pharmacology*, Vol. I, Craker LE, Simon JE, (eds.). Phoenix, Oryx Press,
      1986:233–280.
33    Fuchs J, Milbradt R. Skin anti-inflammatory activity of apigenin-7-glucoside in rats.
      *Arzneimittelforschung.* 1993;43:370–372.
34    Albring M et al. The measuring of the anti-inflammatory effect of a compound on the
      skin of volunteers. *Methods and Findings in Experimental and Clinical Pharmacology.*
      1983;5:75–77.
35    Della Loggia R et al. Depressive effects of *Chamomilla recutita* (L.) Rausch. tubular
      flowers, on central nervous system in mice. *Pharmacol Res Comm.* 1982;14:153–162.
36    Viola H et al. Apigenin, a component of *Matricaria recutita* flowers is a central ben-
      zodiazepine receptors ligand with anxiolytic effects. *Planta Med.* 1995;61:213–216.
37    Merfort I et al. *In vivo* penetration studies of camomile flavones. *Pharmazie.*
      1994;49:509–511.
38    Dstychova E, Zahejsky J. Contact hypersensitivity to camomile. *Ceskoslovenska Der-
      matologie.* 1992;67:14–18.
39    Subiza J et al. Allergic conjunctivitis to chamomile tea. *Ann Allergy.* 1990;65:127–132.
40    Paulsen E, Andersen KE, Hausen BM. Compositae dermatitis in a Danish dermatology
      department in one year. *Contact Dermatitis.* 1993;29:6–10.
41    Rodriguez-Serna M, Sanchez-Motilla JM, Ramon R, Aliaga A. Allergic and systematic
      contact dermatitis from *Matricaria chamomilla* tea. *Contact Dermatitis.* 1998;39:
      192–193.
42    Rudzki E, Rebandel P. Positive patch test with Kamillosan in a patient with hypersen-
      sitivity to camomile. *Contact Dermatitis.* 1998;38:164.

43    Pereira F, Santos R, Pereira A. Contact dermatitis to chamomile tea. *Contact Dermatitis*. 1997;36:307.

44    Subiza J et al. Allergic conjunctivitis to chamomile tea. *Ann Allergy*. 1990;65:127–132.

45    Benner MH, Lee HJ. Anaphylactic reaction to chamomile tea. *J Allergy Clin Immunol*. 1973;52:307–308.

46    Casterline CL. Allergy to chamomile tea, *JAMA*. 1980;244:330–331.

47    Subiza J et al. Anaphylactic reaction after the ingestion of chamomile tea: A study of cross-reactivity with other composite pollens. *J Allergy Clin Immunol*. 1989;84: 353–358.

48    Jensen-Jarolim E, Reider N, Fritsch R, Breiteneder H. Fatal outcome of anaphylaxis to camomile-containing enema during labor: a case study. *J Allergy Clin Immunol*. 1998;102:1041–1042.

49    Hausen BM. Sesquiterpene lactones – *Chamomilla recutita*. Adverse reactions to Herbal Medicines.

50    Habersang S et al. Pharmakologische Untersuchungen von Kamillen-Inhaltsstoffen. IV. *Planta Med*. 1979;37:115–123.

51    Rivera IG et al. Genotoxicity assessment through the Ames test of medicinal plants commonly used in Brazil. *Enviromental Toxicol Water Qual*. 1994;9:87–93.

# 14

# Germander

## Synopsis

Although germander (*Teucrium chamaedrys*) has been used traditionally for the treatment of obesity, there is no scientific or clinical evidence to supporting this claim. Furthermore, in addition to a lack of therapeutic efficacy, there are numerous case reports of serious liver toxicity associated with the chronic ingestion of germander containing products, particularly tablets and capsules. At least 33 cases of hepatitis (one fatal) and one case of cirrhosis have been reported in the medical literature. Hepatitis generally develops within 9 weeks of germander ingestion. However, no direct relationship has been established between liver toxicity, and the dose or duration of treatment. The furano-neoclerodane diterpenes, the main chemical constituents of germander, appear to indirectly cause hepatotoxicity. The mechanism involves the transformation of the furano-neoclerodane diterpenes to hepatotoxic metabolites, whose structure has not yet been elucidated. Recovery from germander-induced hepatitis usually occurs after 6 weeks to 6 months after cessation of treatment. However, one fatal case has been reported in an elderly female patient. Considering the serious potential for hepatotoxicity, and the lack of scientific evidence supporting its efficacy, the therapeutic use of products containing germander cannot be justified.

## Introduction

The blossoms of germander, known scientifically as *Teucrium chamaedrys*, have been used in folk medicine for the treatment of obesity.[1] The flowers and aerial parts were used to prepare herbal teas, which in addition to obesity, were employed as an antiinflammatory for the treatment of rheumatism, and as a diuretic.[2] Ingestion of the plant was generally thought to be safe, until 1991 when an encapsulated product containing germander powder was marketed in France, as an adjuvant to weight control products. Promotion of this product led to its large-scale use in France, which resulted in an epidemic of hepatitis.[3] In 1992, the French Department of Health banned the sale of *Teucrium chamaedrys* in France after 26 cases of acute cytolytic hepatitis were reported.[4] Furthermore, several other cases of fulminant hepatitis, chronic hepatitis and cirrhosis have been associated with ingestion of the plant.[3,4] All products containing germander were banned in France on May 12, 1992.[3] Since 1992, there have been at least 7 cases of hepatitis associated with germander ingestion reported in the medical literature, indicating that these products are still available to the consumer.[1,5–8]

## Quality Information

- The correct Latin name for germander is *Teucrium chamaedrys* (Lamiaceae). No botanical synonyms appear in the scientific literature. Numerous vernacular (common) names for the plant include: calamandrie, camerio, common germander, ground oak, Edelgamander, Germandree petit chene, wall germander, and wild germander.[9]
- Commercial products of germander are prepared from the dried flowers or aerial parts of *Teucrium chamaedrys*.[10,11]
- Germander is native to Eastern Europe and Mediterranean countries.[12]
- The flowers and aerial parts of *Teucrium chamaedrys* contains a number of different chemical constituents, including saponins, glycosides and flavonoids, plus several furano-neoclerodane diterpenes such as teucrins A-G, and teuchamaedryns A-B.[1,10,13]
- Daily dosage: Due to a lack of safety and efficacy there are no dosage recommendations.

## Medical Uses

In spite of the fact that germander has been used in traditional medicine for the treatment of obesity, there are no scientific or clinical data to support its use. Considering the serious potential for hepatotoxicity, and the lack of scientific information supporting its use, there are no justifiable therapeutic applications.

## Summary of Clinical Evidence

No clinical data are available.

## Pharmacokinetics

No pharmacokinetics data have been published in the scientific literature.

## Safety Information

### A. Adverse Reactions
In 1992, the sale of germander was banned in France by the French Department of Health after 26 case reports of acute cytolytic hepatitis and several other cases of fulminant hepatitis, chronic hepatitis and cirrhosis were associated with the

ingestion of the plant.[1,3,14] Hepatitis occurred within 9 weeks of treatment and was characterized by jaundice, and high aspartate and alanine aminotransferase levels.[14] Recovery usually occurred within 6 weeks to 6 months after cessation of use.[14] A retrospective study investigated seven cases of hepatotoxicity, from the period of 1986 to 1991, that were associated with the ingestion of germander products for weight loss.[1] These reports described patients with acute hepatitis associated with germander ingestion, who had no other cause of liver injury. Hepatitis, characterized by jaundice and a marked increase in serum aminotransferase levels, was observed 3 to 18 weeks after germander administration. Liver biopsy specimens in three patients showed hepatocyte necrosis. After discontinuance, the jaundice disappeared within 8 weeks and recovery was complete within 1.5 to 6 months. In three cases, reoccurrence was observed after rechallenge.[1] Since 1992, at least seven additional cases of hepatotoxicity, associated with the ingestion of germander products, have been reported.[5–8] Mostefa-Kara and coworkers describe a case of fatal fulminant hepatic necrosis in a female patient, after the daily ingestion of a germander-containing product for 6 weeks.[7] Three other cases of hepatitis and one case of cirrhosis, associated with the chronic intake of germander, have also been reported.[5,6] The onset of hepatitis occurred after 4 to 7 months of treatment and resolved following the discontinuation of treatment. In two more recent cases, patients ingesting germander products for weight loss presented with asthenia, jaundice and a marked increase in serum aminotransferase levels after 5 to 6 months of use.[8] The jaundice disappeared within 8 weeks after cessation of the product administration. Subsequent readministration of germander by one of the patients led to the reoccurrence of hepatitis.[8]

## C. Drug Interactions
None reported.

## D. Toxicology
The hepatotoxicity of an aqueous lyophilisate (freeze-dried tea) of germander was demonstrated in mice after intragastric administration of 1.25 g/kg body weight and 0.125 mg/kg of the furano-neoclerodane diterpene fraction in a single dose.[12] Midzonal liver cell necrosis was observed after 24 hours and was entirely attributed to the toxicity of the furano-neoclerodane diterpenes.[12,15] The *in vivo* hepatotoxicity induced by the diterpenes of germander was prevented by preadministration of a single dose of troleandomycin, a specific inhibitor of cytochrome P450 3A, and was enhanced by pretreatment with either dexamethasone or clotrimazole, two inducers of cytochrome P450 3A. The mechanism of hepatotoxicity appears to be via the transformation of the furano-neoclerodane diterpenes (FND's) to hepatotoxic metabolites whose structure has not yet been elucidated.[12]

The hepatotoxicity of germander was investigated in isolated rat hepatocytes.[3] A crude extract of the aerial parts of germander containing mainly the FND's, or the purified teucrin A and teuchamaedryn A, covalently bound to hepatocyte proteins at a concentration of 100 μg/ml. The binding resulted in the depletion of cellular glutathione and cytoskeleton-associated protein thiols, and led to the formation of plasma membrane blebs and cell death. Troleandomycin slowed the depletion of glutathione and decreased toxicity of the extract. The study concluded that the furanoditerpenoids of germander are activated by cytochrome P450 3A into electrophilic metabolites that deplete glutathione and cytoskeleton-associated protein thiols causing the formation of plasma membrane blebs.[3]

*E.  Dose and Dosage Forms*
No dosage recommendations.

## References

1     Larrey D et al. Hepatitis after germander (*Teucrium chamaedrys*) administration: another instance of herbal medicine hepatotoxicity. *Ann Int Med.* 1992;117:129–132.
2     Delaveau P. La Germandrée petit-chêne (Wild Germander). *Actual Pharm.* 1986;238:34–39.
3     Lekehal M et al. Hepatoxicity of the herbal medicine germander: metabolic activation of its furano diterpenoids by cytochrome P450 3A depletes cytoskeleton-associated protein thiols and forms plasma membrane blebs in rat hepatocytes. *Hepatology.* 1996;24:212–218.
4     Mattei A et al. Atteinte hepatique associée à la prise d'un produit de phytothérapie contenant de la Germandrée petit-chêne. *Gastroenterol Clin Biol.* 1992;16:798–800.
5     Dao T et al. Hépatite chronique cirrhogène à la germandrée petit-chêne. *Gastroenterol Clin Biol.* 1993;17:609–610.
6     Ben Yahia M et al. Hépatite chronique active et cirrhose induites par la germandrée petit-chêne. *Gastroenterol Clin Biol.* 1993;17:959–962.
7     Mostefa-Kara N et al. Fatal hepatitis after herbal tea. *Lancet.* 1992;340:674.
8     Laliberté L, Villeneuve JP. Hepatitis after the use of germander, a herbal remedy. *Canadian Med Assoc J.* 1996;154:1689–1692.
9     Farnsworth NR, Mahady GB (eds.) *Teucrium chamaedrys.* Napralert[sm] database. Copyright the Board of Trustees, University of Illinois at Chicago, Chicago, IL, 1999.
10    De Vincenzi M, Mancini E. Monographs on botanical flavouring substances used in foods. Part V. *Fitoterapia.* 1996;67:241–251.
11    Delaveau P. La germandrée petit-chêne (wild germander). *Actual Pharm.* 1986;238: 34–39.
12    Loeper J et al. Hepatoxicity of germander in mice. *Gastroenterology.* 1994;106: 464–472.
13    Reinbold AM, Popa DP. Minor diterpenoids of *Teucrium chamaedrys. Khim Prirod Soed.* 1974;5:589–598.
14    Castot A, Larrey D. Hépatite observees au cours d'un traitement par un médicament ou une tisane contenant de la germandrée petit-chêne. *Gastroenterol Clin Biol.* 1992;16: 916–922.
15    Kouzi SA, McMurty RJ, Nelson SD. Hepatotoxicity of germander (*Teucrium chamaedrys* L.) and one of its constituent neoclerodane diterpenes, teucrin A in the mouse. *Chem Res Toxicol.* 1994;7:850–856.

# 15

# Ginger

**Synopsis**

Clinical evidence supports the use of ginger as an antiemetic for the prophylaxis of nausea and vomiting associated with motion sickness, post-operative nausea, hyperemesis gravidarum, and seasickness. While therapeutic doses of ginger rhizome appear to be clinically effective for the treatment of hyperemesis gravidarum, there is significant controversy concerning the mutagenic effects of certain ginger constituents. Therefore until further safety data is available, ginger should only be administered during pregnancy upon the advice of a physician.

Ginger rhizome preparations have also been used traditionally for the treatment of dyspepsia, flatulence, colic, vomiting, diarrhea, spasms and other stomach complaints. Currently there are no data from controlled clinical trials to support these indications. The recommended dose of ginger for the treatment of motion sickness for adults and children $> 6$ years is 0.5 g, 2 to 4 times daily. For treatment of dyspepsia the recommended dose is 2 to 4 g daily, as powdered drug or extracts. While there are no known contraindications, due to possible anticoagulant activity ginger preparations should be used with caution in those patients receiving other anticoagulant or antiplatelet therapy, or having bleeding disorders. Minor adverse reactions such as heartburn have been reported in clinical trials. Ginger preparations should not be administered to children less than 6 years of age, or during pregnancy and lactation without the advice of a physician.

**Introduction**

Ginger (*Zingiber officinale*), a pungent aromatic spice plant commonly used in baking and flavoring beverages, is the 10th most important spice in commerce.[1] Commercial products of ginger, sometimes referred to as ginger root, are actually prepared from the rhizome, which is an underground stem of the plant. The commercial varieties are often named after their place of origin, such as Chinese ginger, African ginger or Jamaican ginger.[2] Ginger has a long history of use throughout the world as a medicinal plant, and has been employed as a digestive aid for upset stomachs and dyspepsia, and also as an antiemetic agent. Originating in the tropical jungles of Asia, it is still important both as a food and medicine in China, Thailand, Malaysia, Vietnam and Cambodia.[3] The medical uses of ginger were recorded in the early Sanskrit and Chinese texts, as well as being documented in the ancient Greek, Roman and Arabic medical literature.[3] It is believed that ginger has been employed in China for over 2,500 years as a medicinal plant, and is an ingredient in many traditional Chinese herbal medicines.[2] The use of ginger in western cultures is more recent, as ginger was introduced to Europe around the 9th century and to North America in the 16th century.[1] In most developing countries ginger is used to treat dyspepsia, stomach ailments, toothache, coughs due to the common cold, rheumatism, and fevers. Much of the modern clinical studies of ginger have focused on its use as an antiemetic for the prevention and treatment of motion and seasickness, as well as morning sickness. In the U.S., ginger is sold as a dietary supplement and is on the Food and Drug Administration's GRAS (Generally Regarded as Safe) list.

## Quality Information

*   The correct Latin name for ginger is *Zingiber officinale* Roscoe (Zingiberaceae). Botanical synonyms that may appear in the scientific literature include *Amomum zingiber* L., and *Zingiber blancoi* Massk. Numerous vernacular (common) names for ginger include: African ginger, allam, baojiang, beuing, chiang, citaraho, Cochin ginger, common ginger, gingembre, ginger root, Zingberwurgel, Jamaica ginger, janzabeil, shengiang, shenjing, shoga, shouhkyoh, Zingiberis rhizoma, zinjabil, and Zingiber.[4]
*   Standardized extracts and other commercial products of ginger are prepared from the dried rhizome of *Zingiber officinale*.[4]
*   Ginger is native to Southeast Asia and is cultivated in the tropical regions in both the Eastern and Western Hemispheres, being commercially grown in India, China, Africa and the West Indies (Jamaica).[4]
*   Ginger rhizomes contain approximately 1 to 4% essential oil and an oleoresin, the composition of which varies depending on the geographical origin of the plant. However, the major constituent sesquiterpene hydrocarbons, that are responsible for the aroma, appear to remain constant. The five major sesquiterpene hydrocarbons in ginger oil include α-zingiberene, (+)-ar-curcumene, (−)-β-sesquiphellandrene, (E,E)-α-farnesene and β-bisabolene. The constituents responsible for the pungent smell and taste of ginger rhizome and may be the active anti-emetic constituents are called the gingerols and their corresponding dehydration products, are named shogaols.[4]
*   Daily dosage: Motion sickness: adults, children > 6 years: 0.5–1.0 g, 2 to 4 times daily. Dyspepsia: 2 to 4 g daily, as powdered drug or extracts.[4]

## Medical Uses

Ginger is employed for the prophylaxis of nausea and vomiting associated with motion sickness,[5-8] post-operative nausea,[9] hyperemesis gravidarum,[10] and sea sickness.[11-12] Although ginger appears to be clinically effective in the treatment of hyperemesis gravidarum, it is currently not recommended for use in morning sickness during pregnancy.[10]

Powdered ginger rhizome is also used traditionally for the treatment of dyspepsia, flatulence, colic, vomiting, diarrhea, the common cold and flu, and as an anti-inflammatory agent in the treatment of migraine headache, and rheumatic and muscular disorders.[4]

## Summary of Clinical Evidence

At least eight clinical trials have been performed to assess the antiemetic effects of ginger rhizome.[7,8,11-16] Five of the investigations have shown that orally administered ginger was effective for prophylactic therapy of nausea and vomiting associated with motion sickness. The other three studies have shown that ginger was no more effective than placebo in treating motion sickness.[8,14,15] The discrepancy between the results of clinical trials appears to be a function of the focus of these investigations. The results from clinical trials that assessed the gastrointestinal reactions involved in motion sickness have recorded better responses to ginger therapy than those clinical trials that focused primarily on responses involving the central nervous system.[14,15]

A placebo controlled clinical trial compared the efficacy of orally administration of powdered ginger rhizome (940 mg) with that of dimenhydrinate (100 mg) for preventing the gastrointestinal symptoms of motion sickness in 36 healthy subjects. Powdered ginger rhizome was superior to dimenhydrinate in preventing the gastrointestinal symptoms of motion sickness. The results of this study further suggested that ginger did not act centrally on the vomiting center, but had a direct effect on the gastrointestinal tract through its aromatic, carminative and absorbent properties, by increasing gastric motility and adsorption of toxins and acids.[7]

It has been hypothesized that the mechanism by which ginger exerts its antiemetic effects may be due to an increase in gastrointestinal motility, thereby increasing gastric emptying. However, two randomized, placebo-controlled clinical trials have demonstrated that oral doses of ginger (1 g) did not affect the gastric emptying rate in healthy volunteers, as measured by sequential gastric scintigraphy[15] or paracetamol absorption technique.[16]

Two randomized, double-blind clinical trials have assessed the effect of powdered ginger root was tested as a prophylactic treatment for seasickness.[11,12] A placebo-controlled clinical trial involving 80 naval cadets in heavy seas, demonstrated that 1 gram of orally administered ginger was statistically better than placebo in decreasing the incidence of vomiting and cold sweating 4 hours after ingestion, however it did not significantly reduce the associated nausea or vertigo.[12] A comparison trial assessed the effects of seven over-the-counter and prescription antiemetic drugs for the prevention of seasickness in 1489 subjects. This study concluded that ginger was as effective as the other antiemetic drugs tested.[11]

A randomized, double-blind cross-over trial assessed the efficacy of orally administration of powdered ginger for the treatment of *hyperemesis gravidarum*.[10] Thirty pregnant women with morning sickness received 250 mg of powdered ginger rhizome or placebo four times daily for four days. After a two-day washout, the patients were crossed over for another four days of treatment. Seventy percent of women in the study preferred the ginger treatments, which significantly reduced the degree of nausea and the number of vomiting attacks ($p = 0.035$).[10]

In a prospective, randomized, double-blind study, there were statistically significant fewer cases of post-operative nausea and vomiting in 60 women who had major gynecological surgery the effects of ginger ($2 \times 0.5$ gm) was compared to placebo.[17] There was a statistically significant reduction in the incidence of nausea in the group treated with ginger as compared with the placebo group. The effect of ginger on post-operative nausea and vomiting was reported to be similar to that of metoclopramide.[17] A similar prospective, randomized double blind trial involving 120 women undergoing laparoscopic surgery found that treatment of the patients with ginger (1 g) was equivalent to metoclopramide, and both treatments were superior to placebo.[18] In contrast, another double-blind randomized study concluded that orally administered ginger BP was ineffective in reducing the incidence of postoperative nausea and vomiting.[19]

In a clinical trial involving 11 patients experiencing nausea after administration of 8-MOP prior to photopheresis demonstrated that ginger administration reduced the incidence of nausea in two-thirds of the patients. The only adverse effect reported was a feeling of heartburn in several patients.[20]

A systematic review of the randomized controlled clinical trials assessing the efficacy of ginger as an antiemetic for nausea and vomiting has been performed.[21] Six controlled trials met the inclusion criteria and were reviewed. The review concluded that while ginger is a promising antiemetic, the clinical data are insufficient to draw firm conclusions and further rigorous studies are needed to establish its efficacy.[21]

### Pharmacokinetics

No pharmacokinetics studies have been performed.

### Mechanism of Action

In animal studies, ginger extracts have been shown to stimulate the motility of the gastrointestinal tract, and enhance bile production and secretion. Intraduodenal administration of an acetone extract (primarily essential oils) of ginger rhizome to rats increased bile secretion for three hours.[4] A water extract of the rhizome was not active. The active chemical constituents of the essential oil were identified as [6]- and [10]-gingerol.[4]

The antiemetic effects of extracts of ginger rhizome were investigated in healthy dogs with emesis induced by 3 mg/kg of cisplatin.[22] An acetone and 50% ethanol extract inhibited cisplatin-induced emesis at an oral dose of 200 mg/kg, but the water extract was not active at any of the tested doses. No ginger extract was effective against apomorphine-induced emesis.[22] The emetic action of the peripherally acting agent, copper sulfate, was inhibited in dogs given an intragastric dose of ginger extract, but emesis in pigeons treated with centrally acting emetics such as apomorphine and digitalis are not inhibited after administration of a ginger extract. These results suggest that ginger's antiemetic activity is peripheral and does not involve the CNS.[4]

Intragastric administration of an acetone extract of ginger rhizome (75 mg/kg), or the isolated active constituents, [6]-shogaol (2.5 mg/kg), or 6-, 8- or 10-gingerol enhances gastrointestinal motility in mice.[23] The effect of an acetone extract on gastrointestinal motility was comparable to or slightly weaker than that of metoclopramide (10 mg/kg) and domperidone. The 6-, 8- or 10-gingerols are reported to have antiserotoninergic activity, and it has been suggested that the effects of ginger on gastrointestinal motility may be due to this activity.[23] The mode of administration plays an important role in the effects of ginger on gastrointestinal motility. Both (6)-gingerol and [6]-shogaol inhibit intestinal motility when administered intravenously, however these compounds accentuate gastrointestinal motility after oral administration.[24] Ginger also has gastroprotective activity and inhibits the development of gastric lesions in mice after treatment with ethanol, hydrochloric acid, sodium hydroxide, indomethacin and aspirin.[25]

Recent studies have demonstrated that methanol extracts of dried ginger rhizome inhibit the growth of 15 clinical strains of *Helicobacter pylori in vitro* (MIC range 12.5 to 50 μg/ml).[26] Inasmuch as dyspepsia, *hyperemesis gravidarum*[27] and peptic ulcer disease[28] are all associated with *H. pylori* infections, these data suggest that ginger may exert its therapeutic effects through inhibition of the growth of this bacterium.

Ginger has also been reported to have antiinflammatory activity. *In vitro* studies have shown that hot water extracts of ginger rhizome inhibit the activities of cyclooxygenase and lipoxygenase in the arachidonic acid cascade, thus exerting an antiinflammatory effect by decreasing in the formation of prostaglandins and leukotrienes.[29,30] Two labdane-type diterpene dialdehydes isolated from ginger extracts have been shown to be inhibitors of human 5-lipoxygenase *in vitro*.[31] Ginger is also a potent inhibitor of thromboxane synthetase, and raised prostacyclin levels without a concomitant rise in $PGE_2$ or $PGF_2$ alpha.[32] *In vivo* studies in rodents have shown that oral administration of ginger extracts decreased rat pedal edema.[33,34] The potency of the extracts was comparable to acetyl salicylic acid. (6)-shogaol inhibited carrageenin-induced pedal edema in rats via inhibition of cyclooxygenase activity.[35]

**Safety Information**

*A. Adverse Reactions*
Minor cases of heartburn are reported as the primary side effect in clinical trials.[20]
Contact dermatitis of the fingertips has been reported in sensitive patients.[36]

*B. Contraindications and Warnings*
Patients taking anticoagulant drugs or those with blood coagulation disorders should
consult their physician prior to self-medication with ginger. Patients with gallstones
should consult their physician before using ginger preparations.[6]

*C. Drug Interactions*
It has been suggested that ginger may alter blood coagulation and immunological
parameters due to its ability to inhibit the activity of thromboxane synthetase, and to
act as a prostacyclin agonist.[37,38] However, a randomized double-blind study
assessed the effects of dried ginger (2 g daily, orally for 14 days) on platelet function, and found that there were no differences in bleeding times in patients receiving
ginger or placebo.[39] Another randomized, placebo-controlled crossover trial involving 18 healthy volunteers found that consumption of 15 grams of raw ginger or 40
grams of cooked ginger per day for two weeks had no effect on platelet thromboxane production.[40] Large doses (10–14 g) of dried ginger rhizome per day may
enhance the hypothrombinemic effects of anticoagulant therapy, however the clinical significance has yet to be evaluated.[4]

*D. Toxicology*
Assessment of the mutagenicity of ginger rhizome extracts is controversial. Hot
aqueous extracts of ginger was reported to be mutagenic in B291I cells and *Salmonella typhimurium* strain TA 100, but not in strain TA98.[41] A number of constituents
of fresh ginger have been identified as mutagens. Both [6]-gingerol and shogaol have
been determined to be mutagenic in a *Salmonella*/microsome assay,[42] and increased
mutagensis was observed in a Hs30 strain of *Escherichia coli* treated with [6]-gin-
gerol.[43] However, the mutagenicity of [6]-gingerol and shogaol is reduced in the
presence of different concentrations of zingerone, an antimutagenic constituent of
ginger.[42] Additionally, ginger juice is anti-mutagenic and suppresses spontaneous
mutations induced by [6]-gingerol, except in cases where the mutagenic chemicals
2-(2-furyl)–3-(5-nitro-2-furyl)acryl amide and N-methyl-N'-nitro-N-nitrosoguani-
dine were added in addition to [6]-gingerol. Other investigations have demonstrated
that ginger juice is antimutagenic.[44,45]

No teratogenic effects have been observed in animal or human studies. In a double-blind randomized crossover clinical trial, ginger (250 mg orally, 4 × daily) was
found to be effective in the treatment of *hyperemesis gravidarum*.[10] No teratogenic
aberrations were observed in infants born during this study, and all newborns had
Apgar scores of 9–10 after 5 min.[10] However, there is limited information concerning

the effect of therapeutic doses of ginger on the newborn or children under the age of 6 years.

*E.  Dose and Dosage Forms*
Motion sickness: adults, children > 6 years: 0.5 g, 2 to 4 times daily.
Dyspepsia: 2 to 4 g daily, as powdered drug or extracts.[4]

## References

1     Awang DVC. Ginger. *Can Pharm J.* 1992;309–311.
2     Robbers JE, Tyler VE. *Herbs of Choice*, Hawthorne Herbal Press, New York 1999.
3     Bone K. Ginger. *Brit J Phytother.* 1997;4:110–120.
4     Anon. Rhizoma Zingiber. *WHO Monographs on Selected Medicinal Plants.* Volume I, WHO, Geneva, Switzerland 1999.
5     Reynolds JEF ed. *Martindale The Extra Pharmacopoeia.* 30th ed. London, The Pharmaceutical Press, 1993, pp. 885.
6     German Commission E monographs, Ginger rhizome. *Bundesanzeiger.* 1990.
7     Mowrey DB, Clayson DE. Motion sickness, ginger, and psychophysics. *Lancet.* 1982: 655–657.
8     Holtmann S et al. The anti-motion sickness mechanism of ginger. A comparative study with placebo and dimenhydrinate. *Acta Otolaryngol.* (Stockholm), 1989;108:168–174.
9     Bone ME et al. Ginger root – a new antiemetic. The effect of ginger root on postoperative nausea and vomiting after major gynaecological surgery. *Anesthesia.* 1990;45: 669–671.
10    Fischer-Rasmussen W et al. Ginger treatment of *hyperemesis gravidarum. European Journal of Obstetrics, Gynecology and Reproductive Biology.* 1991;38:19–24.
11    Schmid R et al. Comparison of seven commonly used agents for prophylaxis of seasickness. *Journal of Travel Medicine.* 1994;1:203–206.
12    Grontved A et al. Ginger root against seasickness. A controlled trial on the open sea. *Acta Otolaryngol.* (Stockholm), 1988;105:45–49.
13    Stott JR, Hubble MP, Spencer MB. A double-blind comparative trial of powdered ginger root, hyosine hydrobromide, and cinnarizine in the prophylaxis of motion sickness induced by cross coupled stimulation. *Advisory Group for Aerospace Research Deveolpment. Conference Proceedings* 1984;39:1–6.
14    Wood CD et al. Comparison of the efficacy of ginger with various antimotion sickness drugs. *Clinical Research Pract Drug Regulatory Affairs.* 1988;6:129–136.
15    Stewart JJ et al. Effects of ginger on motion sickness susceptibility and gastric function. *Pharmacology.* 1991;42:111–120.
16    Phillips S, Hutchinson S, Ruggier R. *Zingiber officinale* does not affect gastric emptying rate. *Anaesthesia.* 1993;48:393–395.
17    Bone ME et al. Ginger root – a new antiemetic. *Anaesthesia.* 1990;45:669–671.
18    Phillips S, Ruggier R, Hutchinson SE. *Zingiber officinale* (ginger): an antiemetic for day case surgery. *Anaesthesia.* 1993;48:715–717.
19    Arfeen Z et al. A double-blind randomized controlled trial of ginger for the prevention of postoperative nausea and vomiting. *Anaesthesia and Intensive Care.* 1995;23: 449–452.
20    Meyer K et al. *Zingiber officinale* (Ginger) used to prevent 8-MOP associated nausea. *Dermatol Nurs.* 1995; 7:242–244.

21    Ernst E, Pittler MH. Efficacy of ginger for nausea and vomiting: a systematic review of randomized clinical trials. *Brit J Anaesthesia.* 2000;84:367–371.

22    Sharma SS et al. Antiemetic efficacy of ginger (*Zingiber officinale*) against cisplatin-induced emesis in dogs. *J Ethnopharmacol.* 1997;57:93–96.

23    Yamahara J et al. Gastrointestinal motility enhancing effect of ginger and its active constituents. *Chem Pharm Bull.* 1991;38:430–431.

24    Suekawa M et al. Pharmacological studies on ginger. I. Pharmacological actions of pungent components, (6)-gingerol and (6)-shogaol. *Journal of Pharmacobio-Dynamics.* 1984;7:836–848.

25    Al Yahya M et al. Gastroprotective activity of ginger *Zingiber officinale* Rosc. in rats. *Am J Chin Med.* 1989;17:51–56.

26    Mahady GB et al. Inhibition of *Helicobacter pylori* by plants used traditionally for the treatment of gastrointestinal disorders. *Phytomedicine.* 2000;7:95.

27    Frigo P et al. *Hyperemesis gravidarum* associated with *Helicobacter pylori* seropositivity. *Obstet Gyneol.* 1998;91:615–617.

28    Laine L, Fendrick AM. *Helicobacter pylori* and peptic ulcer disease. *Postgrad Med.* 1998; 103:231–238.

29    Srivastava KC, Mustafa T. Ginger (*Zingiber officinale*) in rheumatism and musculosketetal disorders. *Med Hypothesis.* 1992;39:342–348.

30    Mustafa T, Srivastava KC, Jensen KB. Drug development report 9. Pharmacology of ginger, *Zingiber officinale. J Drug Develop.* 1993;6:25–39.

31    Srivastava KC. Aqueous extracts of onion, garlic and ginger inhibit platelet aggregation and alter arachidonic acid metabolism. *Biomed Biochim Acta.* 1984;43:335–346.

32    Mascolo N et al. Ethnopharmacologic investigation of ginger (*Zingiber officinale*). *J Pharmacol.* 1989;27:129–140.

33    Sharma, JN, Srivastava, KC, Gan, EK. Suppressive effects of eugenol and ginger oil on arthritic rats. *Pharmacology.* 1994;49:314–318.

34    Suekawa M, Yuasa K, Isono M. Pharmacological studies on ginger: IV. Effects of (6)-shogaol on the arachidonic cascade. *Folia Pharmacologia Japan.* 1986;88:236–270.

35    Kawakishi S, Morimitsu Y, Osawa T. Chemistry of ginger components and inhibitory factors of the arachidonic acid cascade. *American Chemical Society Symposium Series.* 1994;547:244–250.

36    Seetharam KA, Pasricha JS. Condiments and contact dermatitis of the finger tips. *Indian Journal of Dermatology, Venereol Leprol.* 1987;53:325–328.

37    Bordia A et al. Effect of ginger (*Zingiber officinale* Roscoe) and fenugreek (*Trigonella foenum graecum* L.) on blood lipids, blood sugar and platelet aggregation in patients with coronary artery disease. *Prostaglandins, Leukotrienes and Essential Fatty Acids.* 1997;56:379–384.

38    Backon J. Ginger as an antiemetic: possible side effects due to its thromboxane synthetase activity. *Anaesthesia.* 1991;46:05–706.

39    Srivastava KC. Isolation and effects of some ginger components on platelet aggregation and eicosanoid biosynthesis. *Prostaglandins Leukotrienes in Medicine.* 1986;25:187–198.

40    Janssen PLTMK et al. Consumption of ginger (*Zingiber officinale* Roscoe) does not affect *ex vivo* platelet thromboxane production in humans. *Eur J Clin Nutr.* 1996;50:772–774.

41    Yamamoto H, Mizutani T, Nomura H. Studies on the mutagenicity of crude drug extracts. *Yakugaku Zasshi.* 1982;102:596–601.

42    Nagabhushan M, Amonkar AJ, Bhide SV. Mutagenicity of gingerol and shogoal and antimutagenicity of zingerone in *Salmonella*/microsome assay. *Cancer Letters.* 1987;36:221–233.

43    Nakamura H, Yamamoto T. Mutagen and anti-mutagen in ginger, *Zingiber officinale*.
      *Mut Res.* 1982;103:119–126.
44    Kada T, Morita M, Inoue T. Antimutagenic action of vegetable factor(s) on the muta-
      genic principle of tryptophan pyrolysate. *Mut Res.* 1978;53:351–353.
45    Morita K, Hara M, Kada T. Studies on natural desmutagens:screening for vegetable and
      fruit factors active in inactivation of mutagenic pyrolysis products from amino acids. *Agr
      Biol Chem.* 1978;42:1235–1238.

# 16

# *Ginkgo biloba*

## Synopsis

Clinical evidence supports the use of *Ginkgo biloba* for the symptomatic treatment of age-associated memory impairment and dementia, including primary degenerative dementia, vascular dementia (multi-infarct), and Alzheimer disease (early stages), and for the symptomatic treatment of intermittent claudication. The recommended dose of a standardized *Ginkgo biloba* extract (standardized to contain between 24–26% flavonoid glycosides and 5–7% ginkgolides) is 40–60 mg three to four times daily. Treatment for 1 to 3 months may be required before the full therapeutic effects are apparent. While there are no known contraindications, due to potential PAF antagonism and hemorrhagic reactions, Ginkgo preparations should be used with caution in patients receiving anticoagulant or antiplatelet therapy, or having bleeding disorders. Other minor adverse reactions such as nausea, vomiting, headaches and dizziness have been reported in clinical trials. Administration *Ginkgo biloba* preparations to children, or during pregnancy and lactation is not recommended.

## Introduction

*Ginkgo biloba* (Ginkgoaceae) is a tall, monotypic dioecious tree, and the only living representative of the Ginkgoales.[1–4] Known as a living fossil, the tree has a long life span and is remarkably resistant to insects, bacterial and viral infections, and air pollution.[2] In fact, many specimens in the China are thought to be over 1000 years old. The tree, known for its great beauty, has fan-shaped, bilobed leaves, which turn from green to golden yellow in the autumn. *Ginkgo biloba* is commonly referred to as the maidenhair tree because the shape and vein of the leaves are similar to that of the maidenhair fern.[2] The tree is native to China, but is grown as an ornamental shade tree in Europe, Japan, Australia, South East Asia and the USA.[1,3–6] It is commercially cultivated in China, France, Korea and the USA.[3]

Therapy with *Ginkgo* preparations can be traced back approximately 5000 years to the origins of traditional Chinese medicine, and *Ginkgo* was described in the ancient Chinese medical texts such as *Chen Noung Pen T'sao*.[1] In China, the seeds of the

Ginkgo tree are considered to be a tonic, however the fruit and seeds contain large quantities of ginkgolic acids, which are potent allergens.[7,8] The medicinal uses of *Ginkgo* seeds were reported in the *Pen Ts'ao Kang Mu* (Great Herbal, 1596 AD) written by Li Shih-Chen, which is considered to be the treatise of traditional Chinese medicine.[2] *Ginkgo* seeds were used for the treatment of alcohol abuse, asthma, bladder inflammation, coughs, and leukorrhea.[2] The medical use of *Ginkgo biloba* leaves actually dates back to 1550 AD, when they were used for the treatment of asthma and cardiovascular diseases.[2,4] However, the medicinal qualities of *Ginkgo biloba* were rarely mentioned in European herbals, until the middle of the 19th century. In contrast to traditional Chinese medicine, where the seeds are utilized, in Europe only the leaf extracts are employed therapeutically. In 1965, Dr. Willmar Schwabe, a German physician-pharmacist, introduced standardized *Ginkgo biloba* leaf extracts into Western medical practice.[1,2] Today, the standardized extracts have been used in Europe for the treatment of memory deficits associated with aging, dementia, and peripheral arterial occlusive diseases. Over the last 20 years, much of the clinical work with *Ginkgo biloba* extract has been performed in Europe, particularly Germany, France and Italy.

## Quality Information

*   The correct Latin name for the tree is *Ginkgo biloba* L. Botanical synonyms that may appear in the scientific literature include: *Pterophyllus salisburiensis* Nelson, *Salisburia adiantifolia* Smith, and *Salisburia macrophylla* C. Koch.[2–6] Numerous vernacular (common) names for *Ginkgo* include: Eun-Haeng, ginnan, gingko, ginkgo balm, ginkgo leaves, ginkyo, ginan, icho, ityo, kew tree, maidenhair tree, pei-wen, temple balm, yin guo, yinhsing.[1–8]
*   Standardized extracts and other commercial products of *Ginkgo biloba* are prepared from the dried whole leaves.[9,10]
*   *Ginkgo biloba* leaves contain a wide variety of chemical constituents, including alkanes, lipids, sterols, benzenoids, carotenoids, phenylpropanoids, carbohydrates, flavonoids and terpenoids.[11] The active chemical constituents include the flavonoids, which consist of mono-, di-, and tri-glycosides and coumaric acid esters based on the flavonols kaempferol and quercetin predominate, with lesser quantities of glycosides derived from isorhamnetin, myricetin, and 3'-methylmyricetin. Non-glycosidic biflavonoids, catechins and proanthocyanidins are also present.[12] Characteristic constituents of this plant material are the unique diterpene lactones, ginkgolides A, -B, -C, -J and -M; and the sesquiterpene lactone, bilobalide.[13]

## Medical Uses

Symptomatic treatment of mild to moderate age-associated memory impairment and dementia, including primary degenerative dementia, vascular (multi-infarct) dementia, and Alzheimer disease.[14–23] The symptoms associated with progressive degenerative dementia are described in *the International Classification of Diseases* (ICD 10) published by the World Health Organization, or the *Diagnostic and Statistical Manual of Mental Disorders* (DSM-IV) of the American Psychiatric Association.[24,25]

Symptomatic treatment (improvement of pain-free walking distance) of peripheral arterial occlusive disease such as intermittent claudication.[1,14,26–34]

Treatment of inner ear disorders such as acoustic trauma, hearing loss, tinnitus and vertigo of vascular and involutive origin.[36–50]

Treatment of early diabetic retinopathy, and senile macular degeneration.[51,52]

## Summary of Clinical Evidence

*Age Associated Memory Impairment and Dementia*
Numerous clinical trials have assessed the efficacy of *Ginkgo biloba* extracts for the treatment of "cerebral insufficiency" and demential syndromes.[14–23] Over 40 of the clinical trials were published in Europe between 1975–1991, and all evaluated the efficacy of a standardized *Ginkgo biloba* extract for the symptomatic treatment of age-related "cerebral insufficiency".[14–15] Symptoms associated with "cerebral insufficiency" include: memory deficits, distrubances in concentration, depressive emotional conditions, dizziness, tinnitus and headache. While the term "cerebral insufficiency" is widely accepted in Europe, it is not listed in the diagnostic classifications of the ICD 10 or the DSM-IV.[25] Cerebral insufficiency is a poorly defined syndrome that actually includes most of the symptoms associated with progressive degenerative dementia of the ICD-10 and DSM-IV categories.[24,25] Thus, much of the clinical work, performed prior to 1991, does not differentiate between a diagnosis of age-associated memory impairment and early dementia.

A review of the clinical trials published between 1975 and 1991, critically assessed the effects of a standardized, orally administered *Ginkgo biloba* extract for the symptomatic treatment of cerebral insufficiency.[14,15] Methodological evaluation of the 40 published trials, determined that only eight of the studies were of acceptable quality. All other trials suffered from methodological flaws such as small patient numbers; inadequate randomization; and poor description of patient characteristics, effect measurements or data presentation.[14,15] Of the eight clinical trials reviewed, all but one study, demonstrated that treatment with a standardized *Ginkgo biloba* extract (120–160 mg/day for 4 to 6 weeks) reduced the symptoms of cerebral insufficiency such as: difficulties in concentration and memory, confusion, fatigue, decreased physical performance, depression, anxiety, dizziness, tinnitus and headache as compared with placebo.[14,15] Comparison of the best clinical trials of *Ginkgo biloba* (120 mg/d of a standardized extract) with five of the best trials of ergoloid mesylates (Hydergine®, 4.5 mg/d), a drug used for the same purpose, demonstrated that both drugs had similar efficacy after 6 weeks of therapy.[17]

A meta-analysis of 11 randomized, double-blind, placebo-controlled clinical trials in aged patients assessed the efficacy of an orally administered *Ginkgo biloba* standardized extract (LI 1370, containing 25% flavonoid glycosides and 6% ginkgolides) for the symptomatic treatment of cerebral insufficiency.[18] Of the 11 trials, 8 studies were comparable with regard to diagnosis, inclusion and exclusion criteria and methodology, and were included in the meta-analysis.[18] Patients were treated with an average daily dose of 150 mg of the extract or placebo for 6–12 weeks. For all analyzed single symptoms, of *Ginkgo biloba* extract was superior to placebo and significant differences were demonstrate. After analysis of the total score of clinical symptoms, seven studies confirmed the effectiveness of *Ginkgo biloba* extract, while one study was inconclusive.[18]

A meta-analysis of the published clinical trials (1975–1997) assessed the effect of treatment with *Ginkgo biloba* extract on the objective measures of cognitive function in patients with Alzheimer disease (AD).[19] Published trials had to meet the following inclusion criteria: (1) a clear diagnosis of AD by either Diagnostic and Statistical Manual of Mental Disorders, 3rd ed, or National Institutes of Neurological Disorders and Stroke- Alzheimer's Disease Association criteria, (2) description of exclusion criteria, (3) use of a standardized extract in any stated dose; (4) randomized, double-blind, placebo-controlled study design; (5) at least 1 outcome measure was an objective assessment of cognitive function; and (6) sufficient statistical information to allow for meta-analysis. Of the 57 published articles identified, only 4 clinical trials met all the inclusion criteria,[20–23] and only one trial prior to 1991 met the acceptable scientific criteria.[23] The 4 clinical trials included in the meta-analysis reported analyzable data for 424 patients. The number of subjects per individual study ranged from 19 to 104. Patients were treated with 120–240 mg of a standardized *Ginkgo* extract (EGb 761) or placebo for 12 to 26 weeks. All patients in the trials were diagnosed with mild to moderate dementia, and two studies included patients with vascular dementia.[20,21] However, only groups composed of patients solely diagnosed as having AD were included in the final analysis. The meta-analysis concluded that that administration of a standardized *G. biloba* extract has a modest effect on cognitive function in patients with AD. Overall there was a statistically significant effect size of 0.40 (p < 0.0001). The effect size translated into a small (3%), but significant difference in the Alzheimer Disease Assessment Scale-cognitive subtest.[19] The effect size was comparable with that observed in a double-blind, placebo-controlled trial of donepezil, a cholinesterase inhibitor, for treatment of patients with Alzheimer's disease.[53]

*Peripheral Arterial Occlusive Disease*
The efficacy of a standardized *Ginkgo biloba* extract (EGb 761) for the treatment of intermittent claudication (peripheral arterial occlusive disease Fontaine stage II) has been assessed in at least 20 clinical trials.[1,14,26–34] A published review of the earlier 15 clinical trials (prior to 1991),[14] determined that only two of the studies employed acceptable methodology.[26,27] In a 6-month, double-blind, placebo-controlled trial of 79 patients suffering from peripheral arteriopathy (Fontaine's stage II b), an increased walking distance of 112 to 222 meters was observed in patients treated with a standardized *Ginkgo* extract (120 mg/day) as compared with placebo

(145 to 176 meters).[26] In the second trial, a reduction of pain at rest: a decrease on a 100 mm visual analogue scale for pain from 61 to 30 mm was observed in patients with intermittent claudication treated with infusions of 200 mg of a standardized *Ginkgo* extract for eight weeks.[27] A meta-analysis and a systematic review of the earlier placebo-controlled and comparison clinical trials (1975 to 1988) evaluated the efficacy of *Ginkgo biloba* special extract (EGb 761) in the treatment of patients with intermittent claudication, as measured by an increase in walking distance.[28,29] In the meta-analysis, five controlled clinical trials with similar design and inclusion criteria were analyzed.[28] The results of the meta-analysis showed that the effect size in all five trials was homogeneous and the global effect size was estimated as 0.75. The conclusion of this meta-analysis was that a standardized *Ginkgo biloba* extract was more effective than placebo for the symptomatic treatment of intermittent claudication.[28] A systematic review of the clinical trials came to similar conclusions, but suggested that further trials employing more rigorous methodology should be performed.[29]

The methodology and results of the randomized, double-blind, placebo-controlled clinical trials (1977–1990) of *Ginkgo biloba* extract (EGb 761) and pentoxifylline for the treatment of intermittent claudication were reviewed.[30] The review concluded that the trials for both drugs suffered from similar methodological flaws, and that there was no difference between the studies with respect to quality. The averaged results of all studies for each drug showed an increase in pain-free walking distance of 45% for *Ginkgo biloba* extract and 57% for pentoxifylline.[30] Since 1991, 5 randomized, double-blind, placebo-controlled clinical trials have been performed to evaluate the efficacy of a standardized *Ginkgo biloba* extract for the symptomatic treatment of intermittent claudication.[31–35] The total number of patients in these trials was 250 (range 18–111), and the dosage ranged from 120 to 320 mg/day of a standardized *Ginkgo* extract (EGb 761), and 120 mg/day of a *Ginkgo* extract (GB-8, no information on standardization). The length of the trials ranged from 4 to 24 weeks. Three of the studies demonstrated that a standardized extract produced a statistically significant increase in pain-free walking distance as compared with placebo ($p < 0.01$–$0.05$).[31–33] Two of the studies further demonstrated that in addition to physical training, patients treated with a standardized *Ginkgo biloba* extract (120 to 160 mg for 24 weeks) had a statistically significant improvement in walking distance.[31,32] A fourth study demonstrated a 38% decrease in the areas of ischemia (determined by measuring the transcutaneous partial pressure of oxygen during exercise and rest) in those patients treated with a standardized *Ginkgo* extract as compared with placebo.[34] The fifth study, which did not use a standardized *Ginkgo* extract, was negative and no significant change in walking distance or the severity of leg pain was observed as compared with placebo.[35]

### Inner Ear Disorders

Standardized *Ginkgo biloba* leaf extracts have been used clinically in the treatment of inner ear disorders such as hearing loss, vertigo and tinnitus. Numerous open and placebo-controlled, double-blind studies have assessed the efficacy of a standardized *Ginkgo biloba* extract in the treatment of vestibular disorders.[36–45] In a multicenter study, 67 patients with vertiginous syndrome of recent onset without definable etiopathology, were treated with *Ginkgo biloba* extract (120 to 160 mg daily).[36] After 12 weeks of therapy, 47% of the patients treated with the extract were symptom free as compared with 18% in the placebo group.[36] In another study of 33 elderly patients (mean age 59 years old) with existing symptoms of vertigo and ataxia were treated with 120 mg of a standardized *Ginkgo biloba* extract or placebo for 12 weeks.[37] The therapeutic outcome was measured by cranio-corpography (CCG) and by direct questioning of the patients. After treatment with Ginkgo, the CCG showed a statistically significant decrease in the lateral sway amplitude in the stepping on the spot test, as compared with placebo ($p < 0.005$). On direct questioning 20% of the patients in the placebo group and 50% of the patients in the treatment group reported a subjective decrease of their vertigo symptoms.[37] A randomized double-blind study of 35 patients with vertigo symptoms due to Menieré's disease, vestibular neuropathy and post-traumatic vertigo assessed the efficacy of a combination of physical training to improve vestibular and proprioceptive reflexes and *Ginkgo biloba* extract for symptomatic treatment.[38] Amelioration of vertigo and a significant decrease of body sway amplitude were observed after vestibular training and placebo. However, concomitant treatment of vestibular training with 160 mg/day of *Ginkgo biloba* extract for 28 days led to a further decrease in sway amplitude (posturographic body sway measurements in Romberg's test, $p < 0.0001$).[38]

While at least eight clinical trials have assessed the efficacy of *Ginkgo biloba* extract for the treatment of tinnitus, results from these trials are conflicting.[41–48] Five studies reported positive results.[41–45] One multicenter, randomized, double-blind, 13 month study of 103 patients with tinnitus showed improved conditions in all patients, irrespective of the prognostic factor, when treated with *Ginkgo biloba* extract (160 mg/daily for 3 months) as compared with placebo.[41] In a comparison trial, *Ginkgo biloba* extract was administered to 64 patients with tinnitus and hearing loss.[42] The patients were treated for 9 weeks and the efficacy was compared with other vasodilating agents such as flunaricine, nicotinic acid and xanthinol nicotinate. Hearing improved in several patients and in 33 patients with permanent severe tinnitus, the tinnitus disappeared completely in 36% and improved in an additional 15%.[42] In a comparative, randomized multicenter study of 259 patients, the effects of *Ginkgo* extract, almitrine-raubasine combination and nicergoline on tinnitus was assessed.[43] The tinnitus disappeared completely in 35.5% of patients when treated with *Ginkgo* extract, 16.7% when treated with nicergoline, and 15.2% when treated with the almitrine-raubasine combination.[43] Two clinical trials assessed the efficacy of a combination of low power laser and *Ginkgo biloba* extract for the treatment of tinnitus.[44,45] Three other clinical trials reported negative outcomes.[46–48] Treatment of 21 patients with 120 mg/day of a standardized *Ginkgo* extract for 12 weeks had no significant positive outcome on tinnitus.[46] Similarly, no positive effects on tinnitus were observed in a *Ginkgo* extract (240 mg/day) versus placebo-controlled study.[47]

Statistical analysis of an open uncontrolled study (80 patients) coupled with a dou-ble-blind, placebo-controlled part (21 patients), demonstrated that oral administration of an unspecified concentrated *Ginkgo biloba* extract (29.2 mg/day for 2 weeks) had no effect on tinnitus.[48]

Three clinical trials have assessed the efficacy of an infusion of *Ginkgo biloba* extract in conjunction with hyperbaric oxygenation, low-power laser or HAES for the treatment of acoustic trauma, and sensorineural or idiopathic hearing loss.[45,49,50] While treatments improved hearing recovery in all studies, no controls were used, and the safety of intravenous administration of *Ginkgo biloba* extracts has not been established.

*Diabetic Retinopathy and Senile Macular Degeneration*
The therapeutic efficacy of a standardized *Ginkgo biloba* extract was assessed in a randomized double-blind placebo-controlled study of 29 diabetic patients with early diabetic retinopathy confirmed by angiography and associated with a blue-yellow dyschromatopsia.[51] Patients were treated with 160 mg/day of *Ginkgo biloba* extract or placebo for 6 months. Improvements in color vision were measured by the Desat-uration Panel (D-15) and the 100-Hue Farnsworth test. A small, but statistically sig-nificant improvement in the D-15 scores was observed in patients without retinal ischemia who were treated with the *Ginkgo* extract. While the D-15 scores of patients in the placebo group continued to worsen over the 6-month period.[51] A double-blind placebo-controlled trial measured the effects of a standardized *Ginkgo biloba* extract in 10 outpatients patients ($> 55$ years old) with recently ($< 1$ year) diagnosed senile macular degeneration, as confirmed by angiography.[52] Patients received either 160 mg of *Ginkgo biloba* extract or placebo daily for 6 months. Distance visual acuity in the most affected eye gained 2.3/10 ($p < 0.05$) in subjects treated with the extract, while the placebo group gained only 0.6/10.[52] Near vision acuity also improved in the treated group, but did not reach statistical significance.

**Mechanism of Action**

Ischemia due to impaired blood circulation, is a common feature of many vascular disorders such as thrombosis, cerebral vascular disease, or chronic venous insuffi-ciency. A decrease in blood circulation leads to a decrease in the oxygen and nutri-ent supply of the tissues. During this process the release of free radicals and lipid peroxidation causes oxidation of the cell membrane lipids, causing damage that is partly responsible for acute tissue damage and chronic diseases such as arthritis, arte-riosclerosis, CNS disturbances, and diabetes. In addition, oxidation of polyunsatu-rated fatty acids in brain cell membranes results in a change in the membrane structure that is associated with impaired neurotransmitter uptake. In dementia due to degeneration with neuronal loss and impaired neurotransmission, a decline of intellectual function is also associated with disturbances in the supply of oxygen and glucose.[14]

The pharmacological effects of standardized *Ginkgo biloba* extracts (GBE) have been attributed to a combination of the flavonoid glycosides and the diterpene lactones (ginkgolides). Several mechanisms of action of GBE are supported by the *in vitro* and *in vivo* studies: increased blood flow to the arteries, veins (due to the release of EDRF/NO and $PGI_2$), as well as improvements in the microcirculation (capillaries) and rheologiocal effects (decreased blood viscosity, platelet activating factor-antagonism); improvements in cerebral metabolism, such as increased tolerance to anoxia; an inhibition of age-related reduction of muscarinergic choline and adrenergic receptors, an increase in the release of neurotransmitters and an inhibition of biogenic amine uptake; antioxidant activity, prevention of neuronal damage by free radicals; increase in memory and learning capacity; improvements in compensation of disturbed equilibrum; reduction of retinal edema and cellular lesions in the retina.[9,14]

In human pharmacological studies, GBE treatment has been shown to improve global and local cerebral blood flow, skin perfusion and microcirculation;[54–58] to protect against hypoxia;[59] to improve blood rheology, including inhibition of platelet aggregation;[54,56,60–62] to improve tissue metabolism;[58,63] and to reduce capillary permeability.[64] The pharmacodynamic and CNS effects of GBE treatment were assessed in 15 young, healthy volunteers, using electroencephalography (EEG) and event-related potential mapping.[65] Oral administration of GBE in single doses of 80 and 160 mg versus placebo, or 160 mg/day for 5 days enhanced the alpha frequencies, while event-related potentials were only slightly modified in amplitude whereas there was a clear decrease of P300 latency. These results were similar to other nootropic drugs.[65] In elderly patients with mild to moderate dementia, oral administration of GBE or tacrine produced similar EEG profiles that were characteristic of cognition enhancers or anti-dementia drugs.[65]

A double-blind, placebo-controlled, crossover study assessed the effects of three different doses of GBE in 12 healthy male subjects on the CNS.[66] Subjects were administered single oral doses of GBE (40–240 mg) or placebo, and the effects were measured by quantitative pharmaco-electroencephalography. An increase in alpha activity and cognitive activating type response was more apparent at the higher dose (240 mg) of GBE as compared with placebo. The computer-analyzed electroencephalogram profile was similar to that of tacrine.[66] In an uncontrolled trial in 18 elderly subjects diagnosed with probable Alzheimer's, the effects of tacrine (40 mg/day) or GBE 240 mg/day) were assessed by computer-analyzed EEGs (CEEG).[67] The results of this study showed that GBE (240 mg) had a typical cognitive activator CEEG profile.[67] Two double-blind clinical trials of elderly patients with symptoms of cerebral insufficiency, demonstrated that treatment with GBE induced a pronounced reduction in the theta portion of the theta/alpha ratio by comparison with placebo, as measured by EEG.[14]

GBE (30 μg/ml) induced vascular relation in porcine basilar arteries in a concentration- and endothelium-dependent manner.[68] In addition, vasorelaxtion induced by transmural nerve stimulation (TNS) was significantly enhanced by GBE (30 μg/ml) in both endothelial-intact and -denuded basilar arteries.[68] Enhancement of TNS-induced relaxation was abolished by N-L-arginine, indicating that nitric oxide was involved in the direct and indirect vasorelaxant activity of GBE.[68]

GBE exerts both contractile and relaxant effects on rabbit aorta strips.[69,70] GBE (100 µg/ml) did not stimulate isometrically recordable contractions in isolated rabbit aorta, but potentiated the contractile effect of norepinephrine.[69] At higher concentrations ($EC_{50}$ ~ 1.0 mg/ml), GBE produced a concentration-dependent contraction that was antagonized by the α-adrenoceptor blocking agent, phentolamine.[69] Both cocaine and desipramine, inhibitors of catecholamine re-uptake, potentiated the contractile effect of norepinephrine, but inhibited the contractile effects of GBE and tyramine.[69] These results suggest that contractile actions of GBE are due to the release of catecholamines from endogenous tissue reserves.[1,69] The effects of GBE, phentolamine, propranolol, gallopamil, theophylline and papaverine, on the biphasic contractile response of norepinephrine in isolated rat aorta, demonstrated that GBE had musculotropic action that was similar to papaverine.[70] The flavonoids, quercetin, kaempferol and isorhamnetin, isolated from the leaves of *Ginkgo biloba*, appear to be responsible for this activity.[71] The flavonoid compounds and papaverine also inhibited cyclic GMP phosphodiesterase, which in turn induced endothelium-dependent relaxation in isolated rabbit aorta by potentiating the effects of endothelium-derived relaxing factors.[1]

A non-flavonoid containing fraction of GBE protected rat brain tissue from hypoxic damage *in vitro*.[72–76] The diterpene lactones, the ginkgolides and bilobalide were responsible for the antihypoxic activity of the extract.[72,73] Ginkgolides A and B were found to protect rat hippocampal neurons against ischemic damage, which appears to be related to their ability to act as platelet-activating factor receptor antagonists.[74–76] Ginkgolides A and B, and bilobalide reduced hypoxic and glutamate-induced damage in cultured neurons isolated from embryonic chick telencephalons.[73,76] GBE and bilobalide protected endothelial cells from hypoxia-induced ATP depletion.[77]

*Ginkgo biloba* extract (GBE) increased sodium-dependent high-affinity choline uptake into rat hippocampal synaptosomes *in vitro* (50µg/mg protein).[78] Suppression hypoxia-induced, phospholipase A2-dependent release of choline from rat hippocampal slices was observed after treated with bilobalide ($EC_{50}$ 0.38 µM) *in vitro*, indicating a reduction in hypoxia-induced membrane phospholipid degradation.[79]

Free radical oxidation of lipid membranes plays an important role in the pathogenesis of cardiovascular and cerebrovascular disease.[80] Numerous *in vitro* studies have demonstrated that GBE has antioxidant activity and acts as a free-radical scavenger.[81–89] GBE has been shown to scavenge superoxide anions,[86,87] hydroxyl radicals,[84,87] and nitric oxide;[88] protect lipid membranes from peroxidative damage;[87] and prolong the half-life of endothelium-derived relaxing factor by scavenging superoxide anions.[84] Treatment of rat liver microsomes with GBE reduced free radical-lipid peroxidation induced by NADPH-$Fe^{3+}$ systems,[81] and protected human liver microsomes from lipid peroxidation induced by cyclosporin A.[82] Treatment of human leukocytes with GBE inhibited the production of reactive oxygen radicals induced by phorbol myristate acetate.[83] Treatment of human erythrocytes with GBE protected against hydrogen peroxide-induced oxidative membrane damage *in vitro*.[89] Both the flavonoid and terpenoid constituents of GBE appear to play a role in the free-radical scavenging activity.[1,84,87,88]

Intragastric or subcutaneous administration of GBE, ginkgolides, or bilobalide protected rats, mongolian gerbils, and mice against induced cerebral ischemia.[73,76,90–94] Intravenous administration of GBE prevented the development of multiple cerebral infarction in dogs injected with fragments of an autologous clot into a common carotid artery.[95] These data suggest that administered of GBE, after clot formation, may have some beneficial effects on the acute phase cerebral infarction/ischemia caused by embolism.[1] In other experiments, animals treated with GBE survived under hypoxic conditions for longer periods of time than untreated controls.[91,93–98] The results were not only due to significant improvements in cerebral blood flow, but an increase in the level of glucose and ATP as well as a decrease in lactate concentration was also observed.[91,97,98] Intragastric administration of GBE (50 mg/kg/day) or ginkgolide B to rats for 8 days increased the viability of rat hippocampal neurons and decreased 2,2'-azobis-2'-amidinopropane-induced apoptosis.[96] Intragastric administration of bilobalide or GBE to rats (20 mg/kg or 200 mg/kg, respectively) protected hippocampal cells from phospholipid breakdown under anoxic conditions.[79] Administration of GBE (100 mg/kg) to aged male rats prevented oxidative damage in brain and liver mitochondria.[99]

GBE was effective in the treatment of cerebral edema, a condition of excessive hydration of neural tissues due to damage caused by neurotoxic agents (such as triethyltin) or trauma, *in vivo*.[100–102] Bilobalide appears to the chemical constituent responsible for the antiedema effect of GBE.[103] Intragastric or subcutaneous administration of GBE to rats, with acute and chronic phases of adriamycin-induced paw inflammation, partially reversed an increase in brain water, sodium and calcium levels, as well as decreased the brain potassium level associated with sodium arachidonate-induced cerebral infarction.[104]

Chronic administration of GBE (100 mg/kg, i.g. or 50 mg/kg/day, i.p.) improved memory and learning in adult and aged mice.[105,106] Chronic oral administration of GBE (50 mg/kg) to male rats reduced the number of sessions required to reach criterion performance in an eight arm radial maze, as well as increased the life span of the animals.[107] In aged rats, intragastric administration of GBE (50 mg/kg/day for 30 days) led to an increase of high-affinity choline uptake into hippocampal synaptosomes, indicating a functional activation of cholinergic nerve terminals in the hippocampus of treated rats.[78] An increase in the density of muscarinic receptors in the hippocampus, and of the $\alpha_2$-adrenoceptors and $5\text{-HT}_{1A}$-receptors in the adrenal cortex was observed in aged rats treated with GBE.[108–110] Intragastric administration of GBE (50 mg/kg/day, for 14 days) prevented stress-induced $5\text{-HT}_{1A}$-receptor desensitization in aged rats.[111]

Intravenous infusion of GBE to cats increased the pial arteriolar diameter,[112] and improved local cerebral blood flow in rats.[113] The chemical constituents of GBE responsible for increasing cerebral blood flow are the ginkgolides,[114] including ginkgolide B, due to its PAF-antagonist activity.[115,116] Intravenous administration of GBE or ginkgolide B to normal rats showed that the extract, but not ginkgolide B, decreased brain glucose utilization.[117] However, in animals subjected to hypoxic or ischemic conditions, cerebral glucose consumption was enhanced after GBE treatment.[91]

Among the various compounds isolated from *Ginkgo* leaves, the ginkgolides, and in particular ginkgolide B, are known antagonists of platelet activating factor (PAF).[118–123] PAF is a potent inducer of platelet aggregation, neutrophil degranulation, and oxygen radical production, leading to increased microvascular permeability and bronchoconstriction. In animals, intravenous injections of ginkgolide B inhibit PAF-induced thrombocytopenia and bronchoconstriction.[121–123] PAF or ovalbumin-induced bronchospasm in sensitized guinea pigs was inhibited by an intravenous injection of ginkgolide B (1–3 mg/kg) 5 min prior to challenge.[123] The antiplatelet and antithrombotic effects of a combination treatment of ticlopidine (50 mg/kg/day) and GBE (40 mg/kg/day) were investigated in normal and thrombosis-induced rats.[124] ADP-induced platelet aggregation (*ex vivo*) was inhibited by the combination treatment and the results were similar to that observed with a 200 mg/kg/day dose of ticlopidine. The combination treatment also prolonged bleeding times by 150% and decreased thrombus weight in an arterio-venous shunt model.[124]

GBE improved the sum action of potentials in the cochlea and acoustic nerve in cases of acoustically-induced sound trauma, and due to cochlear damage-induced by local gentamicin instillation in guinea pigs.[1,125,126] The mechanism of action appears to involve a reduction in the metabolic damage to the cochlea. Intragastric or intravenous administration of GBE to mice (2 mg/kg) improved the ultrastructure qualities of vestibular sensory epithelia when the tissue was fixed by vascular perfusion.[127] Improvements were due to a decrease in capillary permeability and improvements in microcirculation.[1,127] In unilateral vestibular deficits following labyrinthectomy in rats, or vestibular neurectomy in cats, postoperative treatment with GBE (intraperitoneal) accelerated the compensation of spontaneous ocular nystagmus in light and darkness.[128,129]

Intragastric administration of GBE (100 mg/kg) to rats, for 10 days prior to reperfusion injury, reduced degenerative histological changes of the retina induced by free radical damage.[130–133] Repeated intragastric administration of GBE (50–100 mg/kg) significantly attenuated ischemia/reperfusion-induced retinal ion imbalances ($Na^+$, $K^+$, $Ca^{2+}$ and $Mg^{2+}$), in normal and diabetic rats.[134–136] Treatment of rabbits with 40 mg/kg/day of GBE for 3 weeks decreased the susceptibility of retinal cells to proteolytic enzymes, as determined by the number of Müller cells isolated from isolated retina (5,200 controls vs. 3,050 after GBE treatment).[137]

## Pharmacokinetics

The pharmacokinetics and bioavailability of the ginkgolides A, B, and C were determined in 12 healthy adults of both sexes, after oral administration (in fasting conditions and after a standard meal) of 120 mg of a standardized extract of *Ginkgo biloba* (EGb 761) and intravenous administration (100 mg).[1,2] Plasma and urine samples were collected up to 36 and 48 hours following treatment and analyzed by gas chromatography and mass spectrometry using negative chemical ionisation. The bioavailability of ginkgolides A and B, and bilobalide, when given under fasting condition, was high with coefficients of 0.8, 0.88 and 0.79 respectively. After food intake the bioavailability of these three compounds was unchanged, except for an increase in the time to reach peak concentration. Peak concentrations ranged from 16.5 to 33.3 ng/ml under fasting conditions, and from 11.5 to 21.1 ng/ml after a standard meal. The mean values of elimination half-lives were 9.5–10.6 hours and 3.2–4.5 hours for ginkgolide B and other two compounds, respectively. After intravenous administration, ginkgolide A and bilobalide exhibited elimination half lives comparable to those obtained after oral dosing, while ginkgolide B showed a shorter eliminiation half-life. The urinary excretions of ginkgolides A and B, and bilobalide accounted for 72, 41, and 31%, respectively, of the administered oral dose under fasting conditions.[2] In 2 healthy volunteers, flavonoid glycosides (50, 100, and 300 mg LI 1370) were absorbed in the small intestine, and peak plasma concentrations were observed within 2–3 hours.[14] The half-life of the flavonoid glycosides was between 2 and 4 hours.[14]

## Safety Information

### A. Adverse Reactions
Administration of *Ginkgo biloba* may prolong bleeding times, and five case reports of hemorrhage, associated with the ingestion of *Ginkgo* supplements, have been published in the scientific literature.[138–142] In the first case, a 33-yr-old female who had been taking 120 mg of a *Ginkgo* extract for two years developed bilateral subacute subdural hematomas that were not associated with trauma.[140] Two simultaneously drawn bleeding times were 15 and 9.5 minutes (normal upper limit 9). One month after discontinuing the *Ginkgo* supplements, the bleeding time returned to 6.5.[140] In the second case, a 70-yr-old male, taking aspirin daily for 3 years following coronary artery bypass surgery, developed spontaneous hyphema, one week after starting 80 mg/day a standardized *Ginkgo* extract.[139] A 72-yr-old male developed a small subdural hematoma several months after beginning treatment with *Ginkgo biloba*.[141] A 78-yr-old patient, on warfarin therapy for atrial fibrillation, developed a left parietal intracerebral hemorrhage two months after starting *Ginkgo* therapy.[138] A 61-yr-old male was diagnosed with subarachnoid hemorrhage after taking 120–160 mg of *Ginkgo biloba* extract for 6 months.[142] Minor adverse reactions reported in clinical trials included: headaches, gastrointestinal disturbances and allergic skin reactions, however these cases amounted to less than 0.5% of all cases (51 adverse events from 9772 patients in 44 clinical trials).[143]

### B. Contraindications
The use of GBE use during pregnancy or lactation is contraindicated, due to a lack of safety data.

## C.  Drug Interactions

Two cases of increased bleeding times and hemorrhage were reported in patients who were taking warfarin or aspirin therapy in combination with *Ginkgo biloba* extracts.[138,139]

## D.  Toxicology

Acute toxicity: Mice, intragastric administration $LD_{50}$ 7.73 g/kg; intravenous administration $LD_{50}$ 1.1 g/kg.[143]

Chronic toxicity: No evidence of toxicity was observed in rats at a dose of 500 mg/kg/day for 27 weeks, or in dogs at 400 mg/kg/day for 26 weeks.[143]

Intragastric administration of 100–1600 mg/kg/d to rats or 100–900 mg/kg/d to rabbits did not show any teratogenic effects or affects on reproduction.[143]

## E.  Dose and Dosage Forms

Standardized extracts (dry extracts from dried leaves, extracted with acetone/water, drug ratio 35–67:1) containing 22–27% flavonoid glycosides and 5–7% terpene lactones, of which approximately 2.8–3.4% consists of ginkgolides A, B, and C, as well as approximately 2.6–3.2% bilobalide, as coated tablets or solution for oral administration.[9,10] The daily dose range is 40–60 mg three to four times daily.

## References

1    DeFeudis FV. *Ginkgo biloba Extract (EGb 761). Pharmacological Activities and Clinical Applications*. Editions Scientifique, Paris, Elsevier, 1991:1–187.
2    Van Beek TA, Bombardelli E, Morazzoni P, Peterlongo F. *Ginkgo biloba* L. *Fitoterapia*. 1998;69:195–244.
3    Hänsel R, ed. *Hagers Handbuch der Pharmazeutischen Praxis*, 6th ed., Vol. 6, Berlin, Springer, 1998.
4    Bauer R, Zschocke S. *Z Phytother.* 1996;17:275.
5    Huh H, Staba EJ. The botany and chemistry of *Ginkgo biloba* L. *J Herbs Spices Med Plants.* 1992;1:91–124.
6    Farnsworth NR, ed. *NAPRALERT Database*, University of Illinois at Chicago, IL. An on-line database available directly through the University of Illinois at Chicago or through the Scientific and Technical Network (STN) of Chemical Abstracts Services (1999).
7    Keys JD. *Chinese Herbs, Their Botany, Chemistry and Pharmacodynamics*, Rutland, Vermont: C.E. Tuttle Co., 1976:30–31.
8    *Pharmacopoeia of the People's Republic of China* (English Edition), Guangzhou, China, Guangdong Science and Technology Press, 1992:64.
9    German Commission E monograph, Trockenextrakt (35–67:1) aus *Ginkgo-biloba-Blättern* extrahiert mit Aceton-Wasser. *Bundesanzeiger.* 1994;46:7361–7362.
10   Anon. *WHO Monographs on Selected Medicinal Plants*, World Health Organization, Traditional Medicine Programme, Geneva, Switzerland, 1999.
11   Sticher O. Quality of *Ginkgo* preparations. *Planta Med.* 1993;59:2–11.
12   Hasler A. Complex flavonol glycosides from the leaves of *Ginkgo biloba. Phytochemistry.* 1992;31:1391.
13   Sticher, O. Biochemical, pharmaceutical and medical perspectives of *Ginkgo* preparations. In: *New Drug Development from Herbal Medicines in Neuropsychopharmacology.* Symposium of the XIXth C.I.N.P. Congress, Washington, DC June–July 1994.

14   Kleijnen J, Knipschild P. *Ginkgo biloba. Lancet.* 1992;340:1136–1139.
15   Kleijnen J, Knipschild P. *Ginkgo biloba* for cerebral insufficiency. *Brit J Clin Pharmacol.* 1992;34:352–358.
16   Cochrane Collaboration Review. (http://www.cochrane.co.uk)
17   Gerhardt G, Rogalla K, Jaeger J. Medikamentöse Therapie von Hirnleistungsstörungen. Randomisierte Vergleichsstudie mit Dihydroergotoxin und *Ginkgo biloba*-Extrakt. *Fortschritte Medizin.* 1990;108:384–388.
18   Hopfenmüller W. Nachweis der therapeutischen Wirksamkeit eines *Ginkgo biloba* Spezialextraktes. *Arzneimittelforschung.* 1994;44:1005–1013.
19   Oken BS, Storzbach DM, Kaye JA. The efficacy of *Ginkgo biloba* extract on cognitive function in Alzheimer disease. *Arch of Neurol.* 1998;55:1409–1415.
20   Le Bars PL, Katz MM, Berman N, Itil TM, Freedman AM, Schatzberg AF. A placebo-controlled, double-blind, randomized trial of an extract of *Ginkgo biloba* for dementia. *JAMA.* 1997;278:1327–1332.
21   Kanowski S, Herrmann WM, Stephan K, Wierich W, Horr R. Proof of efficacy of the *Ginkgo biloba* special extract EGb 761 in outpatients suffering from mild to moderate primary degenerative dementia of the Alzheimer type or multi-infarct dementia. *Pharmacopsychiatry.* 1996;29:47–56.
22   Hofferberth B. The efficacy of EGb 761 in patients with senile dementia of the Alzheimer type: a double-blind placebo-controlled study on different levels of investigation. *Human Psychopharmacology.* 1994;9:215–222.
23   Wesnes K, Simmons D, Rook M, Simpson P. A double-blind, placebo-controlled trial of Tanakan in the treatment of idiopathic cognitive impairment in the elderly. *Human Psychopharmacol.* 1987;2:159–169.
24   American Psychiatric Association: *Diagnostic and Statistical Manual of Mental Disorders*, 4th ed. American Psychiatric Association, Washington, DC 1994.
25   *The International Classification of Diseases*, 10th Revision. World Health Organization, Geneva, Switzerland, 1994.
26   Bauer U. Six-month double-blind randomized clinical trial of *Ginkgo biloba* extract versus placebo in two parallel groups in patients suffering from peripheral arterial insufficiency. *Arzneimittelforschung.* 1984;34:716–720.
27   Saudreau F, Serise JM, Pillet J. Efficacité de l'extrait de *Ginkgo biloba* dans le traitement des artériopathies obliterantes chroniques des membres inferieurs au stade III de la classification de Fontaine. *J Mal Vascul.* 1989;14:177–182.
28   Schneider B. *Ginkgo biloba* Extrakt bei peripheren arteriellen Verschlukrankheiten. *Arzneimittelforschung.* 1992;42:428–436.
29   Ernst E. *Ginkgo biloba* in der Behandlung der Claudicatio intermittens. Eine systematische Recherche anhand knotrollierter Studien in der Literatur. *Fortsch Med.* 1996;114:85–87.
30   Letzel H, Schoop W. *Ginkgo biloba*-Extrakt EGb 761 und Pentoxifyllin bei Claudicatio intermittens. Sekundäranalyse zur klinischen Wirksamkeit. *J Vascular Dis* (VASA). 1992;21:403–410.
31   Blume J et al. Placebokontrollierte Doppelblindstudie zur Wirksamkeit von *Ginkgo biloba*-Spezialextrakt EGb 761 bei austrainierten Patienten mit Claudicatio intermittens. *J Vascular Dis* (VASA). 1996;2:1–11.
32   Bulling B, von Bary S. Behandlung der chronischen peripheren arteriellen Verschlubkrankheit mit physikalischem Training und *Ginkgo biloba*-Extrakt EGb 761. *Die Medizinische Welt.* 1991;42:702–708.
33   Peters H, Kieser M, Holscher U. Demonstration of the efficacy of *Ginkgo biloba* special extract EGb 761 on intermittent claudication-a placebo controlled, double-blind multicenter trial. *J Vascular Dis* (VASA). 1998;27:106–110.

34 Mouren X, Caillard P, Schwartz F. Study of the antiischemic action of EGb 761 in the treatment of peripheral arterial occlusive disease by TcPo2 determination. *Angiology.* 1994;45:413–417.

35 Drabaek H, Peterson JR, Wiinberg N, Winther-Hansen KF, Mehlsen J. Effekten af *Ginkgo biloba* ekstrakt hos patienter med claudicatio intermittens. *Ugeskr Læger.* 1996;158:3928–3931.

36 Haguenauer JP et al. Traitement des troubles de l'equilibre par l'extrait de *Ginkgo biloba. La Presse Médicale.* 1986;15:1569–1572.

37 Claussen CF, Kirtane MV. Randomisierte Doppelblindstudie zur Wirkung von Extractum *Ginkgo biloba* bei Schwindel und Gangunsicherheit des älteren Menschen. In: *Presbyvertigo, Presbyataxie, Presbytinnitus,* CF Claussen, ed., Springer Verlag, Berlin, pp. 103–115, 1985.

38 Hamann KF. Physikalische Therapie des vestibulären Schwindels in Verbindung mit *Ginkgo biloba*-Extrakt. *Therapiewoche.* 1985;35:4586–4590.

39 Artieres J. Effets thérapeutiques du Tanakan sur les hypoacousies et les acouphènes. *Lyon Méditerr Méd.* 1978;14:2503–2515.

40 Schwerdtfeger F. Elektronystagmographisch und klinisch dokumentierte Therapieerfanrungen mit rökan bei Schwindelsymptomatik. *Therapiewoche.* 1981;31:8658–8667.

41 Meyer B. Etude multicentrique randomisée a double insu face au placebo du traitement des acouphènes par l'extrait de *Ginkgo biloba. La Presse Médicale.* 1986;15:1562–1564.

42 Sprenger FH. Gute Therapieergebnisse mit *Ginkgo biloba. Ärztliche Praxis.* 1986;12:938–940.

43 Meyer B. Multicenter study of tinnitus. *Ann Oto-Laryng.* 1986;103:185–188.

44 Von Wedel H, Calero L, Walger M, Hoenen S, Rutwalt D. Soft laser/Ginkgo therapy in chronic tinnitus. A placebo controlled study. *Adv Otorhinolaryngol.* 1995;49:105–108.

45 Plath P, Olivier J. Results of combined low-power laser therapy and extracts of *Ginkgo biloba* in cases of sensorineural hearing loss and tinnitus. *Adv Otorhinolaryngol.* 1995;49:101–104.

46 Coles RRA. Trial of an extract of *Ginkgo biloba* (EGb) for tinnitus and hearing loss. *Clin Otolaryngol.* 1988;13:501–504.

47 Fucci JM. Effects of *Ginkgo biloba* extract on tinnitus: A double blind study. St. Petersberg, *Association for Research in Otolaryngology.* 1991.

48 Holgers KM, Axelson A, Pringle I. *Ginkgo biloba* extract for the treatment of tinnitus. *Audiology.* 1994;33:85–92.

49 Vavrina J, Muller W. Therapeutic effect of hyperbaric oxygenation in acute acoustic trauma. *Rev Laryngol Otol Rhinol (Bord).* 1995;116:377–380.

50 Hoffmann F, Beck C, Schultz A, Offermann P. *Ginkgo* extract EGb 761 (Tenobin)/HAES versus naftidrofuryl (Dusodril)/HAES. A randomized study of sudden deafness. *Laryngorhinootologie.* 1994;73:149–152.

51 Lanthony P, Cosson JP. Evolution de la vision des couleurs dans la rétinopathie diabétique débutante traitée par extrait de *Ginkgo biloba. J Fr Ophtalmol.* 1988;11:671–674.

52 Lebuisson DA, Leroy L, Rigal G. Traitement des dégénérescennces "maculaire séniles" par l'extrait de *Ginkgo biloba. Le Presse Médicale.* 1986;15:1556–1558.

53 Rogers SL, Farlow MR, Doody RS, Mohs R, Friedhoff LT. A 24-week, double-blind placebo-controlled trial of donepezil in patients with Alzheimer's disease. *Neurology.* 1998;50:136–145.

54 Raabe A, Raabe M, Ihm P. Therapieverlaufskontrolle mittels automatisierter Perimetrie bei chronischer zerebroretinaler Mangelversorgung älterer Patienten. *Klin Mbl Augenheilk.* 1991;199:432–438.

55  Költringer P, Langsteger W, Eber O. Dose-dependent hemorheological effects and microcirculatory modifications following intravenous administration of *Ginkgo biloba* special extract EGb 761. *Clin Hemorheol.* 1995;15:649–656.

56  Jung F, Mrowietz C, Kieswetter H, Wenzel E. Effect of *Ginkgo biloba* on fluidity of blood and peripheral microcirculation in volunteers. *Arzneimittelforschung.* 1990;40:589–593.

57  Heiss WD, Zeiler K. The influence of drugs on cerebral blood flow. *Pharmakotherapie.* 1978;1:137–144 (English translation).

58  Tea S, Celsis P, Clanet M, Marc-Vergnes, JP, Boeters U. Quantifizierte parameter zum Nachweis von zerebraler Durchblutungs und Stoffwechselsteigerung unter *Ginkgo biloba* Therapie. *Therapiewoche.* 1987;37:2655–2657.

59  Schaffler K, Reeh PW. Doppelblindstudie zur hypoxieprotektiven Wirkung eines standardisierten *Ginkgo-biloba*-Präparates nach Mehrfachverabreichung an gesunden Probanden. *Arzneimittelforschung.* 1985;35:1283–1286.

60  Hofferberth B. Simultanerfassung elektrophysiologischer, psychometrischer and rheologischer Parameter bei Patienten mit hirnorganischem Psychosyndrom und erhöhtem Gefässrisiko – Eine Placebo-kontrollierte Doppelblindstudie mit *Ginkgo biloba*-Extrakt EGB 761. In: Stodtmeister R, Pillunat LE, (eds.). *Mikrozirkulation in Gehirn und Sinnesorganen.* Ferdinand Enke, Stuttgart, 1991:64–74.

61  Witte S. Therapeutical aspects of *Ginkgo biloba* flavone glucosides in the context of increased blood viscosity. *Clin Hemorheol.* 1989;9:323–326.

62  Ernst E, Marshall M. Der Effekt von *Ginkgo-biloba*-Spezialextrakt EGb 761 auf die Leukozytenfilterabilitat-Eine Pilotstudie. *Perfusion.* 1992;8:241–241.

63  Rudofsky G. Wirkung von *Ginkgo-biloba*-extrakt bei arterieller Verschlusskrankheit. *Fortschritte der Medizin.* 1987;105:397–400.

64  Lagrue G, Behar A, Kazandjian M, Rahbar K. Oedèmes cycliques idiopathiques. Rôle de l'hyperperméabilité capillaire et correction par l'extrait de *Ginkgo biloba*. *La Presse Médicale.* 1986;15:1550–1553.

65  Lutheringer R, d'Arbigny P, Macher JP. *Ginkgo-biloba* extract (EGb 761), EEG and event-related potentials mapping profile. In: Advances in *Ginkgo-biloba* extract research. Vol. 4 Effects of *Ginkgo-biloba* extract (EGb 761) on aging and age-related disorders. Y Christen, Y Courtois, MT Droy-Lefaix, (eds.). Elsevier, Paris, 1995, pp. 107–118.

66  Itil TM, Erlap E, Tsambis E, Itil KZ, Stein U. Central nervous system effects of *Ginkgo biloba*, a plant extract. *Amer J Therapeutics.* 1996;3:63–73.

67  Itil T, Erlap E, Ahmed I, Kunitz A, Itil K. The pharmacological effects of *Ginkgo biloba*, a plant extract, on the brain of dementia patients in comparison with tacrine. *Psychopharmacol Bull.* 1998;34:391–397.

68  Chen X, Salwinski S, Lee TJ. Extracts of *Ginkgo biloba* and ginsenosides exert cerebral vasorelaxation via a nitric oxide pathway. *Clin Exp Pharmacol Physiol.* 1997;24:958–959.

69  Auguet M, DeFeudis FV, Clostre F. Effects of *Ginkgo biloba* on arterial smooth muscle responses to vasoactive stimuli. *Gen Pharmacol.* 1982;13:169–171, 225–230.

70  Auguet M, Clostre F. Effects of an extract of *Ginkgo biloba* and diverse substances on the phasic and tonic components of the contraction of an isolated rabbit aorta. *Gen Pharmacol.* 1983;14:277–280.

71  Peter H, Fisel J, Weisser W. Zur Pharmakologie der Wirkstoffe aus *Ginkgo biloba*. *Arzneimittelforschung.* 1966;16:719–725.

72  Oberpichler H et al. Effects of *Ginkgo biloba* constituents related to protection against brain damage caused by hypoxia. *Pharmacol Res Comm.* 1988;20:349–352.

73  Krieglstein J, Ausmeier F, El-Abhar H, Lippert K, Welsch M, Rupalla K, Henrich-Noack P. Neuroprotective effects of *Ginkgo biloba* constituents. *Eur J Pharm Sci.* 1995;3:39–48.

74 Braquet P. The ginkgolides: potent platelet-activating factor antagonists isolated from *Ginkgo biloba* L.: chemistry, pharmacology and clinical application. *Drugs of the Future*. 1987;12:643–648.

75 Oberpichler H. PAF-antagonist ginkgolide B reduces postischemic neuronal damage in rat brain hippocampus. *J Cerebral Blood Flow and Metabolism*. 1990;10:133–135.

76 Prehn JHM, Krieglstein J. Platelet-activating factor antagonists reduce excitotoxic damage in cultured neurons from embryonic chick telencephalon and protect the rat hippocampus and neocortex from ischemic injury *in vivo*. *J Neurosci Res*. 1993;34:179–188.

77 Janssens D, Michiels C, Delaive E, Eliaers F, Drieu K, Remacle J. Protection of hypoxia-induced ATP decrease in endothelial cells by *Ginkgo biloba* extract and bilobalide. *Biochem Pharmacol*. 1995;50:991–999.

78 Kristofikova Z, Benesova O, Tejkalova H. Changes of high-affinity choline uptake in the hippocampus of old rats after long-term administration of two nootropic drugs (Tacrine and *Gingko biloba* extract). *Dementia*. 1992;3:304–307.

79 Klein J, Chatterjee SS, Loffelholz K. Phospholipid breakdown and choline release under hypoxic conditions: inhibition by bilobalide, a constituent of *Ginkgo biloba*. *Brain Res*. 1997;755:347–350.

80 Ames BN, Shigenaga MK, Hagen TM. Oxidants, antioxidants and the degenerative diseases of aging. *Proc Natl Acad Sci*. 1993;90:7915–7922.

81 Barth SA et al. Influences of *Gingko biloba* on Cyclosporin induced lipid peroxidation in human liver microsomes in comparison to Vitamin E, Glutathione and N-Acetylcysteine. *Biochem Pharmacol*. 1991;41:1521–1526.

82 Pincemail J et al. *Gingko biloba* extract inhibits oxygen species production generated by phorbol myristate acetate stimulated human leukocytes. *Experientia*. 1987;43:181–184.

83 Pincemail J, Dupuis M, Nasr C. Superoxide anion scavenging effect and superoxide dismutase activity of *Ginkgo biloba* extract. *Experientia*. 1989;45:708–712.

84 Robak J, Gryglewski RJ. Flavonoids are scavengers of superoxide anions. *Biochem Pharmacol*. 1988;37:837–841.

85 Dumont E, Petit E, Tarrade T, Nouvelot A. UV-C irradiation-induced peroxidative degradation of microsomal fatty acids and proteins: protection by an extract of *Ginkgo biloba* (EGb 761). *Free Rad Biol Med*. 1992;13:197–203.

86 Marcocci L, Packer L, Droy-Lefaix MT, Sekaki AH, Gardes-Albert M. Antioxidant action of *Ginkgo biloba* extract EGb 761. In: *Methods in Enzymology*, L Parker, ed., San Diego, Academic Press, 1994a, pp. 462–475.

87 Droy-Lefaix MT. Effect of the antioxidant action of *Ginkgo biloba* extract (EGb 761) on aging and oxidative stress. *Age*. 1997;20:141–149.

88 Marcocci L, Maguire JJ, Droy-Lefaix MT, Packer L. The nitric oxide scavenging properties of *Ginkgo biloba* extract (EGb 761). *Biochem Biophys Res Comm*. 1994b;201:748–755.

89 Artmann GM, Schikarski C. *Ginkgo biloba* extract (EGb 761) protects red blood cells from oxidative damage. *Clin Hemorheol*. 1993;13:529–539.

90 Larssen RG, Dupeyron JP, Boulu RG. Modèles d'ischémie cérébrale expérimentale par microsphères chez le rat. Étude de l'effet de deux extraits de *Ginkgo biloba* et du naftidrofuryl. *Thérapie*. 1978;33:651–660.

91 Rapin JR, Le Poncin-Lafitte M. Consommation cérébrale du glucose. Effet de l'extrait de *Ginkgo biloba*. *La Presse Médicale*. 1986;15:1494–1497.

92 Le Poncin-Lafitte MC, Rapin J, Rapin JR. Effects of *Ginkgo biloba* on changes induced by quantitative cerebral microembolization in rats. *Arch Int Pharmacodynamics*. 1980;243:236–244.

93   Spinnewyn B, Blavet N, Clostre F. Effets des l'extrait de *Ginkgo biloba* sur un mod-
     èle d'ischémie cérébrale chez la gerbille. *La Presse Médicale.* 1986;15:1511–1515.
94   Spinnewyn B, Blavet N, Clostre F, Bazan N, Braquet P. Involvement of platelet-acti-
     vating factor (PAF) in cerebral post-ischemic phase in mongolian gerbils.
     *Prostaglandins.* 1987;34:337–349.
95   Cahn J. Effects of *Ginkgo biloba* extract (GBE) on the acute phase of cerebral
     ischaemia due to embolisms. In: Agnoli A et al., (eds.). *Effects of Ginkgo biloba Extract
     on Organic Cerebral Impairment.* John Libbey, London, 1985:43–49.
96   Rapin JR, Zaibi M, Drieu K. *In vitro* and *in vivo* effects of an extract of *Ginkgo biloba*
     (EGb 761), Ginkgolide B, and Bilobalide on apoptosis in primary cultures of rat hip-
     pocampal neurons. *Drug Develop Res.* 1998;45:23–29.
97   Karcher L, Zagermann P, Krieglstein J. Effect of an extract of *Ginkgo biloba* on rat
     brain energy metabolism in hypoxia. *Naunyn-Schmiedeberg's Arch Pharmacol.*
     1984;327:31–35.
98   Le Poncin-Lafitte M et al. Ischémie cérébrale après ligature non simultanée des artères
     carotides chez le rat: effet de l'extrait de *Ginkgo biloba. Semaine Hopitale Paris.*
     1982;58:403–406.
99   Sastre J, Millan A, De La Asuncion JG, Pla R, Juan G, Pallardo FV, O'Connor E, Droy-
     Lefaix MT, Vina J. A *Ginkgo biloba* extract (EGb 761) prevents mitochondrial aging
     by protecting against oxidative stress. *Free Rad Biol Med.* 1998;24:298–304.
100  Chatterjee SS, Gabard B. Effect of an extract of *Ginkgo biloba* on experimental neu-
     rotoxicity. *Arch Pharmacol.* 1984;325(Suppl), abstr 327.
101  Otani M et al. Effect of an extract of *Ginkgo biloba* on triethyltin-induced cerebral
     edema. *Acta Neuropathologica.* 1986;69:54–65.
102  Borzeix MG. Effects of *Ginkgo biloba* extract on two types of cerebal edema. In:
     Agnoli A et al., (eds.). *Effects of Ginkgo biloba Extract on Organic Cerebral Impair-
     ment*, John Libbey, London, 1985:51–56.
103  Sancesario G, Kreutzberg GW. Stimulation of astrocytes affects cytotoxic brain edema.
     *Acta Neuropathol.* 1986;72:3–14.
104  DeFeudis FV et al. In: Agnoli A, Rapin JR, Scapagnini V, Weitbrecht WV, (eds.). Some
     *in vitro* and *in vivo* actions of an extract of *Ginkgo biloba* (EGb 761). In: Effects of *Ginkgo
     biloba* extract on Organic Cerebral Impairment, John Libbey, London, 1985:17–29.
105  Blavet N. Effect of *Ginkgo biloba* extract (EGb 761) on learning in the aged rat. In: Y
     Christen, J Costenin, M Lacour, (eds.). *Effects of Ginkgo biloba (EGb 761) on the Cen-
     tral Nervous System,* Elsevier, Paris, 1992:119–127.
106  Cohen-Salmon C, Venault P, Martin B, Raffalli-Sebille MJ, Barkats M, Clostre F, Par-
     don MC, Christen Y, Chapouthier G. Effects of *Ginkgo biloba* (EGb 761) on learning
     and possible actions on aging. *J Physiol. (Paris).* 1997;91:291–300.
107  Winter JC. The effects of an extract of *Ginkgo biloba* (EGb 761), on cognitive behav-
     ior and longevity in the rat. *Physiol Behav.* 1998;63:425–433.
108  Taylor JE. Liaisons des neuromédiateurs à leurs récepteurs dans le cerveau de rats. *La
     Presse Medicale.* 1986;15:1491–1493.
109  Huguet F, Tarrade T. $A_2$-agrenoceptor changes during cerebral aging. The effect of
     *Ginkgo biloba* extract. *J Pharm Pharmacol.* 1992;44:24–27.
110  Huguet F, Drieu K, Piriou A. Decreased cerebral 5-$HT_{1A}$ receptors during aging: rever-
     sal by *Ginkgo biloba* extract (EGb 761). *J Pharm Pharmacol.* 1994;46:316–318.
111  Bolanos-Jimenez F, Manhaes de Castro R, Sarhan H, Prudhomme N, Drieu K, Fillion
     G. Stress-induced 5-$HT_{1A}$ receptor desensitization protective effects of *Ginkgo biloba*
     (EGb 761). *Fundam Clin Pharmacol.* 1995;9:169–174.
112  Iliff LD, Auer LM. The effect of intravenous infusion of Tebonin (*Ginkgo biloba*) on
     pial arteries in cats. *J Neurosurg Sci.* 1982;27:227–231.

113 Krieglstein J, Beck T, Seibert A. Influence of an extract of *Ginkgo biloba* on cerebral blood flow and metabolism. *Life Sci.* 1986;39:2327–2334.

114 Beck T et al. Comparative study on the effects of two extract fractions of *Ginkgo biloba* on local cerebral blood flow and on brain energy metabolism in the rat under hypoxia. In: Krieglstein J, ed. *Pharmacology of Cerebral Ischemia.* Elsevier, Amsterdam, 1986:345–350.

115 Krieglstein J, Oberpichler H. *Ginkgo biloba* und Hirnleistungsstörungen. *Pharmazeutische Zeitung.* 1989;13:2279–2289.

116 Oberpichler H. Effects of *Ginkgo biloba* constituents related to protection against brain damage caused by hypoxia. *Pharmacol Res Comm.* 1988;20:349–352.

117 Lamor Y. Effects of ginkgolide B and *Ginkgo biloba* extract on local cerebral glucose utilization in the awake adult rat. *Drug Development Research.* 1991;23:219–225.

118 Akisu M. Platelet-activating factor is an important mediator in hypoxic ischemic brain injury in the newborne rat. Flunarizine and *Ginkgo biloba* extract reduce PAF concentration in the brain. *Biol Neonate.* 1998;74:439–444.

119 Vargaftig BB et al. Platelet-activating factor induces a platelet-dependent bronchoconstriction unrelated to the formation of prostaglandin derivatives. *Eur J Pharmacol.* 1980;65:185–192.

120 Vargaftig BB, Benveniste J. Platelet-activating factor today. *Trends Pharmacol Sci.* 1983;4:341–343.

121 Desquand S et al. Interference on BN 52021(ginkgolide B) with the bronchopulmonary effects of PAF-acether in the guinea pig. *Eur J Pharmacol.* 1986;127:83–95.

122 Desquand S, Vargaftig BB. Interference of the PAF-acether antagonist BN 52021 bronchopulmonary anaphylaxis. Can a case be made for a role for PAF-acether in bronchopulmonary anaphylaxis in the guinea pig? In: Braquet P, ed. *Ginkgolides.* J. R. Prous, Barcelona, Vol 1, 1988:271–281.

123 Braquet P. Involvement of platelet activating factor in respiratory anaphylaxis, demonstrated by PAF-acether inhibitor BN 52021. *Lancet.* 1985:1501.

124 Kim YS, Pyo MK, Park KM, Park, PH, Hahn BS, Wu SJ, Yun-Choi HS. Antiplatelet and antithrombic effects of a combination of ticlopidine and *Ginkgo biloba* extract (EGb 761). *Thrombosis Res.* 1998;91:33–38.

125 Stange VG et al. Adaptationsverhalten peripherer und zentraler akustischer Reizantworten des Meerschweinchens unter dem Einfluss verschiedener Fraktionen eines Extraktes aus *Ginkgo biloba. Arzneimittelforschung.* 1976;26:367–374.

126 Jung HW. Effects of *Ginkgo biloba* extract on the cochlear damage induced by local gentamicin installation in guinea pigs. *J Korean Med Sci.* 1998;13:525–528.

127 Raymond J. Effets de l'extrait de *Ginkgo biloba* sur la préservation morphologique des épithéliums sensoriels vestibulaires chez la souris. *La Presse Médicale.* 1986;15: 1484–1487.

128 Denise P, Bustany P. The effect of *Ginkgo biloba* (EGb 761) on central compensation of a total unilateral peripheral vestibular deficit in the rat. In: Lacour, M et al. (eds.). *Vestibular Compensation: Facts, Theories and Clinical Perspectives*, Paris, Elsevier, 1989:201–208.

129 Lacour M, Ez-Zaher L, Raymond J. Plasticity mechanisms in vestibular compensation in the cat are improved by an extract of *Ginkgo biloba* (EGb 761). *Pharmacol Biochem Behavior.* 1991;40:367–379.

130 Szabo, ME, Droy-Lefaix MT, Doly M, Braquet P. Free radical-mediated effects in reperfusion injury: a histologic study with superoxide dismutase and EGB 761 in rat retina. *Ophthalmic Res.* 1991;23:225–234.

131 Szabo, ME, Droy-Lefaix MT, Doly M, Carré C, Braquet P. Ischemia and reperfusion-induced histologic changes in the rat retina. *Invest Ophthalmol & Vis Sci.* 1991;32:1471–1478.

132   Droy-Lefaix MT, Menerath JM, Szabo-Tosaki E, Guillaumin D, Doly M. Protective effect of EGb 761 on ischemia-reperfusion damage in the rat retina. *Transplantation Proceedings.* 1995;27:2861–2862.

133   Szabo, ME, Droy-Lefaix MT, Doly M. Direct measurement of free radicals in ischemic/reperfused diabetic rat retina. *Clin Neurosci.* 1997;4:240–245.

134   Szabo, ME, Droy-Lefaix MT, Doly M, Braquet P. Modification of ischemia/ reperfusion-induced ($Na^+$, $K^+$, $Ca^{2+}$ and $Mg^{2+}$) by free radical scavengers in the rat retina. *Ophthalmic Res.* 1993;25:1–9.

135   Droy-Lefaix MT, Szabo, ME, Doly M. Ischemia and reperfusion-induced injury in rat retina obtained from normotensive and spontaneously hypertensive rats: effects of free radical scavengers. *Int J Tiss Reac.* 1993;15:85–91.

136   Szabo, ME, Droy-Lefaix MT, Doly M. EGb 761 and the recovery of ion imbalance in ischemic reperfused diabetic rat retina. *Ophthalmic Res.* 1995;27:102–109.

137   Pritz-Hohmeier S, Chao TI, Krenzlin J, Reichenbach A. Effect of *in vivo* application of *Ginkgo biloba* extract EGb 761 (Rökan®) on the susceptibility of mammalian retinal cells to proteolytic enzymes. *Ophthalmic Res.* 1994;26:80–86.

138   Matthews MK. Association of *Ginkgo biloba* with intracerebral hemorrhage. *Neurology.* 1998;50:1933–1934.

139   Rosenblatt M, Mindel J. (1997) Spontaneous hyphema associated with ingestion *of Ginkgo biloba* extract. *N Engl J Med.* 1997;336:1108.

140   Rowin J, Lewis SL. Spontaneous bilateral subdural hematomas associated with chronic *Ginkgo biloba* ingestion. *Neurology.* 1996;46:1775–1776.

141   Gilbert GJ. *Ginkgo biloba. Neurology.* 1997;48:1137.

142   Vale S. Subarachnoid haemorrhage associated with *Ginkgo biloba. Lancet.* 1998;352:36.

143   DeFeudis FV. (1991) Safety of EGb 761-containing products. In: *Ginkgo biloba* extract (EGb 761): pharmacological activities and clinical applications. Elsevier Science, Paris, 1991, pp. 143–146.

# 17

# Horse Chestnut

## Synopsis

Results from pharmacological studies and controlled clinical trials support the use of standardized horse chestnut seed extracts as an alternative therapy for the symptomatic treatment of chronic venous insufficiency (CVI). In clinical trials, oral administration of a standardized horse chestnut seed extract was more effective than placebo and as effective as compression stockings and reference medications in alleviating the symptoms of CVI. Horse chestnut seed extract exerts its effects through tonic actions on the veins, thereby decreasing their permeability and reducing the associated peripheral edema. The recommended daily dosage is 300 mg of a standardized horse chestnut seed extract (corresponding to 50 mg/day of aescin) twice daily. Treatment for 1 to 3 months may be required before the full therapeutic effects are apparent. Two cases of nephrotoxicity have been reported due to the ingestion of high doses of pure aescin, one of the active constituents of horse chestnut seed extract. Therefore, the extract should not be administered to patients with preexisting renal disease or in combination with other drugs known to cause nephrotoxicity. Adverse reactions, such as pruritis, nausea, vomiting, headaches and dizziness, have been reported in clinical trials. Due to a lack of safety data, the administration of horse chestnut seed extract during pregnancy and nursing, and to children under the age of 18 years old is not recommended.

## Introduction

*Aesculus hippocastanum*, commonly known as horse chestnut, is a tree known worldwide for its majestic beauty, and its overall resistance to adverse environmental conditions.[1] The genus name *Aesculus*, refers to a tree with edible acorns, and may have been derived from the Latin word "esca" meaning nutrient. The seeds of the tree have been compared to horse eyes, and were used as treatment of broken-winded horses, hence the common name horse chestnut. Horse chestnut seeds were first introduced into Europe from Turkey around 1565 AD, and were initially described by Mathiole.[1] By the 16th and 17th centuries, the use of horse chestnut for the treatment of general and hemorrhoidal varicosis was well established throughout Europe. In France, extracts from horse chestnut seeds were employed therapeutically during the early 18th century, however the results of scientific investigations of the extracts were not published until 1896 and 1909.[1] These articles reported successful outcomes in the treatment of hemorrhoids with horse chestnut seed extracts. The presence of saponins, the chemical constituents of horse chestnut, was discovered as early as 1835.[2] Other "traditional" medicine uses for horse chestnut seed extracts included the treatment of varicose veins, phlebitis, diarrhea, fever, and prostate enlargement.[1] Since 1976, clinical trials have focused on the use of standardized horse chestnut seed extracts for the symptomatic treatment of chronic venous insufficiency. Currently, there are pharmacopoeial-style monographs for horse chestnut in the European, French, German, Portuguese and Spanish Pharmacopoeias, and also in Martindale's *The Extra Pharmacopoeia*, 30th edition. A review of the quality, safety and efficacy of standardized horse chestnut seed extracts have also been performed by the World Health Organization's Traditional Medicine Programme.[3]

## Quality Information

- The correct Latin names for horse chestnut is *Aesculus hippocastanum* L. (Hippocastanaceae). Other taxonomic synonyms listed in the scientific literature include *Aesculus castanea* Gilib., *A. procera* Salisb., *Castanea equina*, *Hippocastanum vulgare* Gaertner. Horse chestnut should not be confused with the "common chestnut", which botanically is *Castanea dentata* (Marshall) Burkh. (Fagaceae), or related species.[3]
- The common names for *Aesculus hippocastanum* L include the following: castagna amare, castagna cavallina, castagna di cavalle, castagno d'India, castandas da India, castanheiro da India, castan, castaño de Indias, common horse chestnut, châtaignier de cheval, châtaignier de mer, conqueror tree, custul, gemeine Kastanie, gemeine oder weisse Rosskastanie, horse chestnut, Hippocastani Semen, marronier d'Inde, semen castaneae equinae, shahbalout-e hendi, vadgesztenyemag, wilde kastanje, and wilde kest.[3]
- Standardized extracts and other commercial products of horse chestnut are prepared from the dried ripe seed of *Aesculus hippocastanum* L. (Hippocastanaceae).[3] Native to western Asia, the horse chestnut also grows in Iran, Northern India, and Asia Minor, and is now widely cultivated in Europe and the USA.[3]
- The major chemical constituents isolated from horse chestnut seed include saponins, collectively referred to as aescin (also known in the scientific literature as "escin").[3] These constituents are present in concentrations up to 10%, and are considered the active therapeutic principles of horse chestnut seed and its extracts. Aescin (escin) exists in three forms, α-escin, β-escin and cryptoescin, which are differentiated by their physical properties. β-escin is a mixture of more than 30 different glycosides derived from the triterpene aglycones

protoaescigenin (= protoescigenin) and barringtogenol C. Aescin is considered to be the main active constituent of horse chestnut extract, as isolated aescin has shown comparable clinical efficacy in clinical trials. Other constituents include flavonoids (e.g., quercetin, kaempferol and their glycosyl derivatives).[1,3]

- Average daily oral dose: 100 mg of aescin, corresponding to 250–312.5 mg twice daily of a standardized powdered extract of the seeds containing 16–20% triterpene glycosides, calculated as anhydrous aescin.[3] Topical gels containing 2% aescin.[3]

## Medical Uses

Oral administration for the symptomatic treatment of chronic venous insufficiency, including pain, feeling of heaviness in the legs, nocturnal calf muscle spasms, itching, and edema.[5–16]

Topical applications for the symptomatic treatment of chronic venous insufficiency, sprains and bruises.[17–19]

## Summary of Clinical Evidence

Numerous controlled clinical trials have assessed the efficacy of a standardized horse chestnut seed extract (a sustained-release form corresponding to 100–150 mg aescin per day) for the treatment of chronic venous insufficiency (CVI).[5–16]

A double-blind, placebo-controlled clinical trial assessed the efficacy of the standardized extract for the treatment of 131 patients with CVI, after two treatment periods of 20 days each.[5] Symptomatic improvements in skin color, venous prominence, edema, dermatoses, pain, itching and the feeling of leg heaviness were observed in patients treated with the standardized extract (600 mg/day corresponding to 100 mg/day of aescin).[5] Two randomized, double-blind, placebo-controlled clinical trials, involving 212 and 95 patients respectively, assessed the efficacy of a standardized horse chestnut extract for the symptomatic treatment of CVI, using a 0–3 point scale to rate the severity of symptoms.[9,11] The measured symptoms included edema, calf spasms, pain and itching as well as feeling of heaviness. After two treatment periods of 20 days, a significant improvement in symptomatic improvement in those patients treated with the horse chestnut extract (600 mg/day corresponding to 100 mg/day of aescin) as compared with placebo. Reduction in the symptoms of edema, calf spasms, pain, and feeling of heaviness, was observed ($p < 0.01 - p < 0.05$).[9,11]

A randomized, double-blind, placebo-controlled study assessed the edema reducing effects of a standardized horse chestnut extract in a trial involving 30 outpatients with peripheral venous edema caused by CVI.[15] The patients were treated a standardized horse chestnut seed extract at a dose of 300 mg twice daily (corresponding to 100 mg/day of aescin/triterpene glycosides) or placebo for 20 days. A statistically significant ($p < 0.05$) reduction in the leg circumference measurements was found as compared to placebo.[15]

A randomized, double-blind, placebo-controlled study assessed the efficacy of a standardized horse chestnut extract (600 mg/day corresponding to 100 mg/day aescin) for the treatment of 74 patients with CVI and lower extremity edema.[10] Leg volumes were determined by water plethysmography, and leg circumferences were measured before and after induction of edema. In the patients treated with the standardized extract, provoked leg edema (volume) was reduced from 32 to 27 ml, while in the placebo group it rose from 27 to 31 ml.[10] In a randomized, double-blind, placebo-controlled parallel trial involving 40 patients with venous edema due to chronic deep vein incompetence stage II according to Hach, the edema reducing effects of a standardized extract were assessed.[8] Patients were treated orally with 369 to 412 mg of a standardized extract (corresponding to 75 mg of aescin) or placebo twice daily for six weeks. The reduction in leg edema was measured by hydroplethysmography. A statistically significant reduction in leg volume (after edema provocation), and leg circumference were observed in the treated group as compared with placebo ($p < 0.01$).[8]

A randomized, placebo-controlled, single-blinded parallel study compared the efficacy and safety of class II compression stockings with a standardized horse chestnut seed extract or placebo in 240 patients with CVI.[7] Patients were treated over a period of 12 weeks with 300 mg of a standardized extract twice daily, placebo, or compression stockings. The lower leg volume of the affected limbs decreased by an average of 43.8 ml in those patients treated with the standardized extract and 46.7 ml in those patients treated with compression stockings. However, in the placebo group, the lower leg volume increased by 9.8 ml over 12 weeks. The results of this trial demonstrated that treatment with either the standardized extract or compression stockings resulted in similar reduction in lower-leg volume.[7]

A standardized horse chestnut seed extract (360–412 mg, corresponding to 75 mg aescin, twice daily) was investigated in a randomized, double-blind comparison clinical trial with oxerutins ($O$-($\beta$-hydroxyethyl)-rutinosides 2000 mg/day) in 40 patients with CVI and peripheral venous edema.[21] The results of this trial demonstrated that both treatment groups had a reduction in edema as assessed by measuring the leg circumference.[21] A randomized double-blind comparison study assessed the effect of a standardized seed extract and oxerutins ($O$-($\beta$-hydroxyethyl)-rutinosides) in the treatment of 137 postmenopausal women with CVI stage II. Following one week placebo run-in, the patients were treated with either 1000 mg/day oxerutins, 600 mg/day standardized extract (corresponding to 100 mg aescin) or 1000 mg/day oxerutins for 4 weeks then 500 mg/day oxerutins for 12 weeks and then observation for a further 6 weeks. Patients treated with 1000 mg/day of oxerutins had a greater decrease in leg volume reduction than patients treated with either the standardized extract or 1000 mg/day (4 weeks) and then 500 mg/day oxerutins (12 weeks).[22]

A double blind, placebo-controlled study assessed the efficacy of a standardized horse chestnut extract for the treatment of 20 female subjects, 13 with pregnancy-related varicose veins, and 7 with chronic venous insufficiency.[13] Leg volumes were measured by water plethysmography, and the leg circumferences were measured at three levels. The patients treated with the standardized extract (2 × 14 days) had a significant reduction in leg volume (114 ml and 126 ml, p < 0.01) as compared with placebo.[13] A placebo-controlled double-blind crossover study assessed the effect of a standardized horse chestnut seed extract in the treatment of 52 pregnant women with edema due to venous insufficiency.[23] Patients were treated with one capsule twice daily of the standardized extract (300 mg equivalent to 50 mg aescin) or placebo for 2 weeks. The results of this trial demonstrated that the standardized extract significantly reduced edema and symptoms such as pain, fatigue and itching, as compared to placebo. Patients treated with the extract also showed a greater resistance to edema provocation.[23]

Two clinical trials investigated the effects of a standardized horse chestnut extract on the intravascular volume of the lower extremity veins and on interstitial filtration (measured indirectly by venous-occlusion or water plethysmography) in subjects with CVI.[16,22] The first study was a randomized, placebo-controlled, double-blind crossover trial involving 22 patients with chronic venous insufficiency, assessing the effect of a standardized horse chestnut extract on transcapillary filtration. Three hours after the administration of a single dose of the standardized extract (600 mg, corresponding to 100 mg of aescin), the transcapillary filtration coefficient decreased by 22%, as compared with a slight increase in the placebo group.[6] The results of this study demonstrated that the extract exerts its action primarily by reducing capillary permeability. In the second trial, patients treated with 600 mg/day of the standardized extract (100 mg of aescin) for 28 days had a significant improvement in extravascular volume changes of the foot and ankle (p < 0.01).[12] Other symptoms such as edema, feeling of tension, pain; leg fatigue and itching were also significantly improved (p < 0.05). However, no improvements in venous capacity or calf muscle spasms were observed.[12]

A double-blind placebo-controlled trial investigated the effect of a standardized seed extract (600 mg/day, corresponding to 100 mg of aescin daily) on vascular capacity and filtration in the arms and legs of 12 volunteers with healthy circulation.[24] Using vein plethysmography, the study showed a decrease in both the vascular capacity and filtration coefficients in those patients treated with the extract.[24] The effect of a standardized extract on the flow velocity of venous blood between the instep and the groin was quantitatively determined in 30 patients with varicose veins by the 133-xenon appearance method.[25] Blood flow increased by >30% with a lasting effect observed after 12 days of treatment. Blood viscosity was also reduced and there was a 73% improvement in subjective complaints.[25]

A randomized double-blind study assessed the effect of a standardized extract (600 mg/day, corresponding to 100 mg aescin) on lower leg edema in 10 healthy humans during a 15 hour airflight.[26] A single dose of the extract completely prevented or significantly reduced the increase in ankle and foot edema (p < 0.05, measured by determining the pre-flight circumference of the ankle and heel).[26] A post-marketing surveillance study in over 5000 patients suffering from CVI demonstrated that twice daily treatment with a standardized seed extract (containing 75 mg aescin for 4–10 weeks) reduced the symptoms of leg pain, fatigue, edema and itching.[27] Another postmarketing surveillance study involving over 4000 patients, typical symptoms of CVI were improved in more than 85% of the patients during the treatment with a standardized seed extract (daily dose corresponding to 100 mg of aescin).[28]

A criteria-based systematic review assessed thirteen published randomized, double-blind, controlled clinical trials of oral horse chestnut seed extracts for the treatment of chronic venous insufficiency.[15] The data were extracted from the trials in a standardized method, and the trial outcomes and 2 independent reviewers assessed the methodological quality of each trial. In all trials, horse chestnut seed extract was shown to be superior to placebo for the treatment of chronic venous insufficiency. Treatment with the extract was associated with a decrease in lower-leg volume, and a reduction in leg circumference at the calf and ankle. Other symptoms such as leg pain, itching and a feeling of fatigue were also reduced. Results from five comparative trials demonstrated that the seed extract was as effective as $O$-($\beta$-hydroxyethyl)-10-rutosides, and one trial indicated that the seed extract was as effective as compression therapy.[15]

In an open multicenter clinical study, 71 patients with CVI were treated with a topical gel containing 2% aescin. After 6 weeks of treatment, a significant reduction in ankle volume (0.7 cm, $p < 0.001$) and a significant reduction in the symptom score (60%, $p < 0.001$) was reported.[20] Other clinical trials have assessed the therapeutic effect and safety of a topically applied gel containing 2% aescin for the treatment of bruises and sprains.[17–19] The efficacy of a topically applied gel containing 2% aescin in reducing the tenderness to pressure hematoma (experimentally induced by injection) was demonstrated in a randomized, placebo-controlled, single dose study involving 70 healthy volunteers.[19] Based on tonometyric sensitivity measurements, the aescin gel significantly reduced ($p < 0.001$) tenderness to pressure, that was observed from 1 hour after treatment and lasted for 9 hours.[19]

## Mechanism of Action

Various horse chestnut seed extracts have anti-inflammatory activity and antiexudative effects in animal models. Intragastric administration of a 30% ethanol extract suppresses carrageenan-induced pedal edema and adjuvant-induced arthritis in female rats (doses 0.6 ml/kg and 1.5 ml/kg, respectively).[29] Intraperitoneal administration of a saponin fraction, isolated from a horse chestnut seed extract, exhibited analgesic, anti-inflammatory and antipyretic activities in mice and rats.[30] The same saponin fraction also inhibited prostaglandin synthetase activity *in vitro*.[30] A hydroalcoholic extract of the seeds induced contractions in canine saphenous veins *in vitro*, and increased the venous pressure in perfused canine saphenous vein *in vivo* (25–50 mg i.v. bolus).[31] Administration of a hydroalcoholic extract of the seeds (200–400 mg/kg) suppressed peroxide- and carrageenan-induced pedal edema in rats. Chloroform-, serotonin-, or histamine-induced cutaneous capillary hyperpermeability was also reduced in rats and rabbits after intragastric administration of a hydroalcoholic extract of the seeds (dose range, 50–400 mg/kg).[31]

*In vitro* studies have shown that horse chestnut seed extract and aescin, one of the active chemical constituents of the extract, increases the tension of isolated human saphenous veins and rabbit portal veins (concentration range 5–10 μg/ml).[32] This effect appears to be due to the preferential formation of prostaglandin $F_{2\alpha}$ and can be reversed by treatment with indomethacin.[32] This effect was not blocked by pretreatment with phentolamine, demonstrating that the effects are not mediated by α-adrenergic receptors. The effects of α-aescin on peripheral blood vessels were assessed *in vitro* on isolated arteries and veins, for example in constant-flow perfused cat rear paw, isolated perfused carotis of the guinea pig or isolated porcine iliac vein.[33] The results show that aescin has a biphasic effect on blood vessels. An initial transitory dilation, that is followed by a constricting or tonicizing effect, which lasts for hours in isolated blood vessels, but is transient in feline peripheral blood vessels.[33]

Aescin has also been shown to inhibit the activities of elastase and hyaluronidase *in vitro*.[34] Both of these enzymes are involved in enzymatic proteoglycan degradation, which is a part of capillary endothelium and is one of the primary components of the extravascular matrix.[34] Horse chestnut seed extracts also have antioxidant activity *in vitro*. Hydroalcoholic extracts of horse chestnut seeds (250 μg/ml) reduce lipid peroxidation and scavenge free radicals ($IC_{50}$ 0.24 μg/ml for superoxide dismutase radicals).[35]

*In vivo* studies have demonstrated that aescin has anti-inflammatory activity. Intravenous or oral administration of aescin (0.5 mg/kg to 120 mg/kg) inhibited dextran-induced hind paw edema, cotton-pellet granuloma and formalin-paper-induced granuloma in rodents.[36–38]

## Pharmacokinetics

After oral administration, aescin is rapidly absorbed from the gastrointestinal tract and has a half-life of one hour.[2] There is a pronounced first pass effect; thus the bioavailability is 1.5%. The pharmacokinetics of aescin can be described as compartmentalized, after intravenous administration. At a dose of 5 mg, an infusion rate of 718 μg/min and an infusion rate of 6.9 min, the following elimination half-lives have been observed: α-aescin 6.6 min; β-aescin 1.74 hours and γ of 14.36 hours. The distribution volume in steady state is 100.9 liters, with a total plasma clearance of 21.8 ml/min, renal clearance is 1.7 ml/min, and urine elimination from 0 to 120 hours is 8.2%. Aescin is bound to plasma proteins at a rate of 84%.[2]

The bioavailibility of the enteric-coated preparations is reported to be 100%.[2] After oral administration of a 300 mg dose of a standardized horse chestnut seed extract (corresponding to 50 mg aescin), the maximum plasma concentrations of approximately 25 ng/ml are achieved over 2.4 hours.[2]

## Safety Information

### A. Adverse Reactions
Clinical trials have indicated gastrointestinal side effects such as nausea, stomach discomfort, and allergic reactions.[4,28,39]

### B. Contraindications
While no contraindications have been reported, the use of horse chestnut seed extracts is not recommended during pregnancy or nursing without medical advice. There is no therapeutic rationale for the administration of horse chestnut seed extract to children.[3]

### C. Drug Interactions
Two suspected cases of toxic nephropathy, which appear to be secondary to the use of very high doses of aescin have been reported.[40] Therefore, horse chestnut seed should not be administered with other drugs known to cause nephrotoxicity, such as gentamicin[3] or to patients with preexisting renal disease.

### D. Toxicology
Acute toxicity: Intravenous administration of a standardized horse chestnut extract to rats at a dose of 90 mg/kg for a period of 8 weeks, increased liquid consumption of the animals and resulted in a few sporadic cases of death. The no-effect dose was 30 mg/kg, which is approximately 7 times the individual oral dose for humans.[2]

Chronic toxicity: Treatment of dogs and rats with a standardized extract for 32 weeks resulted in vomiting at a dose of 80 mg/kg and 400 mg/kg, for dogs and rats respectively. Vomiting was reduced when enteric-coated preparations were used.[32]

No embryotoxic or teratogenic effects were observed in rats and rabbits in doses up to 300 mg/kg, 30 times the human dose.[2] A 30% ethanol extract of the seeds was not mutagenic in the Ames test using *Salmonella typhimurium* strains TA98 and TA100 (200 μl).[41] Sodium-aescinate had no toxic effects on the fertility of male rats.[42]

### E. Dose and Dosage Forms
The average daily dose is 100 mg of aescin, corresponding to 250–312.5 mg twice daily of a standardized powdered extract of the seeds containing 16–20% triterpene glycosides, calculated as anhydrous aescin.[4] Topical gels containing 2% aescin.[17–20]

## References

1    Bombardelli E, Morazzoni P. *Aesculus hippocastanum* L. *Fitoterapia*. 1996;67: 483–510.
2    Hiltzenberger G. The therapeutical efficacy of horse chestnut seed extract. *Wiener Medizinische Wochenschrift*. 1989;139:385–389.
3    Anon. Semen Hippocastani. *WHO Monographs on Selected Medicinal Plants*, Volume II, WHO, Geneva Switzerland: WHO Publications, in press.

4   German Commission E monograph, Hippocastani semen (Horse chestnut seed), *Bundesanzeiger*, April 15, 1994.

5   Alter H. Zur medikamentösen Therapie der Varikosis. *Zeitschrift für Allgemeine Medizin*. 1973;49:1301–1304.

6   Bisler H et al. Wirkung von Rosskastaniensamenextrakt auf die transkapilläre Filtration bei chronischer venöser Insuffizienz. *Deutsche Medizinische Wochenschrift*. 1986;111: 1321–1328.

7   Diehm C et al. Comparison of leg compression stocking and oral horse-chestnut seed extract therapy in patients with chronic venous insufficiency. *Lancet*. 1996;347: 292–294.

8   Diehm C et al. Medical edema protection-clinical benefit in patients with chronic deep vein incompetence. *VASA*. 1992;21:188–192.

9   Friederich HC et al. Ein Beitrag zur Bewertung von intern wirksamen Venenpharmaka. *Zeitschrift für Hautkrankheiten*. 1978;53:369–374.

10  Lohr E et al. Ödemprotektive Therapie bei chronischer Veneninsuffizienz mit Ödemneigung. *Münchener Medizinische Wochenschrift*. 1986;128:579–581.

11  Neiss A, Böhm C. Zum Wirksamkeitsnachweis von Rosskastaniensamenextrakt beim varikösen Symptomenkomplex. *Münchener Medizinische Wochenschrift*. 1978:213–216.

12  Rudofsky G et al. Ödemprotektive Wirkung und klinische Wirksamkeit von Roßkastaniensamenextrakt im Doppelblindversuch. *Phlebologie und proktologie*. 1986;15:47–54.

13  Steiner M, Hillemanns HG. Untersuchung zur ödemprotektiven Wirkung eines Venentherapeutikums. *Münchener Medizinische Wochenschrift*. 1986;31:551–552.

14  Pilz E. Ödeme bei Venenerkrankungen. *Die Medizinische Welt*. 1990;41:1143–1144.

15  Pittler MH, Ernst E. Horse-chestnut seed extract for chronic venous insufficiency. A criteria based review. *Archives of Dermatology*. 1998;134:1356–1360.

16  Lange S et al. Practical experience with the design and analysis of a three-armed equivalence study. *Eur J Clin Pharmacol*. 1998;54 (7):535–540.

17  Götz AK, Giannetti BM. Naturstoffe in der Therapie stumpfer Sportverletzungen-heute noch zeitgeass? *Erfahrungsheilkunde*. 1990;6:362–371.

18  Calabrese C, Preston P. Report on the results of a double-blind, randomized, single-dose trial of a topical 2% aescin gel versus placebo in the acute treatment of experimentally induced hematoma in volunteers. *Planta Med*. 1993;59:394–397.

19  Calabrese C, Preston P. Äscin bei der Behandlung von Hämatomen-eine randomisierte Doppelblind-Studie. *Zeitschrift für Phytotherapie*. 1994:112.

20  Geissbühler S, Degenring FH. Behandlung von chronisch venöser insuffizienz mit Aesculaforce Venengel. *Scheiw Zesch Ganzheits Medizin*. 2000, in press.

21  Erler M. Roßkastaniensamenextrakt bei der Therapie peripherer venöser Ödeme. *Die Medizinische Welt*. 1991;42:593–596.

22  Rehn D et al. Comparative clinical efficacy and tolerability of oxerutins and horse chestnut extract in patients with chronic venous insufficiency. *Arzneimittelforschung*. 1996;46:483–487.

23  Steiner M, Hillemanns HG. Venostatin retard in the management of venous problems during pregnancy. *Phlebology*. 1990;5:41–44.

24  Pauschinger P. Neuere Untersuchungen zur Wirkung von Venostasin retard auf die kapilläre Funktion. *Ergebnisse der Angiologie*. 1984;30:129–137.

25  Klemm J. Strömungsgeschwindigkeit von Blut in varikösen Venen der unteren Extremitäten. *Münchener Medizinische Wochenschrift*. 1982;124:579–582.

26  Marshall M, Dormandy JA. Oedema of long distant flights. *Phlebology*. 1987;2:123–124.

27  Greeske K, Pohlmann BK. Roßkastaniensamenextrakt-ein wirksames Therapieprinzip in der Praxis. *Fortschritte der Medizin*. 1996;114:196–200.

28    Masuhr T et al. Nutzen-Risiko-Bewertung von Venoplant® retard, einem auf Aescin standardisierten Präparat aus Roßkastaniensamenextrakt, bei Patienten mit chronischer Veneninsuffizienz. *Top Medizin*. 1994;8:21–24.

29    Leslie GB. A pharmacometric evaluation of nine Bio-Strath herbal remedies. *Medita*. 1978;8:3–19.

30    Cebo B et al. Pharmacological properties of saponin fractions obtained from domestic crude drugs: *Saponaria officinalis*, *Primula officinalis* and *Aesculus hippocastanum*. *Herba Polonica*. 1976;22:154–162.

31    Guillaume M, Padioleau F. Veinotonic effect, vascular protection, antiinflammatory and free radical scavenging properties of horse chestnut extract. *Arzneimittelforschung*. 1994;44:25–35.

32    Longiave D et al. The mode of action of aescin on isolated veins: relationship with $PGF_{2-\alpha}$. *Pharmacological Research Communications*. 1978;10:145–153.

33    Felix W et al. Vasoaktive Wirkungen von α-Aescin. *Ergebnisse der Angiologie*. 1984;30:93–105.

34    Facino RM et al. Anti-elastase and anti-hyaluronidase activities of saponins and sapogenins from *Hedera helix*, *Aesculus hippocastanum*, and *Ruscus aculeatus*: factors contributing to their efficacy in the treatment of venous insufficiency. *Archives der Pharmazie* (Weinheim). 1995;328:720–724.

35    Masaki H et al. Active-oxygen scavenging activity of plant extracts. *Biol Pharm Bull*. 1995;18:162–166.

36    Aizawa Y et al. Antiinflammatory action of aescin. Intravenous injection. *Oyo Yakuri*. 1974;8:211–213.

37    Damas P et al. Antiinflammatory activity of escin. *Bulletin de la Societe Royale des Sciences de Liège*. 1976;45:436–440.

38    Tarayre JP et al. Pharmacological study of some capillary acting substances. *Annales de Pharmacie Française*. 1975;33:467–469.

39    Escribano MM et al. Contact urticaria due to aescin. *Contact Dermatitis*. 1997;37:233–253.

40    Grasso A, Corvaglia E. Due casi di sospetta tubulonefrosi tossica da escina. *Gazzetta Medica Italiana*. 1976;135:581–584.

41    Schimmer O et al. An evaluation of 55 commercial plant extracts in the Ames mutagenicity test. *Pharmazie*. 1994;49:448–451.

42    Kreybig H, Prechtel K. Toxizitäts und Tertilitätsstudien mit Aescin bei der Ratte. *Arzneimittelforschung*. 1977;7:1465–1466.

# 18

# Kava

## Synopsis

Results from controlled clinical trials support the use of kava (*Piper methysticum*) for the symptomatic treatment of mild cases of anxiety and insomnia due to menopause, nervousness, stress or tension. A wide variety of products are available on the market, including standardized extracts containing 15 to 70% kava pyrones, the active chemical constituents. The recommended daily dose is a total of 60 to 210 mg of kava pyrones per day in two or three divided doses. One controlled clinical trial has shown that the anxiolytic activity of 210 mg of kava pyrones per day is similar to that of oxazepam (15 mg/day). Kava preparations should not be taken for more than three months at a time, unless under medical supervision. Adverse reactions to kava range from minor gastrointestinal disturbances, headaches, and dizziness to allergic skin reactions. Chronic excessive ingestion (abuse) of kava may lead to the development of a scaly, eruptive dermopathy, and a transient yellow discoloration of the skin and nails. These conditions are reversible upon discontinuation of the drug. A single case report has suggested a possible interaction has been reported between kava and alprazolam, but the clinical significance of this interaction has not been established. While kava administration does not appear to impair vigilance, patients should be cautioned that motor reflexes, driving ability and the operation of heavy machinery may be adversely affected. Due to a lack of safety data, kava should not be administered to patients with an allergy to the plants in the Piperaceae (black pepper family), during pregnancy or nursing, or to children under the age of twelve.

## Introduction

Kava is a psychoactive beverage prepared from the roots of *Piper methysticum* Forst. (Black pepper family, Piperaceae), a plant that is indigenous to Polynesia, Melanesia and Micronesia.[1] The generic name *Piper* comes from the Latin for "pepper", and the species name *methysticum* from the Greek meaning "intoxicant", thus *Piper methysticum* when translated into English means "intoxicating pepper".[1] Kava is a tall perennial shrub indigenous to nearly all the Pacific islands except New Zealand,

New Caledonia and most of the Solomon Islands.[2] When the plant reaches 2 to 2.5 meters tall, the root is harvested, chewed or ground, and then mixed with water or coconut milk to prepare the kava drink. The indigenous people of the South Pacific have used kava as a ceremonial beverage for centuries, although the origins of the kava ceremony are unknown.[1] Social drinking of kava is denoted as a token of good-will or respect in Oceania, and kava is commonly used at formal gatherings and celebrations, much like alcohol is used in the United States. Kava usage is reported to exert a relaxing effect on the mind and the body, and promotes a sense of sociability. In traditional medicine, kava was used to soothe the nerves, induce relaxation and sleep, for congestion of the urinary tract, for the treatment of asthma and rheumatism, and to reduce weight.[1] The plant was first introduced into Europe around 1620 BC by the Dutch explorers Jacob LeMaire and William Schouten. However, scientific investigations into the pharmacological activities of kava did not begin until the 20th century, and in 1966 the German pharmacologist H.J. Meyer, demonstrated that the kava pyrones (kava lactones) were chemical constituents responsible for the psychotropic properties of the plant.[1] Currently a variety of standardized extracts and other kava products are available. In the United States, these products are regulated as dietary supplements.

**Quality Information**

- The correct Latin name for the plant is *Piper methysticum* Forst. (Piperaceae).[1] Botanical synonyms that may appear in the scientific literature include *Macropiper latifolium* Miq., *M. methysticum* (G. Forst.) Hook. et Arnott, *Piper inebrians* Soland.[2] The vernacular (common) names for the plant include: Ava, ava root, awa, gea, gi, kao, kava, kava-kava, kava-kava root, kavakava, kavapipar, kawa, kawa kawa, kawa pepper, Kawapfeffer, malohu, maluk, meruk, milik, maori kava, racine de poivre enivrant, rhizoma de kava-kava, rhizoma di kava-kava, yagona, and yaqona.[2]
- Commercial products of Kava are prepared from the dried rhizome (root) of *Piper methysticum.*[2]
- *Piper methysticum* is native to and cultivated in the islands of Oceania from Hawaii to New Guinea, with the notable exception of New Zealand, New Caledonia, and most of the Solomon Islands.[1,2]
- The major chemical constituents include the kava lactones, which often occur in excess of 5%. The major kava pyrones (kava lactones) being kawain (1.8%), methysticin (1.2%), dihydromethysticin (0.5%), desmethoxyyangonin (1.0%), yangonin (1.0%), dihydrokawain (1–1.06%).[2]

**Medical Uses**

Short-term (1–3 months maximum) symptomatic treatment of mild states of anxiety or insomnia, due to menopause, nervousness, stress or tension.[3–10]

**Summary of Clinical Evidence**

At least seven randomized, double-blind placebo-controlled clinical trials have assessed the efficacy of two standardized kava extracts for the symptomatic treatment of anxiety.[3–9] Two of the trials used an ethanol-water extract, standardized to contain 15% kava pyrones (kava lactones),[3,7] and five trials used an extract standardized to 70% kava pyrones.[4–6,8,9]

In a randomized, double-blind pilot study, 59 patients undergoing surgical procedures were treated (oral) with a standardized kava extract pre-operatively.[3] Patients treated with the extract showed improvements in mood, however since only two doses of 60 mg/day of the extract was administered, the clinical significance of this study is questionable.

Two placebo-controlled clinical trials involving menopausal women assessed the effects of two different extracts, a standardized extract containing 15% kava pyrones, and one standardized to 70% kava pyrones on climacteric symptoms and anxiety.[7,8] In the first study, 40 women with climacteric symptoms were treated with a kava extract (30–60 mg/day of kava pyrones) for 56–84 days.[7] The outcomes measured were the overall score in the Kuppermann index (KI) and the Anxiety Status Index. The results demonstrated that the extract was superior to placebo for reducing the symptoms associated with menopause. In the second trial, 40 women suffering from climacteric psychosomatic disturbances were treated with a standardized extract (210 mg/day of kava pyrones) for 56 days in a randomized, placebo-controlled, double blind study.[8] The main outcome measured was the overall score of anxiety symptoms using the Hamilton Anxiety Rating Scale (HAMA) as the confirmatory parameter. The Depression Status Inventory (DSI), the KI, and climacteric symptoms were used as secondary outcome measures. The total HAMA score improved after one week of treatment and reached a plateau at four weeks. The therapeutic response was statistically significant as compared with placebo ($p < 0.001$).[8]

A double blind, placebo-controlled clinical trial involving 58 patients with symptoms of anxiety, tension or agitation of non-psychotic origin assessed the efficacy of a kava extract (210 mg of kava pyrones per day) over a 4-week period.[4] The main outcome was assessed using the total score on the HAMA rating scale, and other adjunctive rating scales, the Erlanger scale for anxiety, Clinical Global Impressions and the Fischer Somatic Symptoms. After one week of therapy, the patients treated with the extract showed a significant reduction in the HAMA total score as compared with placebo.[4]

A randomized double-blind comparative trial assessed the efficacy of a kava extract (210 mg/day of kava pyrones), with that of oxazepam (15 mg/day) or bromazepam (9 mg/day), in 172 patients suffering from anxiety, tension and agitation syndromes of non-psychotic origin.[9] The main outcome criterion measured was a decrease in the total score of the HAMA rating scale. After 6 weeks of therapy, no significant difference between treatment groups was observed, and all three groups showed a decline in the total score of the HAMA scale.[9] The efficacy of a standardized kava extract, containing 70% kava pyrones, was assessed in a random-ized, placebo-controlled, double-blind multicenter study, involving 100 outpatients with anxiety of non-psychotic origin (Diagnostic and statistical manual of mental disorders-III-R criteria: agoraphobia, specific phobia, generalized anxiety disorder, and adjustment disorder with anxiety).[6] Patients were treated with a kava extract (210 mg/day of kava pyrones), and the outcome was assessed by a decrease in the HAMA rating scores over 24 weeks. Secondary outcome measures were the rating scales Clinical Global Impressions (CGI) and Von Zerssen mood scale (VZ). After 24 weeks of treatment with the extract, a decrease in the HAMA scores (mean of 30.7 at week 0 to 9.7 at week 24) in the treated groups was statistically significant when compared with the placebo group (p = 0.005). The CGI and VZ also improved over the 24-week period.[6]

A randomized, placebo-controlled, double-blind trial involving 58 patients assessed the efficacy of a kava extract for the treatment of anxiety of non-psychotic origin.[5] Patients were treated with a kava extract (containing 210 mg/day of kava pyrones), and the therapeutic efficacy was measured by a decrease in the HAMA rat-ing scores over 4 weeks. After one week of treatment with the extract, there was a statistically significant reduction in the HAMA scores (mean of 25.6 to 16.2) in the treated group as compared with the placebo group (p = 0.004).[5]

The clinical trials for kava have been criticized due to a lack of sufficiently rigor-ous inclusion criteria. Most trials included a heterogenous population, such as patients with depression with anxious features, panic disorders, phobias, somatoform disorders and generalized anxiety disorders.[10]

In addition to the seven clinical trials with kava extracts, nine double-blind clini-cal trials have been performed with a synthetic ($\pm$) kawain, one of the kava pyrones from kava.[10,11] Two of the trials were comparison studies, and seven were placebo-controlled. The results of these trials showed that therapeutic anxiolytic effects were observed at doses of 200–600 mg/day of kawain.[10] Similar deficiencies in the clini-cal trial methodologies were noted in these studies as well. However, these data sup-port the use of kava extracts for the treatment of anxiety.

Six clinical trials (two single-blind and four placebo-controlled, double-blind) have assessed the effect of a standardized kava extract (containing 70% kava pyrones) on drug-induced electroencephalograph (EEG) changes and psychometric tests of intel-lectual and motor functions in healthy subjects.[12–17] Observed changes in EEG record-ings and psychomotor test results showed no evidence of a decline in vigilance or responsiveness in subjected treated with 600 mg of extract (420 mg of kava pyrones) daily for 5 days.[16,17] Examination of the EEG sleep patterns of health volunteers, given a one-time dose of the extract (105 mg or 210 mg kava pyrones), showed an increased sleep spindle density by 20%, and an increase in slow-wave-sleep (i.e., deep sleep).[12] The rapid eye movement (REM) sleep phase was significantly changed.[12]

Treatment with 300 mg or 600 mg the extract (equivalent to 210 or 420 mg kava pyrones daily) for one week increased the beta/alpha index, which was typical for the pharmaco-EEG profile of anxiolytics.[13] The increased beta activity was most marked in the beta$_2$-range.[13] Treatment of healthy volunteers with a kava extract (300 mg = 210 mg of kava pyrones daily, 8 to 14 days) had no influence on safety-related performance when administered with or without ethyl alcohol in healthy volunteers.[14,15]

In a cross-over study, administration of a single dose of a rhizome extract standardized to 30% kava pyrones was compared with diazepam or placebo in a 7-day trial measuring EEG changes and psychometric testing.[18] No decline in vigilance was observed.[18] Safety-related performance was assessed after oral administration of a standardized extract (30% kava pyrones, 400 mg = 240 mg kava pyrones daily), bromazepam (9 mg daily) or their combination for 14 days.[19] Safety-related performance remained unaffected in those patients treated with the extract, whereas performance was impaired in those patients treated with bromazepam or the combination. No differences were observed between the treatment with bromazepam or a combination of kava and bromazepam, indicating that kava does not have additive effects when given in combination with bromazepam.[19]

**Pharmacokinetics**

A maximal sedating effect was observed 2 to 3 hours after oral administration of a single dose of dihydromethysticin to healthy volunteers.[20] In human studies, seven major and several minor kavalactones have been identified in urine samples.[21] The observed metabolic transformations were a reduction in the 3,4-double bond, and/or demethylation of the 4-methoxyl group of the $\alpha$-pyrone system.[2]

**Mechanism of Action**

The four main pyrones of kava including kawain, dihydrokawain, methysticin and dihydromethysticin act centrally as muscle relaxants and anticonvulsants. An aqueous kava extract, kawain, dihydrokawain, methysticin or dihydromethysticin (DHM) inhibited serotonin and nicotine-induced spasms of guinea pig ileum *in vitro*.[22,23] DHM has been shown to inhibit serotonin-, acetylcholine-, and barium-induced spasms in rat colon and uterus.[23] The antispasmodic activity of the aqueous root extract and its constituent kava pyrones was attributed to a direct musculotropic action.[22,23] Desmethoxyyangonin, DHM and kawain antagonized serotonin-induced contractions in rat uterus *in vitro* at concentrations of 3.2, 7.5 and 10$\mu$g/ml.[24] The effects of an aqueous extract of the rhizome on muscle contractility and neuromuscular transmission were investigated in mouse hemi-diaphragms and frog sartorius muscles *in vitro* using twitch tension and intracellular recording techniques.[25] The extract (2–5 mg/ml) induced muscle relaxation by a direct action on muscle contractility rather than by an inhibition of neuromuscular transmission.[25]

Intraperitoneal administration of an aqueous or dichloromethane kava extract has been shown to decrease spontaneous motor activity in mice, without a loss of muscle tone.[26–29] The aqueous extract, however, was not active orally in doses of 500 mg to 2.5 g/kg in mice or rats. A kava extract produced hypnosis and analgesia in mice after intraperitoneal administration of a dose of 150 mg/kg body weight.[27] Intraperitoneal administration of a lipid-soluble fraction of a rhizome extract decreased the conditioned avoidance response in rats.[28] The aqueous extract was inactive at doses up to 500 mg/kg.[28]

Intraperitoneal administration of (±)-kawain (10–50 mg/kg) or a kava extract (50–100 mg pyrones/kg) to cats reduced muscle tone.[30] Marked effects on the electroencephalogram, inducing high amplitude delta waves, spindle-like formations, and a continuous alpha- or beta-synchronization in the amygdalar recordings ($p <$ 0.001) were observed. Hippocampal responses, following stimulation of the amygdalar nucleus showed an increase in amplitude after intraperitoneal administration of (±)-kawain (50 mg/kg; $p < 0.05$) or the extract (100 mg pyrone/kg; $p < 0.01$).[30]

Intraperitoneal or intragastric administration of an aqueous or lipid soluble kava extract (150–250 mg/kg) induced analgesia in mice, as measured by the tail-flick reaction time test, and acetic acid writhing inhibition test.[31] Both dihydrokawain and dihydromethysticin induced analgesia in rats after intraperitoneal administration (140 mg/kg).[32] Depression of the CNS was observed in rodents after intraperitoneal administration of an aqueous rhizome extract (50–170 mg/kg).[33] Intraperitoneal administration of an aqueous or chloroform extract of the rhizome (140–300 mg/kg) inhibited strychnine-induced convulsions, depressed the central nervous system, and potentiated the effects of barbiturates in mice.[34]

Anticonvulsant activities have been observed in rodents after treatment with a chloroform kava extract, dihydromethysticin, dihydrokawain, methysticin and other kava against strychnine, electroshock- or chemically-induced convulsions.[35–39] More recent studies have demonstrated that both (+) kawain and (±) kawain, a synthetic kava pyrone, inhibited the voltage-dependant calcium and sodium channels of rat cerebrocortical synaptosomes.[40–42] (±) Kawain also inhibited a veratridine- or potassium chloride-induced increase in intracellular calcium levels, and glutamate release in rat cerebrocortical synaptosomes.[40] Both (±) kawain and methysticin inhibit voltage-dependent sodium channels in acutely dissociated rat CA1 hippocampal neurons (1–400 µM).[43]

Kava extracts and purified kava pyrones have neuroprotectant as demonstrated both *in vivo* and *in vitro*. A standardized acetone extract of the rhizome, methysticin and dihydromethysticin protected rodents against hypoxia or ischemia-induced cerebral damage.[44] The extract also protected against neuronal damage in cultured neurons from chick embryo cerebral hemispheres.[45]

Although the pharmacological mechanisms by which kava exerts anxiolytic activity are not well understood, recent investigations have suggested that the kava pyrones effect several neurotransmitter systems such as activation of the mesolimbic dopaminergic neurons,[46] GABAergic,[47] glutamatergic,[39,42] and serotinergic systems.[48] A root extract enriched with 58% kava pyrones enhanced the binding of [$^3$H]-muscimol to GABA-A receptors in a concentration-dependent manner in rat hippocampus, amygdala and medulla oblongata (EC$_{50}$ 200–300 µM).[47] Another study, however, found no significant *in vitro* or *in vivo* interactions with GABA (A and B) or benzodiazepine receptor binding sites for a dichloromethane extract of the root or the kava pyrones.[49] Both kawain and dihdromethysticin (10–100 μM) reduced the field potential changes induced by the serotonin 1A agonist, ipsapirone, in of guinea pig hippocampal slices *in vitro*.[50] These results suggest that both compounds may modulate serotonin-1A receptor activity.[50] Methysticin and kawain inhibited the uptake of noradrenalin but not serotonin in synaptosomes prepared from the cerebral cortex and hippocampus of rats.[51] Intragastric administration of (+)-dihydromethysticin in a single dose (100 mg/kg) or chronic administration of (±)-kawain (10.8 mg/kg/day for 78 days) to rats did not alter dopamine or serotonin levels in the striatal or cortical brain regions.[52]

**Safety Information**

*A. Adverse Reactions*
The incidence of adverse reactions to standardized kava extracts is low and in the range of 1–3%. An observational trial involving 4049 patients reported adverse reactions in only 61 patients (1.5%) after oral administration of a standardized kava extract (105 mg/day, containing 70% kava pyrones) for 7 weeks.[53,54] The major adverse reactions reported were gastrointestinal complaints and allergic skin reactions. A 2.3% incidence of adverse reactions was reported in a four-week trial, involving 3029 patients taking 800 mg/day of a 30% standardized extract (240 mg/day kava pyrones).[53,55] Nine cases of allergic reactions, 31 cases of gastrointestinal complaints and 22 cases of headache or dizziness were reported.[53,55] Four cases of clinical central dopaminergic antagonism have been reported in the medical literature.[55] Two cases of acute dystonic reactions were reported in otherwise young healthy subjects, 90 mintues to four hours after administration of 100 mg of a kava extract. The third patient experienced involuntary oral and lingual dyskinesia after taking a 450 mg of a kava extract for four days. The fourth case involved a 76-year-old woman with idiopathic Parkinson's'disease.[55]

Chronic, excessive ingestion of the rhizome or commercial preparations may cause a transient yellow discoloration of the skin and nails, which is reversible upon discontinuation of the drug.[56] Historically, excessive, long-term abuse of kava tea has been associated with a scaly, eruptive dermopathy, of unknown etiology.[57]Allergic skin reactions and ichthyosis have also been reported.[58–60] Two cases of a drug reaction in sebaceous gland-rich skin areas have been reported after oral administration of a kava extract for 3 weeks.[61] The reaction involved papules and plaques on the face, and ventral and dorsal thorax of these patients.[61] One study in an aboriginal community found that chronic excessive use of kava lead to malnutrition and weight loss, increased levels of γ-glutamyl transferase, depression of plasma protein levels, and a decrease in platelet volume and the number of lymphocytes.[62]

A single case of mydriasis and poor nearsighted accommodation, involving enlargement of the pupils, and disturbances in oculomotor equilibrium was reported following the ingestion of large doses of kava by an otherwise healthy subject.[63] Chronic consumption (6 months) of large quantities of a kava tea (5–6 cups daily) has been reported to cause anorexia, diarrhea, and visual disturbances.[59] A single case report of choreoathetosis involving limbs, trunk, neck and facial musculature, with marked athetosis of the tongue was associated with chronic heavy consumption of kava.[64]

A single case report of a 39-year-old woman experiencing acute hepatitis following ingestion of an unknown kava product has been reported.[65] However, the identity of the plant material in the preparation was never authenticated.

### B. Contraindications
Kava preparations should not be administered during pregnancy and nursing, or to patients with endogenous depression.[56] Administration of kava to children is not recommended.[2]

### C. Drug Interactions
According to the German Commission E, potentiation of the effectiveness of other centrally acting drugs such as alcohol, barbiturates and psychopharmacological agents may be possible.[56] A single case of a possible drug interaction between kava, alprazolam (a benzodiazepine), cimetidine and terazosin has been reported.[66] The clinical significance of this interaction has not yet been established. Kava ingestion may cause extrapyramidal side effects in senstive patients, and should be used with caution in these patients as well as patients with Parkinson's disease as it may worsen the symptoms due to its antagonistic effects on the dopaminergic system.[55]

### D. Toxicology
A dose-dependent decrease in spontaneous motility, ataxia, sedation, unconsciousness and death was observed after administration of a single intragastric (770–2800 mg/kg) or intraperitoneal (280–600 mg/kg) dose of an acetone-water kava extract to mice.[53] Similar results were observed after intragastric or intraperitoneal administration of dihydromethysticin and dihydrokawain.[53] Intragastric administration of an acetone-water extract of kava to rats (doses up to 320 mg/kg) or dogs (doses up to 60 mg/kg daily) for 26 weeks had no toxic effects.[53]

The $LD_{50}$ values of a standardized kava extract are:

Mice and rats: >1500 mg/kg (intragastric) and > 360 mg/kg (intraperitoneal).[53]

Intragastric administration of up to 600 mg/kg of a standardized extract (70% kava pyrones) to rodents did not increase the incidence of micronuclei-containing polychromatic erythrocytes or lead to any changes in the ratio of polychromatic to normochromatic erythrocytes. Extracts were not mutagenic in the Ames test with *Salmonella typhimurium* strains TA 98, 100, 1535, 1537 and 1538 with or without metabolic activation (rat liver S9-mix). There was no increase in the number of revertants up to a dose of 2.5 mg/plate.[53]

## E. Dose and Dosage Forms

Crude drug and extract preparations equivalent to 60–210 mg kava pyrones per day in divided doses.[5–9,56] Kava preparations should not be taken for more than 3 months without medical advice. Clinical trials have suggested that kava does not impair vigilance, or the concomitant ingestion of kava with alcohol does not have additive effects.[67,68] However, as a precaution, patients should be warned that even when administered within the recommended dosage range, motor reflexes, driving ability and the operation of heavy machinery may be adversely affected.[56]

## References

1    Singh YN. Kava: an overview. *J Ethnopharmacol.* 1992;37:13–45.

2    Anon. *WHO Monographs on Selected Medicinal Plants*, Volume II, Rhizoma Piperis Methystici, World Health Organization, Geneva, Switzerland, in press.

3    Bhate H et al. Orale Prämedikation mit Zubereitungen aus *Piper methysticum* bei operativen Eingriffen in Epiduralanästhesie. *Erfahrungsheilkunde.* 1989;6:339–345.

4    Kinzler E et al. Wirksamkeit eines Kava-Spezial Extractes bei Patienten mit Angst-, Spannungs- und Erregungszuständen nict-psychotischer Genese. *Arzneimittelforschung.* 1991;41:584–588.

5    Lehmann E et al. Efficacy of a special Kava extract (*Piper methysticum*) in patients with states of anxiety, tension and excitedness of non-mental origin – A double-blind placebo-controlled study of four weeks treatment. *Phytomedicine.* 1996;3:113–119.

6    Volz HP, Kieser M. Kava-kava extract WS 1490 versus placebo in anxiety disorders- a randomized placebo-controlled 25-week outpatient trial. *Pharmacopsychiatry.* 1997;30:1–5.

7    Warnecke G. Psychosomatische Dysfunktionen im weiblichen Klimakterium. *Fortschritte der Medizin.* 1991;109:119–122.

8    Warnecke G et al. Wirksamkeit von Kawa-kawa-Extrakt beim klimakterischen Syndrom. *Zeitschrift für Phytotherapie.* 1990;11:81–86.

9    Woelk H et al. Behandlung von Angst-Patienten. *Zeitschrift für Allgemeinedizin.* 1993;69:271–277.

10   Volz HP, Hänsel R. Kava-Kava und Kavain in der Psychopharmakotherapie. *Psychopharmakotherapie.* 1994;1:33–39.

11   Klimke A et al. Effectivity of kavain in tranquilizer indication. *Psychopharmacology.* 1988;96 (Supp):34.

12   Emser W, Bartylla K. Verbesserung der Schlafqualität. Zur Wirkung von Kava-Extrakt WS 1490 auf das Schlafmuster bei Gesunden. *Neurologie/Psychiatrie.* 1991;5: 636–642.

13   Johnson D et al. Neurophysiologisches Wirkprofil und Verträglichkeit von Kava-Extrakt WS 1490. *Neurologie/Psychiatrie.* 1991;5:349–354.

14   Heinze HJ et al. Pharmacopsychological effects of oxazepam and kava-extract in a visual search paradigm assessed with event-related potentials. *Pharmacopsychiatry.* 1994;27:224–230.

15   Herberg KW. Fahrtüchtigkeit nach Einnahme von Kava-Spezial Extract WS 1490. *Zeitschrift für Allgemeinmedizin.* 1991;67:842–846.

16   Herberg KW. Zum Einfluß von Kava-Spezialextrakt WS 1490 in Kombination mit Ethylalkohol auf sicherheitsrelevante Leistungsparameter. *Blutalkohol.* 1993;30:65.

17   Münte TF et al. Effects of oxazepam and an extract of kava roots (*Piper methysticum*) on event-related potentials in a word recognition task. *Neuropsychobiology.* 1993;27:46–53.

18    Gessner B, Cnota P. Untersuchung der Vigilanz nach Applikation von Kava-kava Extrakt, Diazepam oder Placebo. *Zeitschrift für Phytotherapie.* 1994;15:30–37.

19    Herberg, KW. Alltagssicherheit unter Kava-Kava-Extrakt, Bromazepam und deren Kombination. *Zeitschrift für Allgemeinmedizin.* 1996;72:973–977.

20    Hänsel R, Woelk H. eds., Unerwünschte Wirkungen des Kavatrinkens. In: *Spektrum Kava-Kava*, 2nd ed., Basel, Aesopus Verlag, 1995.

21    Duffield AM, Jaieson DD, Lidgard RO, Duffield PH, Bourne DJ. Identification of some human urinary metabolites of the intoxicating beverage kava. *J Chromatogr.* 1989;475:273–282.

22    Kretzschmar R et al. Spasmolytische Wirksamkeit von aryl-substituierten alpha-pyronen und Wässrigen extrakten aus *Piper methysticum* Forst. *Archives Internationales de Pharmacodynamie et de Therapie.* 1969;180:475–491.

23    Meyer HJ. Spasmolytische effekte von dihydromethysticin, einem wirkstoff aus *Piper methysticum* Forst. *Archives Internationales de Pharmacodynamie et de Therapie.* 1965;154:449–467.

24    Buckley JP et al. Pharmacology of kava. In: DH Efron, ed. *Ethnopharmacologic Search for Psychoactive Drugs*, U.S. Public Health Service Publication no. 1645, 1967.

25    Singh YN. Effects of kava on neuromuscular transmission and muscle contractility. *J Ethnopharmacol.* 1983;7:267–276.

26    Duffield PH, Jamieson D. Development of tolerance to kava in mice. *Clin Exp Pharmacol Physiol.* 1991;18:571–578.

27    Duffield PH et al. Effect of aqueous and lipid-soluble extracts of kava on the conditioned avoidance response in rats. *Archives Internationales de Pharmacodynamie et de Therapie.* 1989;301:81–90.

28    Jamieson DD et al. Comparison of the central nervous system activity of the aqueous and lipid extract of kava (*Piper methysticum*). *Archives Internationales de Pharmacodynamie et de Therapie.* 1989;301:66–80.

29    O'Hara MJ et al. Preliminary characterization of aqueous extracts of *Piper methysticum* (Kava, Kawa Kawa). *J Pharmaceut Sciences.* 1965;54:1021–1025.

30    Holm E et al. Untersuchungen zum Wirkungsprofil von D,L-Kavain. *Arzneimittelforschung.* 1991;41:673–683.

31    Jamieson DD, Duffield PH. The antinociceptive actions of kava components in mice. *Clin Exp Pharmacol Physiol.* 1990;17:495–507.

32    Brüggemann F, Meyer HJ. Die analgetische Wirkung der Kawa-Inhaltsstoffe Dihydrokawain und Dihydromethysticin. *Arzneimittelforschung.* 1962;12:407–409.

33    Furgiuele AR et al. Central activity of aqueous extracts of *Piper methysticum* (Kava). *J Pharmaceut Sciences.* 1965;54:247–252.

34    Klohs MW et al. A chemical and pharmacological investigation of *Piper methysticum* Forst. *J Med Pharmaceut Chem.* 1959;1:95–103.

35    Keller F, Klohs MW. A review of the chemistry and pharmacology of the constituents of *Piper methysticum. Lloydia.* 1963;26:1–15.

36    Meyer HJ. Spasmolytische effekte von dihydromethysticin, einem wirkstoff aus *Piper methysticum* Forst. *Archives Internationales de Pharmacodynamie et de Therapie.* 1965;154:449–467.

37    Meyer HJ, Meyer-Burg J. Hemmung des elektrokrampfes durch die Kawa-pyrone dihydromethysticin und dihydrokawain. *Archives Internationales de Pharmacodynamie et de Therapie.* 1964;148:97–110.

38    Kretzschmar R et al. Strychnine antagonistic potency of pyrone compounds of the kava root (*Piper methysticum* Forst). *Experientia.* 1970;26:283–284.

39   Schmitz D et al. Effects of methysticin on three different models of seizure like events studied in rat hippocampal and entorhinal cortex slices. *Naunyn-Schmiedeberg's Archives of Pharmacology.* 1995;351:348–355.

40   Gleitz J et al. Anticonvulsant action of (±)-kavain estimated from its properties on stimulated synaptosomes and $Na^+$ channel receptor sites. *Eur J Pharm.* 1996a;315: 89–97.

41   Gleitz J et al. Kavain inhibits non-stereospecifically veratridine-activated $Na^+$ channels. *Planta Med.* 1996b;62:580–581.

42   Gleitz J et al. (±)-Kavain inhibits the veratridine- and KCl-induced increase in intracellular $Ca^{2+}$ and glutamine-release of rat cerebrocortical synaptosomes. *Neuropharmacology.* 1996c;35:179–186.

43   Magura EI et al. Kava extract ingredients, (+)-methysticin and (±)-kavain inhibit voltage-operated Na+-channels in rat CA1 hippocampal neurons. *Neuroscience.* 1997;81:345–351.

44   Backhauß C, Krieglstein J. Extract of kava (*Piper methysticum*) and its methysticin constituents protect brain tissue against ischemic damage in rodents. *Eur J Pharm.* 1992a;215:265–269.

45   Backhauß C, Krieglstein J. Neuroprotectant activity of kava extract (*Piper methysticum*) and its methysticin constituents *in vivo* and *in vitro*. *Pharmacology of Cerebral Ischemia*, 1992b, International symposium, 501–507.

46   Baum SS et al. Effect of Kava and individual kavapyrones on neutransmitter levels in the nucleus accumbens of rats. *Prog Neuropsychopharmacol Biol Psych.* 1998;22: 1105–1120.

47   Jussofie A et al. Kavapyrone enriched extract from *Piper methysticum* as modulator of the GABA binding site in different regions of rat brain. *Psychopharmacology.* 1994;116:469–474.

48   Walden J et al. Effects of kawain and dihydromethysticin on field potential changes in the hippocampus. *Prog Neuropsychopharmacol Biol Psych.* 1997a;21:697–706.

49   Davies LP et al. Kava pyrones and resin: studies on $GABA_A$, $GABA_B$ and benzodiazepine binding sites in rodent brain. *Pharmacol Toxicol.* 1992;71:120–126.

50   Walden J et al. Actions of kavain and dihydromethysticin on ipsapirone-induced field potential changes in the hippocampus. *Human Psychopharmacol.* 1997b;12:265–270.

51   Seitz U et al. [$^3$H]-Monoamine uptake inhibition properties of kava pyrones. *Planta Med.* 1997;63:548–549.

52   Boonen G et al. *In vivo* effects of the kavapyrones (+) dihydromethysticin and (±) kavain on dopamine, 3,4-dihydroxyphenylacetic acid, serotonin, and 5-hydroxyindoleacetic acid levels in striatal and cortical brain regions. *Planta Med.* 1998;64:507–510.

53   Hänsel R et al., eds. *Hagers Handbuch der pharmazeutischen Praxis*, 5th ed. Vol. 6. Berlin, Springer-Verlag, 1994.

54   Seigers SP et al. Ergebnisse der Anwendungsbeobachtung L 1090 mit Laitan Kapseln. *Ärztl Forschung.* 1992;39:6–11.

55   Schelosky L, Raffauf C, Jendroska K et al. Kava and dopamine antagonism. *J Neurol Neurosurg Psych.* 1995;58:639–640.

56   German Commission E monograph, Kava-kava rhizoma, *Bundesanzeiger* Nr. 101, 1.6.1990.

57   Norton SA, Ruze P. Kava dermopathy. *J Amer Acad Dermatol.* 1994;31:89–97.

58   Ruze P. Kava-induced dermopathy: a niacin deficiency? *Lancet.* 1990;335:1442–1445.

59   Siegel R. Herbal intoxication. *JAMA.* 1976;236:473–476.

60   Süss R, Lehmann P. Hämatogenes Kontaktezem durch pflanzliche Medikamente am Beispiel des Kavawurzel-extraktes. *Hautarzt.* 1996;47:459–461.

61  Jappe U et al. Sebotropic drug reaction resulting from kava-kava extract therapy: a new entity. *J Amer Acad Dermatol.* 1998;38:104–106.

62  Mathews JD et al. Effects of the heavy usage of kava on physical health: summary of a pilot survey in an Aboriginal community. *Med J Australia.* 1988;148:548–555.

63  Garner LF, Klinger JD. Some visual effects caused by the beverage kava. *J Ethnopharmacol.* 1985;13:307–311.

64  Spillane PK et al. Neurological manifestations of kava intoxication. *Med J Australia.* 1997;167:172–173.

65  Strahl S et al. Nekrotisierende Hepatitis nach Einnahme pflanzlicher Heilmittel. *Deutsches Mediz-Wochebschr.* 1998;123:1410–1414.

66  Almeida JC, Grimley EW. Coma from the health food store: interaction between kava and alprazolam. *Ann Int Med.* 1996;125:940–941.

67  Herberg KW. Driving ability after the intake of the kava-extract WS 1490. *Z Allg Med.* 1991;67:842–846.

68  Herberg KW. Alltagssicherheit unter Kava-Kava-Ekstrakt, Bromazepam und deren Kombination. *Z Allg Med.* 1996;72:973–977.

# 19

# Milk Thistle

## Synopsis

Silymarin is a complex flavonolignan extract isolated from the fruit of the milk thistle (*Silybum marianum*), and is used for symptomatic treatment of toxic liver damage (primarily drug or alcohol induced), chronic inflammatory liver diseases and hepatic cirrhosis. Administration of a standardized milk thistle extract to patients with cirrhosis leads to symptomatic improvements, such as normalization of liver enzyme levels and a decline in subjective complaints. However, a recent randomized controlled clinical trial has shown that treatment with a standardized silymarin preparation did not increase the survival rate of patients with alcohol-induced cirrhosis. Treatment of chronic liver diseases always requires medical intervention, and patients should be instructed to contact their physician prior to self-medicating with milk thistle products. Based on the clinical trials, the recommended dose is 280 to 420 mg of a standardized silymarin preparation (calculated as silybin). Treatment for 1–3 months may be required before the therapeutic effects are apparent. No drug interactions have been reported, and adverse reactions are limited to gastrointestinal disturbances, such as nausea and diarrhea.

## Introduction

The milk thistle, known scientifically as *Silybum marianum*, is an annual or biennial plant native to Southern Europe, but is also naturalized in North and South America, South Australia, Central Europe, China, and Australia.[1-3] The mature plant large bright purple flowers, and spiked leaves with white veins and many stout spines.[4] Extracts from the milk thistle have been used for over at least 2000 years for the treatment of liver diseases. Teas prepared from milk thistle seeds were used by Dioscorides and Pliney the Elder during the 1st century AD for the treatment of snakebite and "bile disorders".[4] Application of milk thistle extracts to the treatment of gall bladder and liver diseases was first reported in the early 16th century.[4] During the early 20th century, a school of medical herbalists called the "Eclectics" employed milk thistle extracts for the treatment of liver congestion, and menstrual, gall bladder and pancreatic disorders.[3,4] Scientific investigation into the therapeutic effects of milk thistle, its active constituents and its mechanism of action, began 25 years ago. Since then, over 20 clinical trials and hundreds of *in vitro* and *in vivo* investigations have been published describing the pharmacological effects of a standardized milk thistle extract in the treatment of liver diseases.[5]

## Quality Information

- The correct Latin name for the plant is *Silybum marianum* L. (Asteraceae).[1,2,6] Botanical synonyms for the plant include: *Carduus marianus* L., *Carthamus maculatum* Lam., *Cirsium maculatum* Scop., *Mariana mariana* (L) Hill., *Silybum maculatum* Moench. Common names include: blessed thistle, Holy thistle, lady's thistle, Lady's milk, marian thistle, mild marian thistle, milk thistle, silybum, St. Mary's thistle, thistle, thistle of the Blessed Virgin, true thistle, and variegated marian thistle.[1–3,5,6]
- Standardized extracts and other commercial products are prepared from the dried ripe fruit (sometimes called seeds) of *Silybum marianum* (L.) Gaertn., freed from the pappus, Asteraceae.[6]
- The major active constituents of the milk thistle are a group of flavonolignans, collectively known as silymarin, occurring in a concentration of 1.5–3%. Silymarin is an isomeric mixture containing silybin and isosilybin (a 1:1 mixture of diastereoisomers), silychristin and silydianin. The isomeric ratio of silybin to isosilybin to silychristin to silydianin is 3:1:1:1.[7,8] Other flavonolignans identified in milk thistle include 2,3-dehydrosilybin, 2,3-dehydrosilychristin. Taxifolin, a 2,3-dihydroflavonol, which may be regarded as the parent flavonol of the silymarin compounds, is also present.[1,2,4,5]

## Medical Uses

Supportive treatment of alcohol or drug-induced liver damage, chronic inflammatory liver diseases and hepatic cirrhosis.[9–21]

A parenteral preparation, silybin hemisuccinate sodium salt, is available in Germany for the treatment of poisoning due to *Amanita phalloides* (deathcap) mushrooms.[22–26] The drug is administered intravenously in four infusions, for a total dose of 20 mg/kg/day of silybin over a 24-hour period each administered over a 2-hour period.[26]

## Summary of Clinical Evidence

At least 20 clinical trials have assessed the efficacy of a standardized silymarin extract for the treatment of alcohol-induced cirrhosis, toxic liver damage due to alcohol or drugs, and hepatitis. However, the results from these studies are often difficult to interpret due to poor methodology. Many trials suffer from small patient numbers, variability in the severity and etiology of liver disease, inadequately defined outcome measures, lack of proper controls and inconsistent assessment of alcohol use by patients. Furthermore, spontaneous resolution of liver injury, in response to the removal of the hepatotoxin, is not generally taken into consideration.

At least 13 controlled clinical trials have assessed the safety and efficacy of a standardized silymarin extract for the treatment of alcohol-induced liver cirrhosis or hepatitis.[10–14,16–18,20,27–30] In five of the clinical trials, the patient numbers ranged from 50 to 100 patients, and two trials included 170 and 200 patients, respectively.[13,29] In most of the trials, a standardized silymarin extract was administered orally, in a dose ranging from 280–420 mg (140 mg two or three times daily), and the treatment periods ranged from 4 to 24 weeks. The outcomes measured were generally levels of liver enzymes, and subjective symptoms such as fatigue, itching, and jaundice. In the trial by Parés and co-workers, an unspecified extract was used, at a dose of 150 mg three times daily.[29] Two trials, including the Parés study, continued for 2 years and used survival rates as the major outcome measure.[13,29]

A 6-month, double-blind, placebo-controlled trial assessed the efficacy of silymarin for the treatment of 36 chronic alcoholics with fibrous alterations of the liver or micronodular cirrhosis.[12,28] Outcomes measured were liver function tests, serum procollagen III peptide levels, and improvements in liver histology. During the trial alcohol consumption was reduced, and remained comparable in both groups. Silymarin treatment (140 mg twice daily, orally), improved liver histology, increased lymphocyte proliferation and decreased lipid peroxidation.[12] During the treatment, serum bilirubin, aspartate aminotransferase and alanine aminotransferase returned to normal, and γ-glutamyl transferase and procollagen III peptide levels decreased as compared with placebo.[12,28] In a follow-up biopsy performed in 6 patients from the control group and 7 patients from the treatment group, an improvement in liver histology was observed in the treated group.[12] In a randomized, double-blind, placebo-controlled trials, 66 patients with clinically, serologically and biopsy confirmed liver damage (primarily due to alcohol) were treated with a standardized silymarin extract (420 mg/day) for 4 weeks.[14] A decrease in the levels of liver enzymes and improved liver function was observed as early as one week after treatment. In another randomized, double-blind, placebo-controlled trial, the effects of the extract were tested in 106 patients with elevated serum GOT and GPT levels for at least one month, of which 80% will still consuming alcohol. The subjects were treated with 140 mg of the extract three times daily for 4 weeks. The mean serum levels of transaminases decreased by ~65% in the treated group, as compared with an average of 24% in the control group. In patients whose alkaline phosphatase, bilirubin and bromosulphalein tests were abnormal at the beginning of the trial, improvements were observed in the silymarin group, although they were not statistically significant.[18] Comparison of histological evaluations of 15 patients in the treated group and 14 in the placebo group showed definite improvements in 11/15 in the treated group as compared with 4/14 in the placebo group.[18] In another randomized, double-blind, placbeo-controlled trial, patients with alcohol-induced liver disease of varying degrees of severity were treated with a standardized extract (420 mg/day) for two months.[27] Significant (p < 0.05) decreases in the serum levels of GOT, GPT and bilirubin were noted in the treated group, and symptoms such as weakness, anorexia and nausea were markedly improved as compared with placebo.[27] A randomized double-blind, placebo-controlled, multicenter trial assessed the efficacy of a standardized silymarin extract for the treatment of 116 patients with alcohol-induced hepatitis, of which 58 were diagnosed with cirrhosis.[30] The patients were treated orally with 420 mg/day of silymarin or placebo for 3 months. Although a significant decrease in liver enzymes was observed in both groups, treatment with a standardized silymarin extract was not statistically superior to placebo.[30]

Two conflicting randomized, double-blind, placebo-controlled trials have assessed the effects of silymarin extracts on the survival of patients with liver cirrhosis over a long-term treatment period.[13,29] The Ferenci trial (trial 1) treated both alcoholic and non-alcoholic patients, while the Parés trial (trial 2) treated only alcoholic patients. In trial 1 the cumulative survival of 170 patients with hepatic cirrhosis of various origins (87 in the treatment group and 83 in the placebo group) was followed over a mean duration of 41 months of treatment.[13] At two years the mortality in the treatment group was 23% as compared with 33% in the placebo group. After four years of treatment, the cumulative survival in the treated group was 58% as compared with 38% in the placebo group. A separate retrospective review of etiology and severity by means of the Child-Turcotte Index (Child A to C) showed that the subgroup of patients with confirmed alcohol-induced cirrhosis and Child A patients had a significant increase in the survival rate (p = 0.01 and 0.03, respectively). In patients with cirrhosis due to other causes, or those of severity grades Child B or C, there was no significant effect on survival.[13] In the Parés study (trial 2) however, treatment of 200 alcoholics (histologically or laparoscopically proven liver cirrhosis) with a silymarin extract (150 mg three times daily, product not specified) for two years failed to increase the survival time as compared with placebo.[29]

A randomized, double-blind, placebo-controlled clinical trial assessed the efficacy of silymarin extracts on cellular immune parameters, and antioxidant effects in chronic alcoholics with fibrotic changes in the liver or hepatic cirrhosis.[11,16,17] Treatment of the subjects with 420 mg/day of a standardized silymarin preparation for 6 months enhanced the expression and activity of superoxide dismutase in erythrocytes and lymphocytes; increased the serum levels of free-SH groups and glutathione peroxidase activity; increased lectin-induced lymphoblast transformation; normalized serum aminotransferase and bilirubin levels; and reduced serum malondialdehyde concentrations as compared with placebo.[11,16,17]

A double-blind, placebo-controlled trial assessed the efficacy of a silymarin extract in the treatment of 21 patients with various chronic inflammatory liver diseases.[10] Oral administration of 420 mg/day of the extract for 1 year, lead to an improvement in histopathological findings in the treated group as compared with placebo.[10] In a randomized trial (no blinding) of 60 patients with secondary insulin-dependent diabetes caused by alcohol-induced cirrhosis, oral treatments of 600 mg/day of silymarin were compared with untreated controls.[31,32] After 6 months of treatment, blood glucose levels, blood malondialdehyde levels, daily insulin need and fasting insulinemia were all significantly lower in the treated group as compared with untreated patients.[31,32]

Five controlled trials have assessed the efficacy of silymarin or silipide (a complex of silybin and phosphatidylcholine) for the treatment of hepatitis.[33–37] A randomized double-blind clinical study of 57 subjects assessed the efficacy of silymarin for the treatment of acute viral hepatitis A or B.[34] Patients received 420 mg per day of a standardized silymarin extract or matching placebo for 3 weeks. At the end of treatment, 40% of patients in the treatment group had normalized bilirubin levels as compared with 11% of the placebo group, and 82% of the treated patients had a normalized level of aspartamine transaminase as compared with 52% of patients treated with placebo.[34] In a second controlled clinical trial, a significant decrease in the duration of in-patient care (23.3 versus 30.4 days) was observed patients treated with silymarin, as compared with patients treated only by supportive care.[33] In the silymarin treated group, a shorter interval to the development of immunity was observed in those patients infected with hepatitis B (30.4 versus 41.2 days with supportive therapy only).[33] Another double-blind study in subjects with acute viral hepatitis indicated that treatment of the patients with silymarin (140 mg three times daily) decreased the complications associated with the infection.[35] Two combined double-blind clinical trials assessed the efficacy of silymarin for the treatment of chronic hepatitis (with or without cirrhosis) in a total of 36 patients.[36] Patients were treated with 420 mg/day of a silymarin preparation or placebo for 3 months to one year. Assessment of liver function tests (bilirubin levels and liver enzymes) showed no significant differences between silymarin or placebo treatments. However, histological improvements (parenchymatous changes, mesenchymal intralobular reaction $p < 0.05$, portal inflammatory reaction) were noted in those patients treated with the silymarin extract.[36] A randomized, controlled phase II trial evaluated the effects of silipide (a complex of silybin and phosphatidylcholine) in 65 patients with biopsy-proven chronic persistent hepatitis.[37] The patients received 240 mg of the drug or placebo for three months. A statistically significant decrease in mean serum aspartate ($p < 0.05$) and alanine ($p = 0.01$) aminotransferases was observed in the treated group as compared with placebo.[37]

A controlled clinical trial measured the effects of a standardized silymarin extract on liver function tests in 30 patients with chronic occupational exposure to toluene and/or xylene vapours (exposure for 5–20 years).[19] The subjects were treated with 140 mg of the preparation three times daily for 30 days, and the results were compared with 19 untreated matched controls. The results showed that elevated AST and ALT levels were decreased, and low platelet counts were markedly improved in the treated patients as compared with the untreated controls.[19] Oral administration of silymarin (420 mg/day) to 14 subjects who were chronically exposed to organophosphates (Malathion), and a group of 10 healthy volunteers assessed the effects of a silymarin extract on liver function.[38] In those patients with chronic exposure to

Malathion, silymarin increased the activity of serum cholinesterase (p < 0.01), but did not improve liver function in healthy volunteers.[38]

An observational study assessed the efficacy of a silymarin extract (1–2 capsules twice daily, standardized to 100 mg or 40% silymarin) for the inhibition of fibrotic activity in 245 patients with various chronic liver diseases.[39] The outcome measured was the level of procollagen III-peptide (P-III-P), a marker for serum fibrogenetic activity in cases of liver fibrosis. Elevated levels of serum N-terminal propeptide of collagen type III were normalized or improved after 4-weeks in those patients treated with silymarin.[39] In an open drug surveillance study, 108 patients with alcohol-induced hepatotoxicity and liver inflammation were treated with silymarin (200–400 mg/kg/day) for 5 weeks.[40] At the end of treatment, the serum P-III-P and liver enzyme levels were significantly lower in comparison to the initial baseline values. The preparation was generally well tolerated by 98% of patients.[40] The safety and efficacy of silymarin has been evaluated in over 3500 patients in two drug-monitoring studies.[9,15] In one multicenter observational study, 2637 patients with toxic liver damage of various origins and degrees of severity were treated with a standardized silymarin preparation (267 ± 104 mg/day) for 8 weeks.[9] After 8 weeks of therapy, reductions in serum GPT and GOT were observed. The therapy decreased subjective symptoms by 63%, improved clinical findings and reduced elevated serum levels of liver enzymes. Treatment was rated as very good, good or satisfactory by 88% of the attending physicians. Minor gastrointestinal side effects were reported in 1% of patients.[9]

A double-blind, placebo-controlled clinical trial assessed the efficacy of silymarin in the prevention of psychotropic drug-induced hepatic damage.[41] Sixty patients receiving chronic psychotropic drug therapy were treated orally with 800 mg/day of silymarin or placebo for 90 days. Silymarin treatment improved liver function, and reduced lipoperoxidative hepatic damage due to treatment with butyrophenones or phenothiazines, as determined by the measurement of serum malondialdehyde levels (the end product of the oxidation of polyunsaturated fatty acids).[41] A small clinical study found improvements in biochemical parameters in 19 patients using psychotropic drugs after 6 months of therapy with silymarin.[42] A review of hepatitis B infection among renal transplant patients has indicated that silymarin appears to confer some beneficial effects during acute episodes of hepatic dysfunction.[43]

Several published case reports have suggested that silymarin and silybin are effective for the treatment of *Amanita phalloides* poisoning (deathcap mushroom).[21–24,26] *Amanita* toxins inhibit the activity of RNA polymerase in hepatocytes, causing cell death after a period of 12–24 hours. In an uncontrolled clinical trial 60 consecutive patients were treated intravenously with silybin (20 mg/kg/day), initiated 24 to 36 hours after *Amanita* ingestion.[21] The survival rate was reported as 100%.[21] The results of a multicenter study involving 252 cases of human poisoning due to ingestion of *Amanita phalloides* indicate that intravenous infusion of silybin, in combination with the standard management techniques, dramatically reduced mortality, without side effects.[22–24]

## Pharmacokinetics

The dose-linearity of the diastereomers of silybin was assessed in a randomized, uncontrolled, four-way, crossover trial, after oral administration of silymarin.[44] In this study, a standardized silymarin extract was administered orally to 6 healthy male subjects in single doses of 102, 153, 203, and 254 mg.[44] Using high-pressure liquid chromatographic analysis, both silybin and isosilybin were measured in the plasma as unconjugated compounds, as well as total isomers after hydrolysis. Areas under the curve were linear with the dose, and only 10% of total silybin in the plasma was found as the conjugated form. For unconjugated silybin the half-life was less than one hour, and for total silybin the elimination half-life was estimated at 6 hours. Approximately 5% of the dose was excreted into the urine as total silybin, corresponding to a renal clearance of 30 ml/min.[44]

Following a single oral dose of a standardized silymarin preparation, to 10 cholecystectomized patients with T-drainage, approximately 7–15% of silybin was measured in the 24-hour bile samples.[45] In 4 of the female patients, silybin excretion at 24 hours was decreased and elimination was delayed. In two of these patients one with a carcinoma of the stomach with liver metastasis and one with pancreatitis, a reduction in silybin elimination was associated with a reduction in bile output.[45]

Following administration of a single oral dose of a standardized silymarin preparation (140 mg), to 9 cholecystectomized patients with T-drainage, the urinary and biliary excretion of silybin, silydanin and silychristin was measured.[46] The urinary excretion of silybin and silychristin was insignificant. Both silybin and silychristin were excreted in the bile in the form of sulfate and glucuronide conjugates. The total elimination of silybin was estimated at 20–40%, and that of silychristin was 4–10%. Excretion of silymarin occurred over a 24-hour period with the maxiumum excretion between 2 and 9 hours post-administration.[46] In an uncontrolled clinical trial, six subjects were given a single oral dose of silymarin (560 mg = 240 mg silybin), and the serum concentration and urinary excretion of silybin, the principal constituent of silymarin, were measured.[47] The maximum serum concentrations ranged from 0.18–0.62 μg/ml, and renal excretion was 1–2% of the administered silybin after 24 hours. However, after oral administration of 140 mg of silymarin (60 mg of silybin) to cholecystectomized patients, bile collected from the T-drainage was found to contain 11 and 47 μg/ml of silybin.[47] The bioavailability of silymarin products varies considerably and is highly dependent on the product formulation.[48]

## Mechanism of Action

Silymarin, and its major constituent, silybin, inhibit acetaminophen-, amitriptyline-, carbon tetrachloride-, ethanol-, erythromycin estolate-, galactosamine-, nortriptyline- and tert-butyl hydroperoxide-induced hepatotoxicity in rat hepatocytes *in vitro*.[49–52] Silybin reduces ischemic damage to nonparenchymal cells and improves postischemic function in porcine livers.[53] Suppression of allyl alcohol-induced toxicity, associated lipid peroxidation and glutathione depletion in isolated rat hepatocytes was observed after treatment with silymarin and silybin at concentrations of 0.1 and 1.0 mM, respectively.[54] Silybin increases the synthetic rate of the ribosomal RNA by 20% in cultured hepatocytes and in isolated liver nuclei via activation of DNA-dependent-RNA polymerase I.[55] Silybin appears to bind to the regulatory subunit of DNA-dependent RNA polymerase I at a the estrogen binding site, and acts as a natural steroid effector, thereby activating the enzyme and increasing the synthetic rate of ribosomal RNA.[56] The increase of ribosomal RNA synthesis stimulates the formation of mature ribosomes, and protein biosynthesis in the liver.[55] The uptake of $^3$H-dimethyl phalloidin in isolated rat hepatocytes was inhibited by 79% in cells treated with silybin ester (100 μg/ml).[57]

Silybin inhibits *f*-met peptide- and anti-IgE-induced histamine release in human basophil leukocytes.[58] The inhibitory effect was significantly attenuated (p < 0.05) by elevating extracellular calcium concentrations.[58] Silybin inhibits the synthesis of leukotriene B$_4$ (IC$_{50}$ 15 μM/L) in isolated rat Kupffer cells, but did not effect prostaglandin E$_2$ formation.[59] Silymarin, silybin, silydanin and silychristin inhibit the activities of lipoxygenase and prostaglandin synthetase *in vitro*.[60–62] The chemotactic and phagocytic activities of human polymorphonuclear leukocytes are not modified by silybin at concentrations up to 25 μg/ml. However, silybin did inhibit luminol-enhanced chemiluminescence suggesting that silybin exerts its antiinflammatory activity through the inhibition of hydrogen peroxide formation.[63]

Both silymarin and silybin have antioxidant activity *in vitro*. These compounds react with oxygen free radicals such as hydroxyl anions, phenoxy radicals, and hypochlorous acid in human blood platelets, fibroblast cultures, rat liver microsomes and mitochondria, as well as in enzymatic and non-enzymatically generated free inorganic radicals.[59,64–68] Silybin (IC$_{50}$ 80 μM/L) inhibits the production of superoxide anion radical and nitric oxide by 50% in rat Kupffer cells.[59] Both silybin and silymarin inhibit free radical-induced lipid peroxidation in erythrocytes, microsomal and mitochondrial preparations, and thereby stabilize the structure of the cell membrane.[64,69–77] Inhibition of cyclic adenosine monophosphate (cAMP) phosphodiesterase by silybin, silydanin and silychristin has been demonstrated *in vitro* (Koch et al. 1985).[78] Since cAMP is known to stabilize lysosomal membranes, an increase in the concentration of this nucleoside has been proposed as the mechanism by which silymarin exerts its membrane stabilization and antiinflammatory effects.[78] Silybin inhibits phospholipid synthesis and the breakdown in rat liver membranes *in vitro*, and corrects the alteration in phospholipid metabolism due to the administration of ethanol to rats.[79] Both silymarin and silybin are incorporated into the hydrophobic-hydrophilic interface of the microsomal membrane bilayer and alter membrane structure by influencing the packing of the acyl chains.[72]

Intraperitoneal administration of either silybin or silymarin to rodents markedly inhibits experimental liver damage induced by acetaminophen, *Amanita phalloides* toxins (e.g., phalloidin and α-amanitin), ethanol, galactosamine, halothane, polycyclic aromatic hydrocarbons, rare earth metals (e.g., cerium, praseodymium and lanthanum) and thallium.[57,80–89] Intraperitoneal or intragastric administration of sily-marin to dogs, mice or rats prevents carbon tetrachloride-induced liver damage (dosage range 15–800 mg/kg).[71,90–92] The antihepatotoxic activity of silymarin is attributed to its antioxidant effects, its membrane stabilization effects, and a decrease in the metabolic activation of carbon tetrachloride in hepatocytes.[71,90,91,93] Intra-venous administration of silybin hemisuccinate sodium salt (50 mg/kg) to beagles, administered sub-lethal doses of *Amanita phalloides* (85 mg/kg), prevented an increase in serum levels of liver enzymes, and the decrease of clotting factors.[94] Intravenous administration of silybin hemisuccinate sodium salt to mice, preinfected with sublethal doses of Frog virus 3 (FV3), protected the animals from infection with vaccinia virus, attenuated histological changes in hepatocyte nuclei, and increased survival times in animals treated with a lethal dose of FV3.[95–97]

Intragastric administration of silymarin (50 mg/kg) improved the metabolism and disposition of aspirin in rats with carbon-tetrachloride-induced liver toxicity.[98] How-ever, intravenous administration of silybin (50 mg/kg) to rats inhibited the protective effect of ethanol on acetaminophen-induced hepatoxicity.[99] This effect appeared to be due to an inhibition of the microsomal metabolism of acetaminophen by the com-bination of ethanol and silybin.[99] Intragastric administration of silymarin (50 mg/kg) inhibited collagen accumulation in early and advanced biliary fibrosis secondary to complete bile duct occlusion induced by injection of sodium amidotrizoate.[100] Sily-marin increased the redox state and the total glutathione content in the liver, intes-tine and stomach of rats after intraperitoneal administration (200 mg/kg).[68,101] Silybin has been shown to compete with estradiol for binding to estrogen receptors isolated from porcine uteri.[102]

In a liver transplantation experiment, explanted porcine liver was subjected to cold-induced ischaemia by storage of the liver at 4°C for 24 hours, followed by extra-corporeal reperfusion for 4 hours.[53] Intravenous administration of 500 mg of silybin ester, prior to removal of the liver, plus 400 mg/L during cold storage and 100 mg/h during reperfusion, significantly reduced histological liver cell damage and improved liver function during reperfusion by 24 to 66%, as measured by bile production and bile acid excretion, respectively.[53]

Intragastric administration of silymarin reduced carrageenan-induced pedal edema in rats ($ED_{50}$ 62.42 mg/kg).[103] Topical applications of silymarin inhibited xylene-induced ear inflammation in mice, and the activity was similar to that of indomethacin (25 mg/kg). Silymarin also inhibited leukocyte accumulation in inflam-matory exudates following intraperitoneal administration of carrageenan to mice.[103]

Intragastric administration of an acetone extract of the fruit (containing silybin 25–1000 mg/kg) increased the volume and dry mass of the excreted bile in rats.[104] Intragastric administration of silymarin (100 mg/kg) prevented gastric ulceration in rats induced by cold-restraint and pyloric ligation, but was not effective in ethanol-induced ulcers.[105] Intragastric administration of silymarin (100 mg/kg) to rats pre-vented gastric injury induced by ischemia-reperfusion.[106]

## Safety Information

### A. Adverse Reactions

One case of anaphylactic shock was reported in a patient ingesting a tea prepared from milk thistle.[107] A mild laxative effect has also been reported.[108]

### B. Contraindications

Extracts of milk thistle should not be administered to patients with an allergy to plants in the daisy family (Asteraceae).[6] There are no data assessing the effects of milk thistle extracts during pregnancy, nursing or in children.[6]

### C. Drug Interactions

No drug interactions have been reported.

### D. Toxicology

The acute oral and intravenous toxicity of both silymarin and silybin were tested in various animal models. No adverse effects were observed after intragastric administration of 20g/kg of silymarin to mice and 1 g/kg to dogs.[3]

The $LD_{50}$ values of silybin sodium hemisuccinate were 1.01 g/kg in mice and 873 mg/kg in rats after intravenous administration. Long-term toxicity tests showed no adverse reactions after intragastric administration of 100 mg/kg/day to rats for 16–22 weeks.[3]

No evidence of embryotoxicity was observed in rats or rabbits.[3]

### E. Dose and Dosage Forms

Daily dosage: Standardized preparations: based on the clinical trials 280–420 mg of silymarin, calculated as silybin.

## References

1    Blaschek W, Hänsel R, Keller K, Reischling J, Pimpler H, Schneider G, (eds.). *Hägers Handbuch der Pharmazeutischen Praxis*, 6th ed., Berlin, Springer, 1998.

2    Wichtl M. *Herbal drugs and phytopharmaceuticals*, English edition [N.G. Bisset, transl and ed], Boca Raton, FL, CRC Press, 1994.

3    Morazzoni P, Bombardelli E, *Silybum marianum (Carduus marianus)*. *Fitoterapia*. 1995;66:3–42.

4    Flora K et al. Milk thistle (*Silybum marianum*) for the therapy of liver disease. *Amer J Gastroenterol*. 1998;93:139–143

5    Farnsworth NR, ed. Napralert database. University of Illinois at Chicago, IL, February 9;1998 (an on-line database available directly through the University of Illinois at Chicago or through the Scientific and Technical Network (STN) of Chemical Abstracts Services).

6    Anon. *WHO Monographs on Selected Medicinal Plants*, Volume II, Traditional Medicines Programme, WHO, Geneva, Switzerland: WHO Publications, 2001.

7    Leng-Peschlow E, Strenge-Hesse A. The milk thistle (*Silybum marianum*) and silymarin as hepatic therapeutic agents. *Zeitschrift für Phytotherapie*. 1991;12:162–174.

8    Leng-Peschlow E. Properties and medical use of flavonolignans (silymarin) from *Silybum marianum*. *Phytother Res.* 1996;12:162–174.

9    Albrecht M et al. Die Therapie toxischer Leberschäden mit Legalon®. *Zeitschrift für Klinische Medizin.* 1992;47:87–92.

10   Berenguer J, Carrasco D. Ensayo doble ciego de Silimarina frente a placebo en el tratamiento de hepatopatías crónicas de diversa génesis. *Münchener Mediziniche Wochenschrift.* 1977;119:240–260.

11   Déak G. et al. Silymarin kezelés immunmoduláns hatása krónikus alkoholos májbetegségben. *Orvosi Hetilap.* 1990;131:1291–1296.

12   Fehér J, Deák G, Müzes G, Láng I, Niederland V, Nékám K, Káeteszi M. Hepatoprotective activity of silymarin Legalon therapy in patients with chronic alcoholic liver disease. *Orvosi Hetilap.* 1989;130:2723–2727.

13   Ferenci P et al. Randomized controlled trial of silymarin treatment in patients with cirrhosis of the liver. *J Hepatol.* 1989;9:105–113.

14   Fintelmann V, Albert A. Nachweis der therapeutischen Wirksamkeit von Legalon® bei toxishen Lebererkrankungen im Doppelblindversuch. *Therapiewoche.* 1980;30: 5589–5594

15   Grüngreiff K et al. Nutzen der medikamentösen Lebertherapie in der hausärztlichen Praxis. *Die Medizinische Welt.* 1995;46:222–227.

16   Müzes M et al. Silymarin (Legalon®) kezelés hatása idült alkoholos májbetegek antioxidáns védörendszerée és a lipid peroxidációra, (kettos vak protokoll). *Orvosi Hetilap.* 1990c,131:863–866.

17   Láng I, Nékám K, Deák G, Müzes G, Gonzales-Cabello, Gergely P, Csomos G, Fehér J. Immunomodulatory and hepatoprotective effects of *in vivo* treatment with free radical scavengers. *Ital J Gastroenterol.* 1990;22:283–287.

18   Salmi HA, Sarna S. Effect of silymarin on chemical, functional, and morphological alterations of the liver. *Scand J Gastroenterol.* 1982;17:517–521.

19   Szilárd S et al. Protective effect of Legaol® in workers exposed to organic solvents. *Acta Med Hungar.* 1988;45:249–256.

20   Varis K et al. Die Therapie der Lebererkrankung mit Legalon: eine konrtrollierte Doppelbklindstudie. In: *Aktuelle Hepatologie, Third International Symposium*, Cologne, Hanseatisches Verlagskontor, Lubeck, 1978:42–43.

21   Vogel G. Natural substances with effects on the liver. In: *New natural products and plant drugs with pharmacological, biological or therapeutic activity.* New York, Springer-Verlag, 1977:2651–2665.

22   Floersheim GL et al. Poisoning by the deathcap fungus (*Amanita phalloides*): prognostic factors and therapeutic measures. *Schweizerische Medizinische Wochenschrift.* 1982;112:1164–1177

23   Hruby C. Silibinin in the treatment of deathcap fungus poisoning. *Forum.* 1984;6: 23–26

24   Hruby C et al. Pharmakotherapie der Knollenblätterpilzvergiftung mit Silibinin. *Wiener Klinische Wochenschrift.* 1983;95:225–231.

25   Schultz V et al. *Rational phytotherapy. A physician's guide to herbal medicine.* Berlin, Springer, 1997.

26   Vogel G. The anti-*Amanita* effect of silymarin. In: *Amanita toxins and poisonings.* International *Amanita* symposium, H. Faulstich et al., (eds.). Baden-Baden, Verlag, Gerhard and Witzstrock, 1980:180–189.

27   Di Mario F, Farini R, Okolicsanyi R, Naccarato R. Die Wirkung von auf die Leberfunktionsproben bei Patienten mit alkoholbedingter Lebererkrankung, Doppelblindstudie. In: *Der toxisch-metabolische Leberschaden*, De Ritis F, Csomos G, Braatz R, (eds.). Hamburg, Harnsisches Verlagskontor Lubecl, 1981.

28 Fehér J, Lang I. Wirkmechanismen der sogenannten Leberschutzmittel. *Bayer Internist.* 1988;4:3–7.

29 Parés A, Planas R, Torres M, Caballeria J, Viver JM, Acero D, Panes J, Rigau J, Santos, Rodes J. E. Effects of silymarin in alcoholic patients with cirrhosis of the liver: results of a controlled, double-blind, randomized and multicenter trial. *J. Hepatol.* 1998;28:615–621.

30 Trinchet JC et al. Traitement de L'hépatite alcoolique par la silymarine une étude comparative en double insu chez 116 malades. *Gastroenterol Clin Biol.* 1989;13:120–124.

31 Velussi M et al. Silymarin reduces hyperinsulinemia, malondialdehyde levels, and daily insulin need in cirrhotic diabetic patients. *Current Ther Res.* 1993;53:533–544.

32 Velussi M et al. Long-term (12 months) treatment with an anti-oxidant drug (silymarin) is effective on hyperinsulinemia, exogenous insulin need and malondialdehyde levels in cirrhotic diabetic patients. *J Hepatol.* 1997;26:871–879.

33 Cavalieri S. Kontrollierte klinische Pruefung von Legalon. *Gazzette Medica Italiana.* 1974;133:628.

34 Magliulo E et al. Zur Wirkung von Silymarin bei der Behandlung der akuten Virushepatitis. *Medizinische Klinik.* 1978;73:1060–1065.

35 Plomteux G et al. Hepatoprotector action of silymarin in human acute viral hepatitis. *International Research Communications Systems.* 1977;5:259–261.

36 Kiesewetter E et al. Ergebnisse zweier Doppelblindstudien zur Wirksamkeit von Silymarin bei chronischer Hepatitis. *Leber Magen Darm.* 1977;7:318–323.

37 Marcelli R, Bizzoni P, Conte D, Lisena MO, Lampertico M, Marena C, De Marco MF, Del Ninno E. Randomized controlled study of the efficacy and tolerability of a short course of IdB 1016 in the treatment of chronic persistent hepatitis. *Eur Bull Drug Res.* 1992;1:131–135.

38 Boari C et al. Silymarin in the protection against exogenous noxae. *Drugs Exp Clin Res.* 1981;7:115–120.

39 Held C. Fibrose-hemmung unter Praxisbedingungen. *Therapiewoche.* 1992;42:1696–1701.

40 Held C. Fibrose-hemmung unter Praxisbedingungen. *Therapiewoche.* 1993;43:2002.

41 Palasciano G et al. The effect of silymarin on plasma levels of malondialdehyde in patients receiving long-term treatment with psychotrophic drugs. *Current Ther Res.* 1994;55:537–545.

42 Saba P et al. Effetti terapeutici della silimarina nelle epatopatie croniche indotte da psicofarmaci. *Gazzette Medica Italiana.* 1976;135:236–251.

43 Chan MK, Chan PCK, Wong KK, Cheng IKP, Li MK, Chang WK. Hepatitis B infection and renal transplantation: the absence of anti delta antibodies and the possible beneficial effect of silymarin during acute episodes of hepatic dysfunstion. *Nephrol Dial Transplant.* 1989;4:297–301.

44 Weyhenmeyer R et al. Study on dose-linearity of the pharmacokinetics of silibinin diastereomers using a new stereospecific assay. *Int J Clin Pharmacol,* 1992;30:134–138.

45 Lorenz D, Mennicke WH. Elimination von Pharmaka bei cholezystektomierten Patienten. Untersuchungen mit Silymarin bei Patienten mit außerhepatischen Begleiterkrankungen. *Meth Find Exp Clin Pharm.* 1981;3:103S–106S.

46 Flory PJ et al. Studies on elimination of silymarin in cholecystectomized patients. *Planta Med.* 1980;38:227–237.

47 Lorenz D, Lucker PW, Mennicke WH, Wetzelsberger N. Pharmacokinetic studies with silymarin in human serum and bile. *Meth Find Exp Clin Pharm.* 1984;6:655–661.

48 Schultz HU et al. Untersuchungen zum Freisetzungs-verha;ten und zur Bioäquivalenz von Silymarin-Präparaten. *Arzneimittelforschung.* 1995;45:61–64.

49    Davila JC et al. Protective effect of flavonoids on drug-induced hepatotoxicity *in vitro*. *Toxicology.* 1989;57:267–286.

50    Hikino H et al. Antihepatotoxic actions of flavonolignans from *Silybum marianum* fruits. *Planta Med.* 1984;50:248–250

51    Joyeux M et al. Tert-butyl hydroperoxide-induced injury in isolated rat hepatocytes: a model for studing antihepatotoxic crude drugs. *Planta Med.* 1990;56:171–174.

52    Ramellini G, Meldolesi J. Liver protection by silymarin: *in vitro* effect on dissociated rat hepatocytes. *Arzneimittelforschung.* 1976;26:69–73.

53    Blumhardt G et al. Silibinin reduces ischemic damage to nonparenchyymal cells and improves postischemic liver function of UW-preserved porcine livers. *Zeitschrift für Gastroenterologie.* Abstract 1994.

54    Miguez MP et al. Hepatoprotective mechanism of silymarin: no evidence for involvement of cytochrome P450 2E1. *Chemico-Biological Interactions.* 1994;91:51–63.

55    Sonnenbichler J, Zetl I. Stimulating influence of a flavonolignane derivative on proliferation, RNA synthesis and protein synthesis in liver cells. In: *Assessment and management of hepatobiliary disease*, Okolicsanyi L et al. (eds.). Berlin, Springer-Verlag, 1987.

56    Sonnenbichler J, Zetl I. Biochemistry of a liver drug from the thistle *Silybum marianum. Planta Med.* 1992;58(Suppl):A580–A581.

57    Faulstich H et al. Silybin inhibition of amatoxin uptake in the perfused rat liver. *Arzneimittelforschung.* 1980;30:452–454.

58    Miadonna A et al. Effects of silybin on histamine release from human basophil leucocytes. *Brit J Clin Pharmacol.* 1987;24:747–752.

59    Dehmlow C et al. Scavenging of reactive oxygen species and inhibition of arachidonic acid metabolism by silibinin in human cells. *Life Sci.* 1996a;58:1591–1600.

60    Baumann J et al. Hemmung de Prostaglandinsynthetase durch Flavonoide und Phenolderivate im Vergleich mit deren $O_2$-Radikalfängereigenschaften. *Arch Pharmacol* (Weinheim). 1980;313:330–337.

61    Fiebrich F, Koch H. Silymarin, an inhibitor of lipoxygenase. *Experientia.* 1979a;35: 1548–1550.

62    Fiebrich F, Koch H. Silymarin, an inhibitor of prostaglandin synthetase. *Experientia.* 1979b;35:1550–1552.

63    Minonzio F et al. Modulation of human polymorphonuclear leukocyte function by the flavonoid silybin. *Int J Tissue Reactions.* 1988;10:223–231.

64    Cavallini L et al. Comparative evaluation of antiperoxidative action of silymarin and other flavonoids. *Pharmacol Res Comm.* 1978;10:133–136.

65    György I et al. Reactions of inorganic free radicals with liver protecting drugs. *Radiation Physical Chemistry.* 1990;36:165–167.

66    Mira ML et al. The neutralization of hydroxyl radical by silibin, sorbinil and bendazac. *Free Radical Research Communications.* 1987;4:125–129.

67    Pascual C et al. Effect of silymarin and silybinin on oxygen radicals. *Drug Development Res.* 1993;29:73–77.

68    Valenzuela A, Garrido A. Biochemical bases of the pharmacological action of the flavonoid silymarin and of its structural isomer silibinin. *Biol Res.* 1994;27:105–112.

69    Bindoli A et al. Inhibitory action of silymarin of lipid peroxide formation in rat liver mitochondria and microsomes. *Biochem Pharmacol.* 1977;26:2405–2409.

70    Koch HP, Loffler E. Influence of silymarin and some flavonoids on lipid peroxidation in human platelets. *Meth Exp Find Clin Pharmacol.* 1985;7:13–18.

71    Lettéron P et al. Mechanism for the protective effects of silymarin against carbon tetrachloride-induced lipid peroxidation and hepatotoxicity in mice. *Biochem Pharmacol.* 1990;39:2027–2034.

72    Parasassi T et al. Drug-membrane interactions: silymarin, silibyn and microsomal membranes. *Cell Biochemistry and Function*. 1984;2:85–88.

73    Ramellini G, Meldolesi J. Stabilization of isolated rat liver plasma membranes by treatment *in vitro* with silymarin. *Arzneimittelforschung*. 1974;24:806–808.

74    Valenzuela A, Guerra R. Differential effect of silybin on the $Fe^{2+}$-ADP and t-butyl hydroperoxide-induced microsomal lipid peroxidation. *Experientia*. 1986;42:139–141.

75    Valenzuela A et al. Inhibitory effect of the flavonoid silymarin on the erythrocyte hemolysis induced by phenylhydrazine. *Biochem Biophys Res Comm*. 1985a;126:712–718.

76    Valenzula A et al. Silymarin protection against hepatic lipid peroxidation induced by acute ethanol intoxication in the rat. *Biochem Pharmacol*. 1985b;34:2209–2212.

77    Valenzuela A et al. Silybin dihemisuccinate protects rat erythrocytes against phenylhydrazine-induced lipid peroxidation and hemolysis. *Planta Med*. 1987;402–405.

78    Koch HP et al. Silymarin: potent inhibitor of cyclic AMP phosphodiesterase. *Meth Exp Find Clin Pharmacol*. 1985;7:409–413.

79    Castigli E et al. The activity of silybin on phospholipid metabolism of normal and fatty liver *in vivo*. *Pharmacol Res Comm*. 1977;9:59–69.

80    Barbarino F et al. Effect of silymarin on experimental liver lesions. *Revue Roumaine Medecine*. 1981;19:347–357.

81    Campos R et al. Silybin dihemisuccinate protects against glutathione depletion and lipid peroxidation induced by acetominophen on rat liver. *Planta Med*. 1989;55:417–419.

82    Janiak B. Die Hemmung der Lebermikrosomenaktivität bei der Maus nach einmaliger Halothannarkose und seine Beeinflußbarkeit durch Silybin (Silymarin). *Anaesthesist*. 1974;23:389–393.

83    Meiß R et al. Effect of silybin on hepatic cell membranes after damage by polycyclic aromatic hydrocarbons (PAH). *Agents and Actions*. 1982;12:254–257.

84    Mourelle M et al. Protection against thallium hepatoxicity by silymarin. *J Applied Toxicol*. 1988;8:351–354.

85    Strubelt O et al. The influence of silybin on the hepatotoxic and hypoglycemic effects of praseodymium and other lanthanides. *Arzneimittelforschung*. 1980;30:1690–1694.

86    Trost W, Lang W. Effect of thioctic acid and silibinin on the survival rate in amanitin and phalloidin poisoned mice. *IRCS Medical Science*. 1984;12:1079–1080.

87    Tuchweber B et al. Prevention of praseodymium-induced hepatoxicity by silybin. *Toxicol Appl Pharmacol*. 1976;38:559–570.

88    Tyutyulkova N et al. Hepatoprotective effect of silymarin (Carsil) on liver of D-galatosamine treated rats. Biochemical and morphological investigations. *Meth Find Exper Clin Pharmacol*. 1981;3:71–77.

89    Wang M et al. Hepatoprotective properties of *Silybum marianum* herbal preparation on ethanol-induced liver damage. *Fitoterapia*. 1996;67:166–171.

90    Martin R et al. Hepatic regeneration drugs in dogs: effect of choline and silibin in dogs with liver damage. *Vet Med*. 1984;April:504–510.

91    Mourelle M et al. Prevention of $CCL_4$-induced liver cirrhosis by silymarin. *Fundamentals Clin Pharmacol*. 1989;3:183–191.

92    Muriel P, Mourelle M. Prevention by silymarin of membrane alterations in acute $CCL_4$-induced liver damage. *J Appl Toxicol*. 1990a;10:275–279.

93    Muriel P, Mourelle M. The role of membrane composition in ATPase activities of cirrhotic rat liver: effect of silymarin. *J Appl Toxicol*. 1990b;10:281–284.

94    Floersheim GL et al. Effects of penicillin and silymarin on liver enzymes and blood clotting factors in dogs given a boiled preparation of *Amanita phalloides*. *Toxicol Appl Pharmacol*. 1978;46:455–462

95    Elharrar M et al. Ein neues Modell der experimentellen toxischen Hepatitis. *Arzneimittelforschung*. 1975;25:1586–1591.

96    Gendrault JL et al. Wirkung eines wasserlöslichen Derivates von Silymarin auf die durch Frog-Virus 3 an Mäusehepatozyten hervorgerufenen morphologischen und funktionellen Veränderungen. *Arzneimittelforschung*. 1979;29:786–791.

97    Steffan AM, Kirn A. Multiplication of vaccinia virus in the livers of mice after frog virus 3-induced damage to sinusoidal cells. *Journal of the Reticuloendothelial Society*. 1979;26:531–538.

98    Mourelle M, Favari L. Silymarin improves metabolism and disposition of aspirin in cirrhotic rats. *Life Sci*. 1989;43:201–207.

99    Garrido A et al. The flavonoid silybin ameliorates the protective effect of ethanol on acetaminophen hepatotoxicity. *Research Communications in Substances of Abuse*. 1989;10:193–196

100   Boigk G et al. Silymarin retards collagen accumulation in early and advanced biliary fibrosis secondary to complete bile duct obliteration in rats. *Hepatology*. 1997;26:643–649.

101   Valenzuela A et al. Selectivity of silymarin on the increase of the glutathione content in different tissues of the rat. *Planta Med*. 1989;55:420–422.

102   Sonnenbichler J, Zetl I. Specific binding of a flavonolignane derivative to an estradiol receptor. In: *Plant flavonoids in biology and medicine II: biochemical, cellular and medicinal properties*, Alan R. Liss, 1988:369–374.

103   De La Puerta R et al. Effect of silymarin of different acute inflammation models and on leukocyte migration. *J Pharm Pharmacol*. 1996;48:969–970.

104   Danieluk R et al. The preparation of vegetable products containing isofraxidin, silibin and glaucium alkaloids and evaluation of their choleretic actions. *Polish Pharmacol Pharm*. 1973;25:271–283.

105   Alarcón de la Lastra C et al. Gastric anti-ulcer activity of silymarin, a lipoxygenase inhibitor, in rats. *J Pharm Pharmacol*. 1992;44:929–931.

106   Alarcón de la Lastra C et al. Gastroprotection induced by silymarin, the hepatoprotective principle of *Silybum marianum* in ischemia-reperfusion mucosal injury: role of neutrophils. *Planta Med*. 1995;61:116–119.

107   Geier J et al. Anaphylakischer Schock durch einen Mariendistel-Extrakt bei Soforttyp Allergie auf Kiwi. *Allergologie*. 1990;13:387–388.

108   German Commission E Monograph: Cardui mariae fructus, *Bundesanzeiger*. 1986;50:13.03.

# 20

# Nettle

## Synopsis

Results obtained from clinical trials support the use of stinging nettle root extracts for the symptomatic treatment of urinary symptoms secondary to benign prostatic hyperplasia stages I and II. Patients should be instructed to obtain a diagnosis of BPH and rule out prostate cancer prior to self-medicating with nettle root extract. The recommended daily dosage is 4–6 g of the crude drug or equivalent preparations as an infusion, or 600–1200 mg of a dried extract preparation (5:1, 20% methanol). Treatment for 1 to 3 months may be required before the full therapeutic effects are apparent. There are no drug interactions reported. Adverse reactions, such as nausea, vomiting, headaches and dizziness, have been reported in clinical trials. Nettle root should not be administered to patients with an allergy to stinging nettle or to plants of the Urticaceae. Due to the effects of nettle on androgen and estrogen metabolism, administration of nettle root is contraindicated during pregnancy and nursing, and in children under the age of 18 years old.

## Introduction

The genus *Urtica*, to which nettle belongs, is reported to have at least 14 distinct species and numerous varieties, many of which have been used medicinally throughout the world.[1] The nettle is a perennial herbaceous plant with stinging hairs on their leaves and stems. The Latin name for the genus *Urtica*, was derived from the Latin *uro* or *urere*, meaning "I burn".[2] Before its popularity as an herbal medicine, nettle plants were used in fabric weaving, and nettle fabric burial shrouds have been found in Bronze Age sites in Denmark.[1] The medicinal use of nettle dates back to the 1st century BC when Greek physicians applied nettle juice externally to treat snakebite and other insect stings. Hippocrates was reported to use nettle juice internally as a diuretic, a laxative, and an antidote for plant poisons such as hemlock and henbane.[1] The Native American Indians employed a tea prepared from nettle tea to stop uterine contractions after childbirth, as well as increase milk production. In the 19th century, the Eclectic physicians recommended nettle root as a diuretic to treat urinary, bladder and kidney ailments, and as a topical treatment for hemorrhoids.[1] A decoction of the roots combined with honey was also used to treat the symptoms of asthma. Case studies published in 1950 reported symptomatic improvements in patients suffering from benign prostatic hyperplasia, after oral administration of a nettle root tea.[2] Since then, much of the scientific research has focused on the use of nettle root extracts for the treatment of urinary symptoms associated with benign prostatic hyperplasia (BPH). Currently, hydroalcoholic extracts of the nettle root are used clinically in Europe for the treatment of BPH.[2]

## Quality Information

- The correct Latin names for nettle include *Urtica dioica* L. and *U. urens* L.[3] Botanical synonyms that may appear in the scientific literature for *Urtica dioica* include *Urtica gracilis* Ait., *U. major* Kanitz. Botanical synonyms for *U. urens* include *U. urens* maxima Blackw. Numerous common names for each of the species. For *Urtica dioica* they include: Brennesselwurzel, common nettle, gazaneh, grande ortie, greater nettle, grosse Brennessel, Haarnesselwurzel, Hanfnesselwurzel, ortica, ortie, ortiga, pokrzywa, Nesselwurzel, nettle root, pokrzywa, qurrays, racine d'ortie, raiz de ortiga, and stinging nettle. Common names for *Urtica urens* are lesser nettle and small nettle.[2,3]
- Standardized extracts and other commercial products of nettle are prepared from the dried root and rhizomes of *Urtica dioica* L., *U. urens* L. (Urticaceae) their hybrids.[3]
- Native to Eurasia and Africa, but have become naturalized in temperate regions of the world: Europe, Africa, Asia, Australia, North and South America.[2,3]
- The chemical constituents isolated from nettle root belong to various chemical classes, including fatty acids, terpenes, phenylpropanes, lignans, coumarins, triterpenes, ceramides, sterols, and lectins. Among these compounds are oxalic acid, linoleic acid, 14-octacosanol, 9-hydroxy-10-*trans*-12-*cis*-octadecadienoic acid, 13-hydroxy-9-*cis*,11-*trans*-octadecadienoic acid, 9-hydroxy-10-*trans*,12-*cis*-octadecadienoic acid, scopoletin, β-sitosterol, stigmasterol, campesterol, β-sitosterol-3-O-glucoside, secoisolariciresinol-9-O-β-D-glucoside, neo-olivil, oleanolic acid, ursolic acid, the lectin *Urtica dioica* agglutinin, and polysaccharides, RP1-RP5.[2,3]

## Medical Uses

Treatment of lower urinary tract symptoms (nocturia, polyuria, urinary retention) secondary to benign prostatic hyperplasia (BPH) stages I and II,[4–17] in cases where a diagnosis of prostate cancer is negative.[3] Nettle root has also been used in traditional medicine as a diuretic, and for the treatment of rheumatism and sciatica.[1] However, there are no clinical data to support these claims.

## Summary of Clinical Evidence

Three double-blind, placebo-controlled clinical trials assessed the efficacy of nettle root extracts for the treatment of lower urinary tract symptoms secondary to BPH.[6,9,17] All three trials employed a 20% methanol extract (dried) of nettle root. A double-blind, placebo-controlled clinical trial assessed the efficacy of a nettle root extract for the symptomatic treatment of BPH stages I and II in 50 male subjects.[17] A significant increase in urine volume (43.7% increase, $p = 0.027$), and a significant decrease in the levels of serum sex hormone binding globulin (SHBG, $p = 0.0005$) was observed in the patients treated orally with 600 mg/day of the extract for 9 weeks as compared with placebo. A modest increase in the maximum urine flow rate (8%) was also observed in the treated group, however it was not statistically significant.[17]

The second double-blind placebo-controlled clinical trial assessed the efficacy of a nettle root extract for the symptomatic treatment of BPH in 40 male subjects.[9] Oral administration of 1200 mg of the extract per day for 6 weeks produced a statistically significant decrease in urinary frequency ($p \leq 0.05\%$) and serum levels of SHBG.[9] The third double-blind, placebo-controlled study assessed the efficacy of a nettle root extract (600 mg/day, oral) for the treatment of 32 male subjects with BPH stage I.[6] A 4–14% increase in average urinary flow and a 40–53% decrease in post-void residual volume were observed in those patients treated with the extract for 4–6 weeks. The placebo group exhibited no improvements in residual volume.[6]

Twelve uncontrolled clinical trials assessed the efficacy of hydroalcoholic nettle root extracts (20% methanol or 30–45% ethanol) for the treatment of lower urinary tract symptoms (nocturia, polyuria, dysuria, urine retention) secondary to benign prostatic hyperplasia.[4,5,7,8,10–14,16,18,19] Although all of the 12 trials demonstrated positive effects on symptoms of BPH, the results should be viewed with caution, as no controls were used, and observed placebo effects for BPH treatments may be as high as 60%.

The efficacy of a 40% ethanol extract of the roots (1:1) for the treatment of urinary symptoms was assessed in 67 men (> 60 yr.) with benign prostatic hyperplasia.[5] Treatment of the patients with 5 ml of the extract daily for six months decreased nocturia and post-void volume, but did not reduce prostate enlargement.[5] The effect of a 20% methanol extract of nettle roots on the urinary symptoms of BPH was assessed in 89 male subjects.[7] Oral administration of 600 mg of the extract per day decreased the post-void volume in 67 patients after 3–24 months of treatment.[7] In an open study, 26 men with BPH stage I or II, were treated with 1200 mg of a nettle root extract for 3 to 24 months.[8] Treatment produced a decrease in prostate volume in 54% of patients, and a decrease in post-void residual volume in 75% of the subjects.[8] Ten men with symptoms of BPH were treated with 30–150 drops daily of a 45% ethanol extract of the root for 30 days.[11] After treatment, the post-void residual volume by decreased by 66%.[11] In an open study of 39 patients with BPH stages I-III, 37 of the patients exhibited an improvement in urine flow, and a reduction in post-void residual volume, nocturia, and polyuria after 6 months of treatment with a 20%

methanol extract per day (600–1200 mg).[18] Twenty-seven patients with BPH stages I and II were treated with a 20% methanol extract of the roots for 3.5 months.[19] Residual urine volume was decreased in 75% of cases ($p < 0.001$), and the maximal urinary flow increased in 50% of cases ($p < 0.002$).[19]

Three large-scale multicenter studies involving 14,033 men assessed the efficacy of a 20% methanol extract of the roots (600–1200 mg/day) for symptomatic treatment of BPH.[10,13,14] One study demonstrated a decrease in nocturia and polyuria in 90.6% of the patients after 6 months of treatment.[10] A 50% decrease in nocturia was observed in the patients treated with 1200 mg/day for 10 weeks.[13] Significant improvements in both urinary flow and post-void residual volume were observed in 4480 patients treated with 600–1200 mg/day for 20 weeks ($p < 0.01$).[14]

An uncontrolled study involving 31 male subjects with BPH stages I and II assessed the effects of treatment with 1200 mg of a 20% methanol root extract (oral) for 20 weeks on prostate morphology.[20] Prostate cells were obtained from the subjects by needle biopsy and analyzed every four weeks for morphological changes. After 4–16 weeks of treatment, an enlargement of the nuclear volume, as well as hydropic swelling and vacuolization of the cytoplasm in prostate cells was observed.[20] In an open study involving 33 patients with BPH, prostate cells were obtained by needle biopsy and analyzed by fluorescent microscopy.[21] Incubation of prostate glandular epithelium cells with a 20% methanol extract of the root caused an enlargement of the nuclear volume with a loosening of the chromatin, and a hydropic swelling of the cytoplasm. In addition, a distinct reduction in the number of homogenous secretory granules was observed, indicating a reduction in the biological activity of these cells.[21]

Morphological changes in prostate tissues, obtained by needle biopsy from patients with BPH (stage II), before and six-months after treatment with a 20% methanol extract (1200 mg/day), were examined by electron microscopy.[22] After six months of treatment, changes in the smooth muscle cells and glandular epithelium cells of the prostate were detected. A reduction in the activity of the muscle cells was reversed, and an increase in the secretory activity of the glandular epithelium cells was observed.[22]

**Mechanism of Action**

Ethanol extracts and isolated polysaccharides from nettle root have anti-inflammatory activity both *in vitro* and *in vivo*. Intragastric administration of a polysaccharide fraction, isolated from nettle roots, to rats (40 mg/kg) suppressed carrageenan-induced pedal edema for up to 22 hours.[23] The activity of the polysaccharide fraction was comparable to that of indomethacin 10 mg/kg.[23] An ethanol extract of the roots inhibited the activity of human leukocyte elastase (HLE) and reduced the amount of HLE released by activated polymorphonuclear granulocytes during the inflammatory response.[24] The ethanol extract inhibited the degradation of a peptide substrate by HLE ($IC_{50}$ 3.6 μg/ml) and bovine elastin ($IC_{50}$ 68 μg/ml).[24]

Polysaccharides isolated from a root extract induced lymphocyte proliferation *in vitro* in a concentration range of 10–100 μg/ml.[25] An aqueous lyophilized extract (10 μg/ml), and a 40% alcohol extract of the roots (100 μg/ml) stimulated lymphocyte proliferation by 63 and 100%, respectively *in vitro*.[25] An ethyl acetate extract of the roots induced cell differentiation in human promyelocytic leukemia HL-60 cells *in vitro* (ED$_{50}$ 4 μg/ml).[26] *Urtica dioica* agglutinin (UDA, 500 ng/ml) inhibited lymphocyte proliferation and the binding of epidermal growth factor to its receptor on A431 epidermoid cancer cells *in vitro*.[27] UDA also exhibited immune stimulant effects on T lymphocytes in a dose dependent manner.[25]

Steroid hormone binding globulin (SHBG) is a plasma protein that binds to circulating androgens and estrogens, and thereby regulates their free concentration in the blood plasma.[2] The plasma membrane of the human prostate contains specific SHBG receptors, and SHBG appears to play a role in the development of BPH. A hydroalcoholic extract of the root (10%) reduced the binding capacity of SHBG for 5-α-dihydrotestosterone by 67% *in vitro* (blood plasma).[28] An aqueous extract of the root inhibited the binding of $^{125}$I-SHBG to human prostate membranes *in vitro* at concentrations of 0.6–10 mg/ml.[29] The lignan, secoisolariciresinol, and an isomeric mixture of the compounds 13-hydroxy-9-*cis*,11-*trans*-octadecadienoic acid and 9-hydroxy-10-*trans*,12-*cis*-octadecadienoic acid isolated from a methanol root extract, reduced the binding of SHBG to 5-α-dihydrotestosterone.[30] Both secoisolariciresinol, and its main metabolites (−)-3,4-divanillyltetrahydrofuran and enterofuran, displaced the binding of 5-α-dihydrotestosterone to SHBG by 60, 95, and 73%, respectively *in vitro*.[31]

Intragastric administration of a 30% ethanol extract of the root to male mice inhibited the activities of 5α-reductase and aromatase *in vivo* (ED$_{50}$ 14.7 and 3.5 mg/ml, respectively).[32] However, a hydroalcoholic extract of the root dissolved in DMSO did not inhibit the activity of 5α-reductase from human prostate cells *in vitro* at a concentration of up to 500 μg/ml.[33] Both ursolic acid and 14-octacosanol isolated from a methanol extract of the roots inhibited the activity of aromatase *in vitro*.[34] The compound 9-hydroxy-10-*trans*,12-*cis*-octadecadienoic acid isolated from the roots inhibited the activity of aromatase *in vitro*.[35] A standardized hydroalcoholic extract of the roots (IC$_{50}$ 338 μg/ml) inhibited the conversion of androstenedione into estradiol by aromatase isolated from placenta microsomes in the presence of NADPH.[24] The heptane-soluble fraction of the extract was the most active for inhibiting the activity of aromatase (IC$_{50}$ 9 μg/ml).[24] A butanol, ether, ethyl acetate or hexane extract of the roots inhibited the activity of Na$^+$, K$^+$-ATPase isolated from BPH tissues by 27.6–81.5%.[36] In addition, stigmast-4-en-3-one, stigmasterol, and campesterol, steroidal components of the roots, inhibited enzyme activity by 23–67% at concentrations ranging from 1 mM to 1 μM.[36]

Intragastric administration of a hexane extract of nettle root (1.28 g per day) to castrated rats did not inhibit testosterone or dihydrotestosterone stimulated prostate growth.[33] Intraperitoneal administration of a hydroalcoholic extract of the roots (20 mg/kg) suppressed testosterone-stimulated increases in prostate weight and intraprostatic ornithine decarboxylase activity in castrated rats.[37] Oral administration of a hydroalcoholic extract of the root (30 mg/kg) decreased prostate volume by 30% in dogs with benign prostatic hyperplasia, after 100 days of treatment.[38] The effect of various root extracts was assessed in adult mice after implantation of fetal uro-genital sinus into the prostate gland.[39] Intragastric administration of a butanol, cyclo-hexane, or ethyl acetate extract of the root (0.25 ml/day) had no effect on the development of benign prostatic hyperplasia in mice models. However, a 20% methanol extract of the root (0.25 ml/day, gavage) reduced the development of benign prostatic hyperplasia in male mice by 51.4%.[39]

## Pharmacokinetics

Intragastric administration of [$^{125}$I]-labeled *Urtica dioica* agglutinin (UDA) to mice (2 µg–10 mg/animal) showed that UDA was distributed to the blood, gut, liver and kidneys.[2] The uptake and excretion of UDA was assessed by enzyme-linked immunoassay in human subjects after oral administration of a 20 mg dose.[2] Approx-imately 30–50% of the oral UDA dose[25] was excreted in the feces, and the total amount of UDA excreted in the urine was less than 1%.[2]

## Safety Information

### A. Adverse Reactions
Results from clinical trials show that nettle root extracts are well tolerated in humans. A few cases of minor transient gastrointestinal side effects such as diarrhea, gastric pain and nausea have been reported.[14,17] as well as allergic skin reactions.[14]

### B. Warnings
Nettle root extracts relieve the symptoms associated with an enlarged prostate without reducing the size of an enlarged prostate. If symptoms do not improve or become wors-ened, or in cases of blood in the urine or acute urinary retention, contact a physician.[3]

### C. Contraindications
Nettle root should not be administered to patients with an allergy to stinging nettle or to plants of the Urticaceae. Due to the effects on androgen and estrogen metabo-lism, the use of nettle root during pregnancy and nursing is contraindicated.[3] There is no therapeutic rationale for the administration of nettle root extracts to children under the age of 18 years old.

### D. Drug Interactions
No drug interactions have been reported.

### E. Toxicology
Intravenous administration of an aqueous extract or infusion of the roots to rats had an LD$_{50}$ of 1721 mg/kg and 1929 mg/kg, respectively.[2] Chronic oral administration of an infusion of the roots to rats was well tolerated at doses up to 1310 mg/kg.[2]

*F. Dose and Dosage Forms*
Daily dose: Oral administration of 600–1200 mg of a dried extract preparation (5:1, 20% methanol); 1.5–7.5 ml fluid extract (1:1, 45% ethanol), or 5 ml ethanol extract (1:5, 40% ethanol).[3] The German Commission E has also approved a dose of 4–6 g of the crude drug or equivalent preparations as an infusion.[40]

## References

1 Patten G. *Urtica. Aust J Med Herbalism.* 1993;5:5–13.
2 Bombardelli E, Morazzoni P. *Urtica dioica* L. *Fitoterapia.* 1997;68:387–402.
3 Mahady GB, Fong HHS, Farnsworth NR. Radix Urticae. *WHO Monographs on Selected Medicinal Plants*, Volume II, WHO, Geneva Switzerland: WHO Publications.
4 Bauer HW et al. Endokrine Parameter während der Behandlung der benignen Prostatahyperplasie mit ERU. *Benigne Prostatahyperplasie II, Klinische und Experimentelle Urologie* 19, H.W. Bauer R, ed., Munich, Zuckschwerdt, 1988.
5 Belaiche P, Lievoux O. Clinical studies on the palliative treatment of prostatic adenoma with extract of *Urtica* root. *Phytother Res.* 1991;5:267–269.
6 Dathe G, Schmid H. Phytotherapie der benignen Prostatahyperplasie (BPH). *Urologie* [B]. 1987;27:223–226.
7 Djulepa J. Zweijährige Erfahrung in der Therapie des Prosta-Syndroms. *Ärztliche Praxis.* 1982;63:2199–2202.
8 Feiber H Sonographische Verlaufsbeobachtungen zum Einfluß der medikamentösen Therapie der benignen Prostatahyperplasie (BPH). *Benigne Prostatahyperplasie II, Klinische und experimentelle Urologie* 19, HW Bauer, ed., Munich, Zuckschwerdt, 1988.
9 Fisher M, Wilbert D. Wirkprüfung eines Phytopharmakons zur Behandlung der benignen Prostatahyperplasie (BPH). *Benigne Prostatahyperplasie III, Klinische und experimentelle Urologie* 22, G Rutishauser, ed., Munich, Zuckschwerdt, 1992.
10 Friesen A. Statistische Analyse einer Multizenter-Langzeitstudie mit ERU. *Benigne Prostatahyperplasie II, Klinische und Experimentelle Urologie* 19, H.W. Bauer, ed., Munich, Zuckschwerdt, 1988.
11 Goetz P. Die Behandlung der benignen Prostatahyoerplasie mit Brennesselwurzeln. *Zeitschrift für Phytotherapie.* 1989;10:175–178.
12 Sonnenschein R. Untersuchung der Wirksamkeit eines prostatotropen Phytotherapeutikums (*Urtica* plus) bei benigner Prostatahyperplasie und Prostatitis-eine prospektive multizentrische Studie. *Urologie* [B]. 1987;27:232–237.
13 Stahl HP. Die Therapie prostatischer Nykturie. *Zeitschrift für Allgemeine Medizin.* 1984;60:128–132.
14 Tosch U et al. Medikamentöse Behandlung der benignen Prostatahyperplasie. *Euromed.* 1983;(6):1–3.
15 Vandierendounck EJ, Burkhardt P. Extractum radicis urticae bei Fibromyoadenom der Prostata mit nächtlicher Pollakisurie. *Therapiewoche Schweiz.* 1986;2:892–895.
16 Vahlensieck W. Konservative Behandlung der benignen Prostatahyperplasie. *Therapiewoche Schweiz.* 1986;2:619–624.
17 Vontobel HP et al. Ergebnisse einer Doppelblindstudie über die Wirksamkeit von ERU-Kapseln in der konservativen Behandlung der benignen Prostatahyperplasie. *Urologie* [A]. 1985;24:49–51.
18 Maar K. Rückbildung der Symptomatik von Prostataadenomen. *Fortschritte der Medizin.* 1987;105:50–52.
19 Romics I. Observations with Bazoton in the management of prostatic hyperplasia. *International Urol Nephrol.* 1987;19:293–297.

20    Ziegler H. Cytomorphological study of benign prostatic hyperplasia under treatment with Radix Urticae extract (ERU) preliminary results. *Fortschritte der Medizin.* 1982;39:1823–1824.

21    Ziegler H. Investigations of prostate cells under the effect of Radix Urticae extract (ERU) by Fluorescent microscopy. *Fortschritte der Medizin.* 1983;45:2112–2114.

22    Oberholzer M et al. Elektronenmikroskopische Ergebnisse bei medikamentös behandelter benigner Prostatahyperplasie. In: *Benigne Prostatahyperplasie*, Bauer HW ed., Munich, Zuckschwerdt, 1986.

23    Wagner H at al. Search for the antiprostatic principle of stinging nettle (*Urtica dioica*) roots, *Phytomedicine.* 1994a;1:213–224.

24    Koch E. Pharmacology and modes of action of extracts of Palmetto fruits (Sabal Fructus), stinging nettle roots (Urticae Radix) and pumpkin seed (Cucurbitae Peponis Semen) in the treatment of benign prostatic hyperplasia. *Phytopharmaka in Forschung und Klinischer Anwendung*, D. Loew, N. Rietbrock, (eds.). Darmstadt, Verlag Dietrich Steinkopf, 1995.

25    Wagner H et al. Lektine und Polysaccharide-die Wirkprinzipien der *Urtica dioica* Wurzel. In: *Benigne Prostatahyperplasie*, Boos G, ed., Frankfurt, PMI, 1994b.

26    Suh N et al. Discovery of natural product chemopreventive agents utilizing HL-60 cell differentiation as a model. *Anticancer Res.* 1995;15:233–239.

27    Wagner H et al. Studies on the binding of *Urtica dioica* agglutinin (UDA) and other lectins in an *in vitro* epidermal growth factor receptor test. *Phytomedicine.* 1995;4: 287–290.

28    Schmidt K. The effect of an extract of Radix Urticae and various secondary extracts on the SHBG of blood plasma in benign prostatic hyperplasia. *Fortschritte der Medizin.* 1983;101:713–716.

29    Hryb DJ et al. The effects of extracts of the roots of the stinging nettle (*Urtica dioica*) on the interaction of SHBG with its receptor on human prostatic membranes. *Planta Med.* 1995;61:31–32.

30    Ganßer D, Spiteller G. Plant constituents interfering with human sex hormone-binding globulin evaluation of a test method and its application to *Urtica dioica* root extracts. *Zeitschrift für Naturforschung.* 1995a;50c:98–104.

31    Schöttner M et al. Lignans from the roots of *Urtica dioica* and their metabolites bind to human sex hormone binding globulin (SHBG). *Planta Med.* 1997;63:529–532.

32    Hartmann RW et al. Inhibition of 5 alpha-reductase and aromatase by PHL-00801 (Prostatonin), a combination of PY 102 (*Pygeum africanum*) and UR 102 (*Urtica dioica*) extracts. *Phytomedicine.* 1996;3:121–128.

33    Rhodes L et al. Comparison of finasteride (Proscar), a 5 $\alpha$-reductase inhibitor, and various commercial plant extracts in *in vitro* and *in vivo* 5 $\alpha$-reductase inhibition. *Prostate.* 1993;22:43–51.

34    Ganßer D, Spiteller G. Aromatase inhibitors from *Urtica dioica* roots. *Planta Med.* 1995b;61:133–140.

35    Kraus R, Spiteller G, Bartsch W. (10E,12Z)-9-hydroxy-10,12-octadecadiensäure, ein Aromatase-Hemmstoff aus dem Wurzelextrakt von *Urtica dioica*. *Liebigs Annales d'Chemie.* 1991:335–339.

36    Hirano T et al. Effects of stinging nettle root extracts and their steroidal components on the $Na^+$, $K^+$-ATPase of the benign prostatic hyperplasia. *Planta Med.* 1994;60:30–33.

37    Scapagnini U, Friesen A. *Urtica dioica*-Extrakt und Folgesubstanzen im Tierversuch. *Klinische und Experimentelle Urologie.* 1992;22:138–144.

38    Daube G. Pilotstudie zur behandlung der benignen prostatahyperplasie bei hunden mit extractum Radicis Urticae. *Benigne Prostatahyperplasie II, Klinische und Experimentelle Urologie* 19, H.W. Bauer, ed., Munich, Zuckschwerdt, 1988.

39   Lichius, JJ, Muth C. The inhibiting effects of *Urtica dioica* root extracts on experimentally induced prostatic hyperplasia in the mouse. *Planta Med.* 1997;63:307–310.

40   German Commission E monograph, Urticae radix (Brennesselwurzel). *Bundesanzeiger* No. 173;18.09.1986; No. 43;2.03.1989, No. 50;13.03.1990; No. 11, 17.01.1991.

# 21

# *Panax ginseng*

**Synopsis**

Although *Panax ginseng* has been used extensively as a tonic, for the enhancement of physical and mental performance, results from recent randomized, controlled clinical trials do not substantiate its purported ergogenic effects in healthy individuals. However, results from both human and animal studies suggest that *Panax ginseng* maybe of some value for the treatment of erectile dysfunction and age associated memory impairment, through a mechanism involving nitric oxide-induced vasodilation. Furthermore, there is an increasing body of evidence to suggest that *Panax ginseng* may have health promoting effects by acting as both an antioxidant and an immune stimulant. The daily dose is 0.5 to 2 grams of dried root, or 100 to 300 mg per day of a standardized extract containing 1.5 to 7% ginsenosides. Patients should be cautioned not to exceed the recommended dose, as adverse reactions such as hypertension, diarrhea, nervousness and insomnia are associated with excessive ingestion. Drug interactions have been reported with both phenelzine and warfarin. While there are no contraindications, *Panax ginseng* should be used with caution during pregnancy, nursing, and in children under the age of 12 years old, as safety data is currently unavailable.

**Introduction**

One of the most popular and well-known herbal products is a group of plants known generically as "ginseng". The plants in this group are not taxonomically uniform, as there are at least 30 different species of plants that are collectively referred to as "ginseng". One species, known scientifically as *Panax ginseng*, is commonly referred to as Korean or Asian ginseng. In China, the root or rhizome (underground stem) and the leaves of *Panax ginseng* have been used medically.[1] However, the root is certainly the most prominent part and dominates the commercial market. The plant is a slow-growing perennial herb, and the roots are usually not harvested until the 5th or 6th year of growth, when the ginsenosides are at the highest concentration.

*Panax ginseng* is a member of a class of "seng" herbs, which are fleshy roots used as tonics. The term "gin" is the Chinese word for "man", as the shape of the roots can resemble the human body.[1] The two major forms of *Panax ginseng* root, called white and red type, are dependent on the way the ginseng root is processed to prevent rotting and microbial contamination. White ginseng is the dried, bleached (sulfur dioxide) root of *Panax ginseng*, sometimes peeled to remove the outer coating (skin) of the root. Red ginseng, on the other hand, is prepared by steaming the root for three hours, and then air-drying. The steamed root turns a caramel-like color and is resistant to invasion by fungi and pests.[1,2] Most commercial ginseng is now cultivated, as wild *Panax ginseng* is a protected species in both Russia and China.[3]

One of the first written records of the medicinal use of *Panax ginseng* is in an ancient text of traditional Chinese herbal medicine *entitled Shen Nung Pen Ts'ao Ching*, also known as the Shen-Nung Pharmacopoeia 196 AD, and later revised around 452–536 AD.[4] Thus, *Panax ginseng* has been used as an herbal medicine for at least two thousand years, and perhaps as long as four thousand years.[4] In the year 1596, Li-Shih-Chen's herbal *Pen-Tsao Kang-Mu* described *Panax ginseng* as a "superior tonic", or an herb that was given to persons suffering from weakness.[1] While the name Panax is derived from the Greek panacea (cure-all), the concept that ginseng is a general "cure-all" is not correct. The traditional Chinese use of *Panax ginseng* is very specific, in that it used to treat patients with chronic illness, especially during periods of convalescence, to restore the person to a normal state of good health. Up to 1937, *Panax ginseng* was an official compendial drug the US, and listed in the US Dispensatory (US Dispensatory). While *Panax ginseng* is not included in the Generally Recognized as Safe (GRAS) list, the US Food and Drug Administration regards ginseng as a food, and it is currently regulated as a dietary supplement in the United States.

## Quality Information

- The correct Latin name for the plant is *Panax ginseng* C.A. Meyer (Araliaceae),[5] while *Panax schinseng* Nees is a botanical synonym.[5] Common names for *Panax ginseng* include the following: Asian ginseng, Ginsengwürzel, hongshen, jenseng, jenshen, Korean ginseng, and Oriental ginseng.[5]
- Standardized extracts and other commercial products are prepared from the dried root of *Panax ginseng*.[5] The plant is indigenous to the mountain regions of Korea, Japan, China (Manchuria), and Russia (eastern Siberia).[1] However, commercial products are prepared from cultivated ginseng, imported from China, Japan, Korea and Russia.[1,5]
- The main chemical constituents of *Panax ginseng* are the triterpene saponins, known as the ginsenosides. More than 30 are based on the dammarane structure, with one, ginsenoside Ro, being an oleanolic acid derivative.[2,6,7] The ginsenosides are derivatives of either protopanaxadiol or protopanaxatriol. Members of the former group include the ginsenosides $Ra_{1,-2,-3}$; ginsenosides $Rb_{1,-2,-3}$; ginsenosides Rc, $Rc_2$; ginsenosides Rd, $Rd_2$; ginsenoside $Rh_2$; 20(S)-ginsenoside $Rg_3$; malonyl-ginsenosides Rb-1; malonyl-ginsenoside $Rb_2$;

malonyl-ginsenosides Rc; and malonyl-ginsenoside Rd. Examples of pro-topanaxatriol saponins are: ginsenosides $Re_{2,\,-3}$; ginsenoside Rf; 20-gluco-gin-senoside Rf; ginsenosides $Rg_{-1,-2}$; 20(R)-ginsenoside Rg-2; 20(R)-ginsenoside $Rh_1$; and ginsenoside $Rh_1$.[5] The most important constituents are the ginseno-sides $Rb_{1,-2}$, Rc, Rd, Rf, $Rg_1$, and $Rg_2$, with $Rb_{1,-2}$ and, $Rg_1$ being the most pre-dominant.[5]

- Daily dose (taken in the morning): dried root 0.5–2.0 grams; or 100–300 mg of a standardized extract containing 1.5 to 7.0% ginsenosides calculated as gin-senoside $Rg_1$.[8–18]

## Medical Uses

*Panax ginseng* is used a tonic or immune stimulant for enhancement of mental and physical capacity during fatigue, chronic illness, and convalescence.[5,8,25–26] The use of *Panax ginseng* as an ergogenic agent in healthy subjects[10–17] has not been sub-stantiated in recent clinical trials.[18–24]

*Panax ginseng* has been used for the treatment of diabetes, erectile dysfunction, and gastritis, however further randomized controlled clinical trials are needed before any therapeutic recommendations can be made.[5]

## Summary of Clinical Evidence

*Ergogenic Activity*
While numerous clinical studies have suggested that ginseng has an ergogenic effect,[10–17] results from recent randomized, controlled clinical trials do not support these data.[19–24] Furthermore, results from clinical studies published prior to 1990 measuring increased performance and anti-fatigue activity in healthy subjects are dif-ficult to interpret due to conflicting data, poor methodology, lack of proper controls, and a lack of standardization of the ginseng extracts used in the trials.

A randomized, double-blind, placebo-controlled trial assessed the effects of a standardized ginseng extract (SGE) on the work performance and energy metabo-lism of healthy adult females.[19] Nineteen women were treated with 200 mg of gin-seng extract (equivalent to 1 g of ginseng root) or placebo for eight weeks. Consumption of other dietary supplement such as vitamins and minerals was not per-mitted during the trial and the subjects maintained their normal diet and levels of physical activity. Before and after the trial intervention, each subject performed a graded maximal cycle ergometry test to exhaustion, and completed a standardized habitual physical activity questionnaire. The results of this trial demonstrated that chronic ginseng supplementation had no effect on maximal work performance; on resting, exercise and recovery exercise uptake, respiratory exchange ratio, minute ventilation, heart rate or the concentration of blood lactic acid levels ($p > 0.05$). Habitual physical activity scores of the subjects were similar in both the placebo and ginseng groups at the beginning and end of the 8-week trial period.[19] In another randomized, double-blind, placebo-controlled clinical trial by the same investigators assessed the effects of chronic ginseng administration (two different dosages) on the physiological and psychological responses during graded maximal aerobic exercise.[20] Thirty-six healthy men were treated with either 200 or 400 mg of a standardized ginseng extract (4% ginsenosides) or placebo for 8 weeks. The main outcomes measured were submaximal and maximal aerobic exercise responses, before and after treatment. Ginseng supplementation had no effect on oxygen consumption, respiratory exchange ratio, minute ventilation, heart rate, the concentration of blood lactic acid levels, or perceived exertion ($p > 0.05$).[20]

A randomized, double-blind, placebo-controlled clinical trial assessed the effects of ginseng on the peak exercise performance in healthy young adults.[21] Twenty men and eight women (mean age 23.2 years) were administered a standardized ginseng extract (7% ginsenosides) 200 mg daily or placebo for three weeks. The subjects performed a symptom limited graded exercise cycle ergometry test before and after treatment. Ginseng supplementation had no effect on maximal exercise capacity, exercise time, workload, plasma lactate and hematocrit, heart rate or perceived exertion.[21]

A placebo-controlled clinical trial with 8 participants reported no significant effect after 7 days of ginseng or placebo administration during a pedal to exhaustion bicycle test.[23] A randomized, placebo-controlled trial assessed the efficacy of a SGE (300 mg daily) on maximum aerobic and anaerobic exercise capacities in 41 male students with aerobic training or placebo.[24] Oral administration of ginseng alone, or in combination with aerobic exercise for 8 weeks increased the maximum aerobic capacity as compared with placebo. However, the combination of ginseng and aerobic exercise was not synergistic, and the combination did not enhance maximum aerobic capacity as compared with aerobic exercise alone. No effect on anaerobic parameters was observed.[24] The effect of chronic ginseng administration (2 g/day orally for 4 weeks) on substrate utilization, hormone production, endurance, metabolism and perception of effort during consecutive days of exhaustive exercise, was assessed in 11 healthy naval cadets. No significant differences were observed between the control group and the group receiving the ginseng supplementation.[24]

A randomized, double-blind, cross-over study assessed the effects of a combination SGE and vitamin supplement on circulatory, respiratory and metabolic functions during maximal exercise in 50 male sports teachers (21 to 47 years old).[11] The total tolerated workload and maximal oxygen uptake was significantly higher following ginseng administration (17880 kg/m and 63 ml/kg/min) as compared with placebo (14,464 kg/m and 59 ml/kg/min). In addition, workload, oxygen consumption, plasma lactate levels, ventilation, carbon dioxide production and heart rate during exercise were all significantly lower in the group taking ginseng supplements. The results of this trial suggest that the ginseng/vitamin preparations effectively increased the work capacity of the participants by improving oxygen utilization.[11] A placebo-controlled, crossover study determined the effects of ginseng on the physical fitness of 43 male triathletes.[12] The participants received a SGE (200 mg twice daily) or placebo for two consecutive training periods of 10 weeks. While no significant changes were observed during the first 10-weeks of treatment, ginseng supplementation appeared to prevent the loss of physical fitness (as measured by oxygen uptake and oxygen pulse) during the second ten week period.[12] The physical performance of 20 elite athletes (18–31 years of age) was assessed after supplementation with a standardized ginseng extract (100 mg twice daily) for nine weeks.[13] Treatment significantly improved aerobic capacity, and reduced serum lactate levels and heart rate during and after physical exertion.[13] A subsequent investigation in elite athletes assessed the effects of two dosages of ginseng (200 mg per day of a SGE containing 4% total ginsenosides or 200 mg per day of an extract containing 7% total ginsenosides) on athletic performance.[14,15] A significant improvement in aerobic capacity and a reduction in blood lactate levels and heart rate was observed following 9 weeks of treatment.[14,15] However, neither of these trials employed placebo or controls. Results from a further extension of the two trials in a randomized, placebo-controlled, double-blind clinical trial of 30 elite athletes reported ginseng treatment significantly increased oxygen uptake, and decreased exercise-induced elevation of blood lactate levels and heart rates as compared with placebo.[16] An assessment of the duration of the effects of a SGE (100 mg twice daily) was evaluated in a double-blind study of 28 athletes before, immediately following, and 11 weeks after cessation of ginseng treatment.[17] After nine weeks of treatment, a significant increase in oxygen uptake and forced expiratory volume, and decrease in exercise heart rate was observed for the ginseng versus placebo group. The differences persisted for approximately three weeks after the last ginseng dose and were no longer observable at 11 weeks.[17]

A placebo-controlled, double-blind crossover study assessed the effects of 1200 mg per day of *Panax ginseng* root in 12, fatigued, night-nurses.[9] The study used self-rating scales for mood, sleep quality, degree of lethargy and bodily feeling; objective tests of psychophysiological performance and hematological and biochemical tests. The results were compared with placebo and daytime work. Ginseng restored ratings on tests 11 of the 16 of mood variables, and 8 of the 14 somatic symptoms as compared with placebo. However, none of the differences were statistically significant, and no alterations were observed in either the biochemical or hematological tests. General performance improved in one of the tests (tapping), and the study concluded that ginseng had antifatigue activity.[9]

**Immune Stimulation**

An uncontrolled study assessed the effects of a SGE on the pulmonary function of patients with severe chronic respiratory diseases (not stated) who required home oxygen treatment.[25] Fifteen patients (mean age 67 years) were evaluated at the beginning of the trial for pulmonary function, respiratory muscle strength and endurance, arterial blood gases and walking distance in 6 minutes. Each subject was treated with 200 mg/day of a SGE for three months, and all functions were reevaluated. The 6-minute walking distance increased from an average of 600 meters to 1123 meters after 3 months of treatment. The study concluded that treatment with SGE improved pulmonary functions and oxygenation capacity in those patients with severe chronic pulmonary disease.[25]

The effects of an aqueous ginseng extract on immune function were compared with that of a standardized ginseng extract in a placebo-controlled, double-blind trial.[26] Sixty healthy patients were divided into three groups of 20 patients each and were treated with a placebo or 100 mg of aqueous ginseng extract or 100 mg of an SGE every 12 hours for eight weeks. Blood samples drawn from the patients showed an increase in the chemotaxis of polymorphonuclear leukocytes, the phagocytic index, the total number of $T_3$ and $T_4$ lymphocytes after 4 and 8 weeks of ginseng treatment, as compared with the placebo group. The group receiving the standardized ginseng extract also had an increase in the $T_4/T_8$ ratio and the activity of natural killer cells. The study concluded that ginseng extracts have immune stimulant activity, and that the standardized extract is more effective than the aqueous extract.[26]

A randomized, placebo-controlled, single-blind trial assessed the effects of a standardized ginseng extract on the activity of alveolar macrophages of patients suffering from chronic bronchitis.[27] Forty volunteers were treated with 100 mg of a ginseng extract or placebo twice daily for 8 weeks and the function of the bronchoalveolar macrophages was determined by measuring phagocytic activity and cytotoxicity against *Candida albicans*. Ginseng treatments enhanced the phagocytic index and increased the phagocytic fraction of macrophages in chronically compromised patients. The results were statistically significant at the end of week eight.[27]

A randomized, double-blind, placebo-controlled clinical trial assessed the immune stimulating effects of a standardized ginseng extract in combination with vaccination against influenza.[28] A total of 227 volunteers received 100 mg of ginseng extract or placebo daily for 12 weeks, in addition to an anti-influenza polyvalent vaccination at week 4. The results showed that the incidence of influenza or common cold between weeks 4 and 12 was 42 in the placebo group, but only 15 in the treated group ($p <$ 0.0001). Antibody titres by week 8 rose to an average of 171 units in the placebo group and 272 in the treated group ($p < 0.0001$), and natural killer cell activity levels were twice as high in the treated group as compared with placebo ($p < 0.0001$).[28] A randomized, placebo-controlled clinical trial assessed the effects of a standardized ginseng extract on peripheral blood leukocytes and lymphocyte subsets in 20 young healthy males.[29] The subjects received 150 mg of a SGE twice daily or placebo for 8 weeks. No significant differences in the total and differential leukocyte counts or subpopulations: T cells, B cells, T-helper cells, T-suppressor cells, T4/T8 ratio or interleukin-2-receptor cells were observed in either group.[29]

*Memory Impairment*
A double-blind, placebo-controlled trial in 38 healthy students assessed the effects of 8 to 14 treatments of ginseng or placebo on puzzle solving and performance on final exams.[30] Treatment did not significantly enhance mathematical or final exam performance, however it did improve proofreading error detection and mood-fatigue.[30]

The affective status and memory functions were assessed in a randomized, double-blind trial of 60 subjects with age-associated memory impairment.[31] Patients were treated with a SGE containing vitamins or placebo for 8 weeks and affective status and memory were assessed using the Symptom Rating Test, the Randt Memory Test and the Life Satisfaction in the Elderly Scale. A significant improvement in the Randt Memory Test was observed in the treated group as compared with placebo.

A double-blind, placebo-controlled clinical trial assessed the effects of a standardized ginseng extract on psychomotor performance in 16 healthy volunteers.[10] Patients were treated with ginseng (100 mg twice daily) or matching placebo for 12 weeks) and various tests of pyschomotor performance were performed. While a favorable effect was found in attention, processing, integrated sensory-motor function and auditory reaction time, performance on the mental arithmetic test was the only case where the ginseng treated group differed significantly from the placebo group.[10]

A double-blind, placebo-controlled clinical reported a statistically significant improvement in psychomotor (spiral trace and letter cancellation tests) performance after supplementation with two ginseng preparations.[32] A double-blind, placebo-controlled trial assessed the effects of ginseng supplementation on reaction time, mental alertness, two-hand coordination and recovery following physical exercise.[33] Sixty subjects (22 to 80 years old) were tested before and after 12 weeks of treatment. A statistically significant improvement was observed in reaction times as well as recovery following exercise.[33] A double-blind placebo-controlled trial assessed the effects of a SGE (200 mg/day) with that of Hydergine (3.0 mg/day) in 45 patients with rheographically detected cerebrovascular deficits.[34] Treatment for 90 days increased cerebral blood flow in both treatment groups as compared with placebo. However, Hydergine was superior to ginseng, 58.43% compared with 34.36%, respectively.[34] A randomized, double-blind, placebo-controlled trial assessed the effects of ginseng on a variety of cognitive functions in 112 healthy volunteers older than 40 years of age.[35] The subjects were treated with 400 mg of a standardized ginseng extract or placebo for 8 to 9 weeks. The results of this trial showed that the ginseng group did not differ significantly with placebo in regard to concentration, memory or subjective experience.[35]

*Antidiabetic Activity*
Clinical trials have suggested that ginseng supplementation may have beneficial effects for both insulin-dependent and non-insulin-dependent diabetic patients.[36,37] A randomized, double-blind, placebo-controlled clinical trial assessed the effects of a SGE (100 or 200 mg daily for 8 weeks) on psychophysical performance, glucose homeostais, serum lipids and amino-terminal-propeptide concentrations and body weight in 36 non-insulin dependent patients.[37] Ginseng supplementation (200 mg) resulted in elevated mood, improved physical performance, and reduced fasting blood glucose and body weight. Treatment also improved serum amino-terminal-propeptide concentrations, and glycosylated hemoglobin as compared with placebo.[37]

*Erectile Dysfunction and Infertility*

An open clinical trial assessed the effect of a SGE on 66 patients, 30 with oligoast-enospermic sine causa, 16 with oligoastenospermic with idiopathic varicocele and 20 age-matched controls.[38] All patients were treated with 4 grams of ginseng extract for 3 months. Treatment moderately increased spermatozoa number/ml and progressive oscillating motility. An increase in testosterone, dihydrotestosterone, follicle stimulating hormone and luteinizing hormone was observed, along with a decrease in prolactin levels.[38]

A placebo-controlled trial assessed the effects of ginseng on psychogenic impotence in 35 male volunteers.[39] The men treated with the ginseng extract had a higher frequency of coitus, morning erection, penile rigidity and tumescence, as compared with the placebo-treated group. The overall therapeutic effect on erectile function was 67% in the ginseng treated group as compared with 28% in the placebo group.[39] A placebo-controlled clinical study assessed the effects of a SGE (1800 mg/day), 25 mg of trazodone or placebo in 90 patients with erectile dysfunction.[40] Subjects treated with ginseng for three months showed improvements in penile rigidity, changes in early detumescence and libido, as compared with the other groups. The therapeutic efficacy on erectile dysfunction was 60% for the ginseng group, and 30% for the placebo and trazodone groups. No changes were noted in the frequency of intercourse, premature ejaculation and morning erections in any of the groups.[40]

**Pharmacokinetics**

The pharmacokinetics of ginsenosides $Rg_1$ and $Rb_1$ were investigated in rats after oral and intravenous administration.[41,42] Ginsenoside $Rg_1$ was rapidly absorbed after oral administration ($t_{max}$ 30 min), and approximately 60% of the intravenous dose was excreted in the bile within 4 hours, and 24% of the dose was excreted in the urine after 12 hours.[41] Ginsenoside $Rb_1$ is poorly absorbed after oral administration.[42] The excretion of $Rb_1$ is biphasic with a half-life of 11.6 min for the $\alpha$-phase, and 14.6 hours for the $\beta$-phase after intravenous administration. Excretion occurs primarily through the urine (44% of the dose within 12 hours).[42]

The ginsenosides, 20(S)-protopanaxatriol and 20(S)-protopanaxadiol were quantified in the urine of athletes consuming ginseng preparations (total ginsenosides varied from 2.1 to 13.3 mg) for at least 10 days prior to urine sample collection.[43] Concentrations of 2 to 35 ng/ml of 20(S)-protopanaxatriol were found in 60 of 65 of the urine samples tested.[43]

**Mechanism of Action**

Pharmacological studies have shown that *Panax ginseng*, or its active constituents, the ginsenosides, can reduce cellular damage *in vitro* and prolong survival times in response to physical or chemical stress *in vivo*. Treatment of cultured mammalian cells, isolated organs and various animal models (primarily mice and rats) with root extracts before or during exposure to physical, chemical or psychological stress results in an increased ability of the respective model systems to resist the damaging effects of various stressors.[4] Treatment of the animals with ginseng reduced injuries caused by radiation poisoning,[44,45] viral infection and tumor load,[46,47] alcohol or,[50,51] light or temperature stress, emotional stress, and electrical shock or restricted movement.[52,53]

In animal models, ginseng exerts its protective effects through a stimulation of the hypothalamus-pituitary-adrenal axis and the immune response.[4] In rats, intraperitoneal administration of a ginseng extract (saponin fractions or the ginsenosides $Rb_1$, $Rb_2$, Rc, Rd, Re), elevated serum levels of adrenocorticotrophic hormone (ACTH) and corticosterone.[54,55] Pretreatment of the animals with dexamethasone, which blocks hypothalamus and pituitary functions, prevented the ginseng saponin-induced release of ACTH and corticosterone. These data suggest that ginseng increases serum corticosterone levels indirectly via release of ACTH from the pituitary.[54,55]

The immune stimulating effects of ginseng is at least partly responsible for its restorative activity.[56,57] Ginseng root extracts have been shown to significantly ($p <$ 0.01) enhance natural killer cell (NKC) activity and antibody-dependent cell cytotoxicity of peripheral blood mononuclear cells obtained from healthy individuals, as well as patients with chronic fatigue syndrome and acquired immunodeficiency syndrome.[58] Addition of the ginsenosides to the culture medium, induced T lymphocyte transformation and proliferation of interleukin-2 in cultured T lymphocytes obtained from healthy and chronic renal failure patients.[59] Alcohol extracts of ginseng root stimulate phagocytosis *in vitro*; are mitogenic in cultured human lymphocytes; increased the production of interferon; and stimulated the activity of natural killer cells.[60,61] Intraperitoneal administration of an aqueous ginseng extract to mice increased interferon production, stimulated cell-mediated immunity against Semliki Forest virus, elevated antibody levels against sheep red blood cells and natural killer cells,[62] and stimulates the production of interferon.[63] The extract inhibited spontaneous and stimulated lymphocyte proliferation and enhanced interferon production *in vitro*.[63] The extracts also enhance graft-to-host reactivity, the immune-mediated expulsion of *Trichinella*, and inhibited macrophage migration.[62,63] A reduction in immune function, marked by inhibition of NKC activity and T-cell mitogenesis induced by exogenous ACTH, was reversed by ginsenosides.[64] The addition of ginsenoside $Rg_1$ to the culture medium stimulated the proliferative response and immune function of lymphocytes obtained from elderly patients.[64]

In rodent models, an improvement in physical and mental performance has been observed after oral or intraperitoneal administration of ginseng extracts.[65,66] However, two studies concluded that ginseng did not enhance physical performance in rodents.[67,68] Intragastric administration of a ginseng extract, or the ginsenosides $Rb_1$ and $Rg_2$, to mice, during passive avoidance response tests, reversed stress-induced memory loss,[69] and improved the retention of learned behavior in rats.[70] Ginsenoside $Rg_1$ and $Rb_1$ are the active nootropic constituents of ginseng, and have been shown to improve memory and learning in normal, as well as cognition-impaired animals. The nootropic effects of ginseng are due to an increase in the synthesis and release of acetylcholine, and a decrease of brain serotonin levels.[71] In cerebral and coronary blood vessels, SGE produces vasodilatation, which resulted in improved brain and coronary blood flow.[72]

One polypeptides and five glycans, isolated from ginseng root, named GP and panaxans A-E respectively, have been shown to reduce blood glucose levels in mice after intraperitoneal administration.[73,74] Two of the glycans, panaxans A and B, stimulate hepatic glucose utilization by increasing the activity of glucose-6-phosphate dehydrogenase, phosphorylase-A, and phosphofructokinase.[73] Panaxan A does not affect plasma insulin levels or insulin sensitivity, but panaxan B elevates the plasma insulin level by stimulating insulin secretion from pancreatic islets, and further enhances insulin sensitivity by increasing insulin binding to receptors.[73] The panaxans are not active after oral administration. Administration of GP (intravenous or subcutaneous) to mice and rats decreased blood glucose and liver glycogen levels.[74] Ginseng root also contains a number of other chemical constituents with reported hypoglycemic activity.[73] Adenosine, isolated from an aqueous extract of ginseng root enhanced lipogenesis and cyclic AMP-induced accumulation of adipocytes *in vivo*.[73] Various ginsenosides have been reported to inhibit adrenocorticotropin-induced lipolysis, suppress insulin-stimulated lipogenesis and stimulate the release of insulin from cultured islet cells.[73]

Intravenous administration of ginseng extracts (10–40 mg/kg) to anaesthetized dogs produced a transient vasodilation, which resulted in reduce blood pressure.[75,76] The vasodilator effects of ginseng were associated with the ginsenosides, which induce relaxation of vascular smooth musculature. Ginsenosides $Rg_1$, $Rb_1$, and Re induced endothelium-dependent relation in rat aorta and increased the synthesis of cyclic GMP.[77] The ginsenosides also inhibited the constricting effects of norepinephrine in isolated aorta strips, and inhibit the uptake of $^{45}Ca^{2+}$ in the membrane and sarcolemma of rabbit heart tissue. Inhibition of $Ca^{2+}$ uptake in the muscle membrane contributes to the mechanism of vasodilatation.[78] Both SGE and purified ginsenosides have been shown to induced pulmonary vasodilation and prevent free radical damage to perfused rabbit lungs.[79,80] This vasodilator effect of ginseng also appears to be mediated by a stimulation of nitric oxide release from the vascular endothelium as well.[72] Both ginsenosides $Rb_1$ and $Rb_2$ enhance the expression of a copper and zinc-dependent superoxide dismutase gene, which was mediated by the AP2 transcription factor.[81] These data suggest ginseng may prevent oxygen free radical damage by promoting the release of nitric oxide.[72]

Subcutaneous administration of a ginseng extract enhanced the mating behavior of male rats.[82] Ginseng extracts have also been shown to stimulate spermatogenesis in rat [83] and rabbit testes, and increased the motility and survival of rabbit sperm outside the body.[82] The mechanism by which ginseng exerts its aphrodisiac effects appears to involve nitric oxide release. *In vitro* studies have shown that ginseng extracts (1 mg/ml) exert a relaxation effect on rabbit corpus cavernosal smooth muscle strips, which is mediated by increasing the release of nitric oxide from the corporal sinusoids, increasing intracellular calcium sequestration, and a hyperpolarizing effect.[84] Furthermore, the ginsenosides are reported to act as a nitric oxide donor and induce the relaxation of rabbit corpus carvernosal smooth muscle, through the L-arginine/nitric oxide pathway.[85]

## Safety Information

### A. *Adverse Reactions*

In a 2-year open study of 133 patients chronically ingesting large doses of ginseng (up to 15 grams per day), 14 subjects presented with symptoms of hypertension, nervousness, irritability, diarrhea, skin eruptions, and insomnia, which were collectively called Ginseng Abuse Syndrome (GAS) by the author of the study.[86] Critical analysis of this report has shown that there were no controls or analysis to determine the botanical identity of the ginseng being ingested, the constituents of the preparation taken. In a latter study, when the ginseng dose was decreased to 1.7 g/day the symptoms of the "GAS" were rare.[87] Thus excessive and uncontrolled intake of ginseng products should be avoided. One case of ginseng-associated cerebral arteritis has been reported in a patient consuming high dose of a rice wine extract of ginseng root [approximately 6 grams in one dose].[88] Two cases of mydriasis and disturbance in accommodation, as well as dizziness have been reported after ingestion of large doses (3–9 g) of an unspecified type of ginseng preparation.[89]

Ginseng supplementation has also been reported to cause estrogenic-like adverse effects in both pre- and post-menopausal women. Seven cases of mastalgia,[90–92] and one case of vaginal bleeding in a post-menopausal woman,[93] have been reported after ingestion of unspecified ginseng products. An increased libido in premenopausal women has also been reported.[92] However, in specific *in vitro* studies assessing the possible hormonal effects of a standardized ginseng extract, there was no interaction of the ginseng extract with either cytosolic estrogen receptors isolated from mature rat uterus, or progesterone receptors from human myometrium.[94] Furthermore, no changes in the male or female hormone status were observed in clinical trials after administration of a standardized ginseng extract.[95,96]

Ginseng supplementation was associated with the development of Stevens-Johnson syndrome (SJS) in one patient.[97] However, the type of the ginseng was not identified, and the patient been taking both aspirin and unspecified antibiotics six days prior to the development of SJS.[97]

### B. *Contraindications*

While no contraindications are reported,[5,8] there are no data to substantiate the use or safety of *Panax ginseng* during pregnancy or nursing, and or in children under 12 years of age.

### C. *Drug Interactions*

Two cases of a manic-like reaction, which included headache and tremulousness, have been reported in female patients taking a ginseng supplement in combination with phenelzine, a monoamine oxidase inhibitor.[98,99] However, the clinical significance of this interaction has not been substantiated. There is one case report of a decrease in the International Normalized Ratio (INR) of warfarin in a patient, two weeks after ingestion of 300 mg per day of a standardized ginseng extract.[100] The INR returned to normal two weeks after discontinuation of the ginseng product.[100]

### D. Toxicology

The $LD_{50}$ values of ginseng root were 10–30 g/kg in mice[101] and 910 mg/kg after intraperitoneal administration to mice.[101] Chronic toxicity tests showed no adverse reactions after intragastric administration in a dosage range of 1.5, 5 or 15 mg/kg/day to male or female beagle dogs for 90 days.[102] No evidence of embryotoxicity was observed in rats or rabbits.[5]

### E. Dose and Dosage Forms

*Panax ginseng* is available as crude drug (dried root), capsules, tablets, standardized extracts, The daily dose (taken in the morning): dried root 0.5 to 2 grams by decoction;[5,8] standardized extracts (containing 1.5 to 7% ginsenosides), 100 to 300 mg per day.[10–18]

### References

1     Hu, SY. The genus *Panax* (*ginseng*) in Chinese medicine. *Econ Bot*. 1976;30:11–28.
2     Shibata, S, Tanaka, O, Shoji, J and Saito, H, Chemistry and Pharmacology of *Panax*. In: Wagner H, Farnsworth NR, (eds.). *Economic and Medicinal Plants Research*. Vol. 1 Academic Press, London, San Diego, New York, pp. 217–284 (1985).
3     Carlson AW. Ginseng: America's Botanical Drug Connection to the Orient. *Econ Bot*. 1986;233–249.
4     Sonnenborn, U, Proppert, Y. Ginseng (*Panax ginseng* C.A. Meyer). *Brit J Phytother*. 1991;2:3–14.
5     Mahady GB, Fong HHS, Farnsworth NR. *WHO Monographs on Selected Medicinal Plants*, Volume 1, World Health Organization, Geneva, Switzerland, WHO Publications 1999.
6     Cui JF. Identification and quantification of ginsenosides in various commercial ginseng preparations. *Eur J Pharmaceut Sci*. 1995;3:77–85.
7     Sprecher E. Ginseng: Miracle drug or phytopharmacon? *Deutsche Apotheker Zeitung*. 1987;9:52–61.
8     Blumenthal M et al. (eds.). *The Complete German Commission E Monographs*. Ginseng monograph. American Botanical Counsel, Austin 1998.
9     Hallstrom C, Fulder S, Carruthers M. Effect of ginseng on the performance of nurses on night duty. *Comparative Medicine East and West*. 1982;6:277–282.
10    D'Angelo L et al. Double-blind, placebo-controlled clinical study on the effect of a standardized ginseng extract on psychomotor performance in healthy volunteers. *J Ethnopharmacol*. 1986;16:15–22.
11    Pieralisi G, Ripari P, Vecchiet L. Effects of a standardized ginseng extract combined with dimethylaminoethanol bitartrate, vitamins, minerals, and trace elements on physical performance during exercise. *Clinical Therapeutics*. 1991;13:373–382.

12   Van Schepdael P. Les effects du ginseng G115 sur la capacité physique de sportifs d'endurance. *Acta Ther.* 1993;19:337–347.

13   Forgo I, Kirchdorfer AM. The effect of different ginsenoside concentrations on physical work capacity. *Notabene Med.* 1982;12:721–727.

14   Forgo I, Kirchdorfer AM. On the question of influencing the performance of top sportsmen by means of biologically active substances. *Ärztliche Praxis.* 1981;33:1784–1786.

15   Forgo I. Effect of drugs on physical performance and hormone system of sportsmen. *Münchener Medizinische Wochenschrift.* 1983;125:822–824.

16   Forgo I. Schimert, G. The duration of effect of the standardized ginseng extract in healthy competitive athletes. *Notabene Med.* 1985;15:636–640.

17   Forgo I, Kayasseh L, Staub JJ. Einfluss eines standardisierten Ginseng-Extraktes auf das Allgemeinbefinden, die Reaktionsfähigkeit, Lungenfunktion und die gonadalen Hormone. *Med Welt.* 1981;32:751–756.

18   Bahrke MS, Morgan WP Evaluation of the ergogenic properties of ginseng. *Sports Med.* 1994;18:229–248.

19   Engels HJ, Said JM, Wirth JC. Failure of chronic ginseng supplementation to affect work performance and energy metabolism in healthy adult females. *Nutr Res.* 1996;16:1295–1305.

20   Engels HJ, Wirth JC. No ergogenic effects of ginseng (*Panax ginseng* C.A. Meyer) during graded maximal aerobic exercise. *J Am Diet Assoc.* 1997;97:1110–1115.

21   Allen JD, McLung J, Nelson AG, Welsch M. Ginseng supplementation does not enhance healthy young adults' peak aerobic exercise performance. *J Am Coll Nutr.* 1998;17:462–466.

22   Cherdrungsi P, Rungroeng K. Effects of standardized ginseng extracts and exercise training on aerobic and anaerobic exercise capacities in humans. *Korean J Ginseng Sci.* 1995;19:93–100.

23   Morris AC, Jacobs I, Klugerman A. No ergogenic effect of ginseng ingestion. *Med Sci Sports Exercise.* 1994;26:S6.

24   Knapik JJ, Wright JE, Welch MJ. The influence of *Panax ginseng* on indices of substrate utilization during repeated, exhaustive exercise in man. *Fed Proceed.* 1983;42:336.

25   Gross D, Krieger D, Efrat R, Dayan M. Ginseng extrakt G115 for the treatment of chronic respiratory diseases. *Schweiz Zeit Ganzheits Med.* 1995;1:29–33.

26   Scaglione F et al. Immunomodulatory effects of two extracts of *Panax ginseng* C.A. Meyer (G115). *Drugs Exp Clin Res.* 1990;16:537–542.

27   Scaglione F et al. Immunomodulatory effects of *Panax ginseng* C.A. Meyer (G115) on alveolar macrophages from patients suffering with chronic respiratory diseases. *Int J Immunother.* 1994;10:21–24.

28   Scaglione F, Cattaneo G, Alessandria M, Cogo R. Efficacy and safety of the standardized ginseng extract G115 for potentiating vaccination against common cold and/or influenza syndrome. *Drugs Exp Clin Res.* 1996;22:65–72.

29   Srisurapanon S, Apibal S, Siripol R, Runroeng K, Cherdrugsi P, Vanich-Angkul V, Timvipark C. The effects of standardized ginseng extract on peripheral blood leukocytes and Lymphocyte subsets: a preliminary study in young healthy adults. *J Med Assn Thailand.* 1997;80:S81–S85.

30   Johnson A, Jiang NS, Staba EJ. Whole ginseng effects on human response to demands for performance. In: *Proceedings of the Third International Ginseng Symposium*, Seoul, Korea, 1980;pp. 237–244.

31   Neri M, Andermarcher E, Pradelli JM, Salvioloi G. Influence of a double blind pharmacological trial on two domains of well-being in subjects with age associated memory impairment. *Arch Gerontol Geriatrics.* 1995;21:241–252.

32    Sandberg F. Vitalitet och senilitet-effekten av ginsengglykosider pa prestationsfor-
      magan. *Svensk Farmaceut Tidskrift.* 1980;84:499–502.
33    Dorling E, Korchdorfer AM, Ruckert KH. Do ginsenosides influence the perform-
      ance? Results of a double-blind study. *Notabene Med.* 1980;10:241–246.
34    Quiroga HA. Comparative double-blind study of the effect of Ginsana G115 and
      Hydergine on cerebrovascular deficits. *Orientación Médica.* 1982;31:201–202.
35    Sorensen H, Sonne J. A double-masked study of the effects of ginseng on cognitive
      functions. *Curr Ther Res.* 1996;57:959–968.
36    Kwan HJ, Wan JK. Clinical study of treatment of diabetes with powder of the
      steamed insam (Ginseng) produced in Kaesong, Korea. *Technical Information.*
      1994;6:33–35.
37    Sotaniemi EA, Haapakoski E, Rautio A. Ginseng therapy in non-insulin-dependent
      diabetic patients. *Diabetes Care.* 1995;18:1373–1375.
38    Salvati G, Genovesi G, Marcellini L, Paolini P, De Nuccio I, Pepe M, Re M. Effects
      of *Panax ginseng* C.A. Meyer saponins on male fertility. *Panminerva Med.*
      1996;38:249–254.
39    Kim YC, Hong KH, Shin JS, Kang MS, Sung DH, Choi HK. Effect of Korean red gin-
      seng on sexual dysfunction and serum lipid levels in old aged men. *Korean J Ginseng
      Sci.* 1996;20:125–132.
40    Choi HK, Seong DH. Effectiveness for erectile dysfunction after the administration
      of Korean red ginseng. *Korean J Ginseng Sci.* 1996;19:17–21.
41    Odani T et al. The absorption, distribution and excretion of ginsenoside Rg1 in the
      rat. *Chem Pharm Bull.* 1983;31:292–298.
42    Odani T et al. The absorption, distribution and excretion of ginsenoside Rb1 in the
      rat. *Chem Pharm Bull.* 1983;31:1059–1066.
43    Cui JF, Garle M, Bjorkhem I, Eneroth P. Determination of aglycones of ginsenosides
      in ginseng preparations sold in Sweden and in urine samples from Swedish athletes
      consuming ginseng. *Scand J Clin Lab Invest.* 1996;56:151–160.
44    Takeda A, Yonezawa M, Katoh N. Restoration of radiation injury by ginseng. I.
      Responses of X-irradiated mice to ginseng extracts. *Journal of Radiation Research.*
      1981;22:323–335.
45    Yonezawa M, Katoh N, Takeda A. Restoration of radiation injury by ginseng. IV.
      Stimulation of recoveries in CFUs and megakaryocyte counts related to the preven-
      tion of occult blood appearance in X-irradiated mice. *Journal of Radiation Research.*
      1985;26:436–442.
46    Wang BX, Cui JC, Liu AJ. The effects of ginseng on immune function. In: *Advances
      in Chinese Medicinal Materials Research*, Chang HM et al. (eds.). Sinapore, *World
      Scientific Publications.* 1985:519–527.
47    Yun TK, Yun YS, Han IW. An experimental study on tumor inhibitory effect of red
      ginseng in mice and rats exposed to various chemical carcinogens. In: *Proceedings
      of the 3rd International Ginseng Symposium.* Seoul, Korea, 1980:87–113.
48    Choi CW, Lee SI, Huk K. Effect of ginseng on hepatic alcohol metabolizing enzyme
      system activity in chronic alcohol-treated mouse. *Korean Journal of Pharmacognosy.*
      1984;20:13–21.
49    Hikino H et al. Antihepatotoxic actions of ginsenosides from *Panax ginseng* roots.
      *Planta Med.* 1985;51:62–64.
50    Chen X et al. Protective effects of ginsenosides on anoxia/reoxygenation of cultured
      rat monocytes and on reperfusion injuries against lipid peroxidation. *Biomed Biochim
      Acta.* 1987;46:646–649.
51    Lu G, Cheng XJ, Yuan WX. Protective action of ginseng root saponins on hypobaric
      hypoxia animals. *Yao Hsueh Hsueh Pao.* 1988;23:391–394.

52   Cheng XJ et al. Protective effects of ginsenosides on anoxia/reoxygenation of cultured rat myocytes and on reperfusion injuries against lipid peroxidation. *Biomed Biochim Acta.* 1987;46:646–649.

53   Saito H. Neuropharmacological studies on *Panax ginseng.* In: Chang HM et al., eds. *Advances in Chinese Medicinal Materials Research*, Singapore,World Scientific Publ. Co., 1985:509–518.

54   Hiai S et al. Stimulation of pituitary-adrenocortical system in ginseng saponin. *Endocrinology Japan.* 1979;26:661.

55   Hiai S, Sasaki S, Oura H. Effects of ginseng saponin on rat adrenal cyclic AMP. *Planta Med.* 1979;37:15–19.

56   Singh VK, Agarwal SS, Gupta BM. Immunomodulatory activity of *Panax ginseng* extract. *Planta Med.* 1984;50:462–465.

57   Sonnenborn U. Ginseng neuere Untersuchungen immunologischer, und endokrinologischer Aktivitäten einer alten Arzneipflanze. *Deutsche Apotheker Zeitung.* 1987;125:2052–2055.

58   See DM, Broumand N, Sahl L, Tilles JG. *In vitro* effect of echinacea and ginseng on natural killer and antibody-dependent cell cytotoxicity in healthy subjects and chronic fatigue syndrome or acquired immunodeficiency syndrome patients. *Immunopharmacology.* 1997;35:229–235.

59   Ma L, Hou GX, Zhou ZL, Yang C, Yan Y. *In vitro* study of ginsenoside effect on cellular immune function in patients with chronic renal failure. *Zhongguo Zhongyao Zazhi.* 1995;20:307–309.

60   Fulder S. The growth of cultured human fibroblasts treated with hydrocortisone and extracts of the medicinal plant *Panax ginseng. Exp Gerontol.* 1977;12:125–131.

61   Gupta S et al. A new mitogen and interferon inducer. *Clin Res.* 1980;28:504A.

62   Singh VK, Agarwal SS, Gupta BM. Immunomodulatory effects of *Panax ginseng* extract. *Planta Med.* 1984:50:459.

63   Jie YH, Cammisuli S, Baggiolini M. Immunomodulatory effects of *Panax ginseng* C. A. Meyer in the mouse. *Agents and Actions.* 1984;15:386–391.

64   Liu J et al. Stimulatory effect of saponin from *Panax ginseng* on immune function of lymphocytes in the elderly. *Mech Ageing Dev.* 1995;83:43–53.

65   Avakian EV et al. Effect of *Panax ginseng* on energy metabolism during exercise in rats. *Planta Med.* 1984;50:151–154.

66   Brekhman II, Dardymov IV. Pharmacological investigation of glycosides from ginseng and *Eleutherococcus. J Nat Prod.* 1969;32:46–51.

67   Lewis WH, Zenger VE, Lynch RG. No adaptogen response of mice to ginseng and Eleutherococcus infusions. *J Ethnopharmacol.* 1983;8:209–214.

68   Martinez B, Staba EJ. The physiological effects of *Aralia, Panax* and *Eleutherococcus* on exercised rats. *Jpn J Pharmacol.* 1984;35:79–85.

69   Wagner H, Norr H, Winterhoff H. Plant Adaptogens. *Phytomedicine.* 1994;1:63–76.

70   Petkov VD et al. Memory effects of standardized extracts of *Panax ginseng* (G115), *Ginkgo biloba* (GK501) and their combination Gincosan (PHL00701). *Planta Med.* 1993;59:106–114.

71   Liu CX, Xiao PG. Recent advances on ginseng research in China. *J Ethnopharmacol.* 1992;36:27–38.

72   Gillis N. *Panax ginseng* pharmacology: a nitric oxide link? *Biochem Pharmacol.* 1997;54:1–8.

73   Marles R, Farnsworth NR. Antidiabetic plants and their active constituents. *Phytomedicine.* 1995;2:137–189.

74   Wang BX et al. Studies on the mechanism of ginseng polypeptide induced hypoglycemia. *Yao Hsueh Hsueh Pao.* 1989;25:727–731.

75    Lee DC, Lee MO, Kim CY. Effect of ether, ethanol and aqueous extracts of ginseng on cardiovascular functions in dogs. *Can J Comparat Med.* 1981;45:182–187.

76    Wood WB, Roh BL, White RP. Cardiovascular actions of *Panax ginseng* in dogs. *Jpn J Pharmacol.* 1964;14:284–294.

77    Kang SY, Schini-Kerth VB, Kim ND. Ginsenosides of the protopanxatriol group cause endothelium-dependent relaxation in the rat aorta. *Life Sci.* 1995;56:1577–1586.

78    Huang KC. Herbs with multiple actions. In: *The Pharmacology of Chinese Herbs*, Boca Raton, CRC Press, Inc., 1993:21–48.

79    Kim H, Chen X, Gillis CN. Ginsenosides protect pulmonary vascular endothelium against free radical-induced injury. *Biochem Biophys Res Commun.* 1992;189:670–676.

80    Rimar S, Lee-Mengel M, Gillis CN. Pulmonary protective and vasodilator effects of a standard *Panax ginseng* preparation following artificial gastric digestion. *Pulm Pharmacol.* 1996;9:205–209.

81    Kim YH, Park KH, Rho HM. Transcriptional activation of the Cu, Zn-superoxide dismutase gene through the AP2 site by ginsenoside Rb2 extracted from a medicinal plant, *Panax ginseng. J Biol Chem.* 1996;271:24539–24543.

82    Kim C. Influence of ginseng on mating behavior in male rats. *Amer J Chinese Med.* 1976;4:163–168.

83    Yamamoto M. Stimulatory effect of *Panax ginseng* principals on DNA and protein synthesis in rat testes. *Arzneimittelforschung.* 1977;27:1404–1405.

84    Choi YD, Xin ZC, Choi HK. Effect of Korean red ginseng on the rabbit corpus cavernosal smooth muscle. *Int J Impotence Res.* 1998;10:37–43.

85    Kim HJ, Woo DS, Lee G, Kim JJ. The relaxation effects of ginseng saponin in rabbit corporal smooth muscle: is it a nitric oxide donor? *Brit J Urol.* 1998;82:744–748.

86    Siegel, RK. Ginseng Abuse Syndrome – problems with the panacea. *JAMA.* 1979;241:1614–1615.

87    Siegel RK. Ginseng use amoung two populations in the United States. In: Korean Ginseng and Tobacco Research Institute, *Proceedings of the Third International Ginseng Symposium*, 1980:229–236.

88    Ryu SJ, Chien YY. Ginseng-associated cerebral arteritis. *Neurology.* 1995;45:829–830.

89    Lou BY et al. Eye symptoms due to ginseng poisoning. *Yen Ko Hsueh Pao.* 1989;5:96–97.

90    Palmer BV, Montgomery AC, Monteiro JC. Gin Seng and mastalgia. *Brit Med J.* 1978;279:1284.

91    Koriech OM. Ginseng and mastalgia. *Brit Med J.* 1978;297:1556.

92    Punnonen R, Lukola A. Oestrogen-like effect of ginseng. *Brit Med J.* 1980;281:1110.

93    Hopkins MP, Androff L, Benninghoff AS. Ginseng face cream and unexplained vaginal bleeding. *Amer J Obstet Gynecol.* 1988;159:1121–1122.

94    Buchi K, Jenny E. On the interference of the standardized ginseng extract G115 and pure ginsenosides with agonists of the progesterone receptor of the human myometrium. *Phytopharm*, internal report 1984 (available through the Napralert database).

95    Forgo I, Kayasseh L, Staub JJ. Effect of a standardized ginseng extract on general well-being, reaction capacity, pulmonary function and gonadal hormones. *Medizinische Welt.* 1981;19:751–756.

96    Reinhold E. Der Einsatz von Ginseng in der Gynäkologie. *Natur- und Ganzheits Medizin.* 1990;4:131–134.

97    Dega H. et al. Ginseng as a cause for Stevens-Johnson syndrome? *Lancet.* 1996;347:1344.

98    Jones BD, Runikis AM. Interaction of ginseng with phenelzine. *J Clin Psychopharmacol.* 1987;7:201–202.

99    Shader RI, Greenblatt DJ. Phenelzine and the dream machine-ramblings and reflections. *J Clin Psychopharmacol.* 1985;5:67.

100  Janetzky K, Morreale AP. Probable interaction between warfarin and ginseng. *Amer J Health-Sys Pharm.* 1997;54;692–693.

101  Brekhman II, Dardymov IV. New substances of plant origin which increase nonspecific resistance. *Annu Rev Pharmacol.* 1969;9:419–430.

102  Hess FG, Parent RA, Stevens KR, Cox GE, Becci PJ. Effect of subchronic feeding of ginseng extract G115 in beagle dogs. *Food Chem Toxicol.* 1983;21:95–97.

# 22

# Siberian Ginseng

## Synopsis

Although Siberian ginseng has been used traditionally as a tonic to enhance mental and physical performance, there is an absence of compelling clinical evidence to support this claim. The results from open clinical trials published prior to 1985 are difficult to interpret due to numerous methodological flaws, and statistical shortcomings. Furthermore, the results from recent randomized, controlled clinical trials are conflicting, and further clinical trials are needed before any definitive statement on therapeutic efficacy can be made. The average daily dose of Siberian ginseng is 2 to 3 grams of the dried root. Currently, no standardized products, with proven clinical efficacy, are available on the U.S. market. Due to safety concerns, women who are pregnant or nursing should not use Siberian ginseng products. Siberian ginseng has been shown to increase blood pressure in clinical trials, therefore it should not be administered to patients with a blood pressure in excess of 180/90 mm Hg. A possible interaction between Siberian ginseng and digitalis has been reported, however the clinical significance of this interaction has not been determined.

## Introduction

*Eleutherococcus senticosus*, also known as Eleuthero or Siberian ginseng, is a slender, thorny hardy shrub, indigenous to the southeastern part of the former USSR, northern China, Korea, and Japan.[1] The root of *Eleutherococcus senticosus* is used in Russian medicine as an adaptogenic or tonic herb in a similar manner to that of *Panax ginseng* in Chinese medicine. Interestingly, *Eleutherococcus* has also been used in Traditional Chinese Medicine for over 2,000 years as a stimulant, tonic and diuretic.[2,3] The plant is official in the Russian Pharmacopoeia and production of an *Eleutherococcus* extract is supervised by the Ministry of Health in Russia. Siberian ginseng belongs to the same plant family as *Panax ginseng*, the Araliaceae and *Eleutherococcus* preparations were popularized in the 1950s in former USSR as an inexpensive substitute for *Panax ginseng*. In Russia, *Eleutherococcus* is used extensively by the Olympic athletes, soldiers, and cosmonauts.[2,3] Unlike *Panax ginseng*,

Siberian ginseng does not contain ginsenosides, but a series of structurally unrelated compounds known as the eleutherosides A-G.[4] Because of the diverse nature of the chemical constituents, and the variability of their occurrence in the plant, analysis and standardization of Siberian ginseng is a challenge.

## Quality Information

- The correct Latin name for Siberian ginseng is *Eleutherococcus senticosus* (Rupr. & Maxim.) Maxim. (Araliaceae).[5] Botanical synonyms include *Acanthopanax senticosus* (Rupr. et Maxim.) Harms., *Hedera senticosa*.[4,6] Common names for the plant include: devil's bush, devil's shrub, eleuthero, eleutherococc, eleutherococoque, many prickle Acanthopanax, prickly Eleutherococcus, Siberian ginseng, stachelpanax, taiga root, taigawurzel, thorny ginseng, thorny Russian pepperbush, touch-me-not, wild pepper, wu cha sang and wu cha seng.[4]
- Commerical products of Siberian ginseng are prepared from the dried roots and rhizomes of *Eleutherococcus senticosus*.[4] The plant is indigenous to Southeast Asia, the southeastern part of Russia, northern China, Japan and Korea.[1,6]
- Unlike *Panax ginseng*, Siberian ginseng does not contain ginsenosides, but a group of unrelated phenylpropane derivatives of diverse structure and various sugar polymers, known as the eleutherosides.[6,7] The principal components of the phenylpropane group are the lignans, sesamin (eleutheroside $B_4$), syringaresinol, and its monoglucoside (eleutheroside $E_1$) and diglucoside (eleutheroside E); the simple phenylpropanes sinapic alcohol and its monoglucoside syringin (eleutheroside B), and the coumarins, isofraxidin and its monoglucoside (eleutheroside $B_1$). A polysaccharide complex and a series of glycans (eleutherans A-G) have also been isolated from the roots.[4] Eleutheroside E has been found in all root samples regardless of geographical origin, whereas eleutheroside B is present in all samples, except those plants grown in Korea.[7]
- The recommended daily dose of Siberian ginseng is 2–3 grams of the root, or equivalent preparations in divided doses.[8]

## Medical Uses

Siberian ginseng is used as a prophylactic and restorative tonic for the enhancement of mental and physical capacities in cases of weakness, exhaustion, and during convalescence.[6,8–10] However, the results from recent clinical trials are conflicting,[11,12] and the quality of the published clinical work is generally poor.

Although the Chinese Pharmacopoeia states that Siberian ginseng may be used for the treatment of rheumatoid arthritis, insomnia and dream-disturbed sleep,[13] there are no clinical data to support these claims.

## Summary of Clinical Evidence

Much of the clinical work on Siberian ginseng was performed in the 1960s and 1970s in Russia, and suffers from lack of proper controls, poor methodology and lack of well-defined outcome measures.[6] The clinical trials were designed to measure the adaptogenic or "stress reducing" effects of a 33% ethanol extract of the roots by measuring the ability of healthy subjects to withstand adverse conditions such as heat; to improve athletic performance; or to increase mental performance and improve work output.[4] The results of these trials were published in Russian, but an English review of these publications is available.[6] In one group of trials, over 2100 healthy subjects were treated with the ethanol extract at a dose of 2 to 16 ml, 1 to 3 times daily for up to 60 days. The overall results showed an improvement in physical and mental performance, without significant adverse effects. Two of the trials, however, noted an increase in blood pressure in some patients.[6]

In another group of 35 open clinical trials, the effects of a 33% ethanol extract of the roots was assessed in 2200 patients (range 5–1200 subjects per study) with various ailments, such as arteriosclerosis, acute pyelonephritis, diabetes, hypertension, hypotension, chronic bronchitis and rheumatic heart disease.[6] An average of 0.5 to 6 ml of the extract was administered orally, one to three times daily, for up to 8 courses of 35 days each (each course was separated by 2–3 weeks of no therapy). The overall results were generally positive, with improvements such as normalization of blood pressure, reduction in serum prothrombin and cholesterol, as well as an overall improvement in well being and physical performance.[6] However, the results from these open clinical trials are difficult to interpret due to poor methodology.

During the period of 1985 to 1999, the results of four small clinical trials have been published.[9–12] These trials assessed the effects of Siberian ginseng on physical performance and various immune parameters. Unfortunately, although the methodology in these trials was better than the previous studies, all of the trials still suffer from one or more of the following methodological flaws: small patient numbers, short follow-up period, lack of double blinding, statistical analysis and/or well-defined outcome measures.

A single-blind placebo-controlled clinical trial in six male baseball players assessed the effects of a 33% ethanol root extract on maximal working capacity.[9] The subjects were administered 2 ml of the extract (orally) twice a day before meals, for eight days. The outcomes measured were maximal oxygen uptake, oxygen pulse, total work and time to exhaustion before and after the treatments. The results of this trial showed a significant improvement in all four parameters, including a 23.3% increase in total work capacity as compared with only 7.5% with placebo.[9]

A randomized, placebo-controlled, double-blind study assessed the effect of an ethanol extract of the root (standardized to contain 0.2% syringin) on cellular immunity in 36 healthy volunteers.[10] This trial utilized quantitative multi-parameter flow cytometry, with monoclonal antibodies directed against specific surface markers of human lymphocyte subsets. The volunteers were treated with 10 ml of the ethanol extract or a placebo preparation (orally, three times daily) for 4 weeks, and the cellular immune status was assessed. Subjects treated with the ethanol extract had a significant increase in the number of immune competent cells, including an increase in the total number of lymphocytes (predominantly total T cells and T helper/inducer cells), as well as natural killer cells. In addition, an enhancement of the activity of T lymphocytes was also observed in treated patients as compared with placebo.[10]

A randomized, placebo-controlled, crossover study of 30 healthy volunteers compared the effects of Siberian ginseng with *Panax ginseng* or placebo on physical performance.[12] The primary outcome measured was cycle ergometer $VO_{2max}$. After 6 weeks of treatment, *Panax ginseng* significantly increased the $VO_{2max}$ as compared with placebo. However, treatment with Siberian ginseng did not enhance physical performance as compared with placebo.[12]

A randomized, double-blind placebo-controlled clinical trial assessed the effects of Siberian ginseng on submaximal and maximal exercise performance.[11] Twenty marathon runners were treated with either 3.4 ml of a 30–34% ethanol extract of the root or placebo for 8 weeks, during which they completed 5 trials of a 10-minute treadmill run, and a maximal treadmill test. The heart rate, consumption of oxygen, expired minute volume, ventilatory equivalent for oxygen, respiratory exchange ratio, and rating of perceived exertion were all measured during the trial. Blood samples were also analyzed for serum lactate. No significant differences were observed for any of the measured parameters in either the placebo or treatment groups.[11]

A comparative trial assessed the efficacy of tinctures of Siberian ginseng or *Leuzea carthamoides*, containing eleutherosides and ecdysone, respectively, to reduce the increase in blood coagulation observed in athletes after intensive training.[13] Highly trained athletes were treated with a 20-day course of the one of the two tinctures and blood coagulation was measured after training. The results of this study demonstrated that treatment of the athletes with a tincture of Siberian ginseng decreased both blood coagulation, and the activity of blood coagulation factors induced by intensive training.[13]

## Mechanism of Action

Numerous *in vivo* studies have assessed the pharmacological activity of a 33% ethanol extract of the roots in a variety of animal models, primarily rodents.[4,6] Much of this work was designed to demonstrate prolonged survival or an increase in physical resistance in response to a variety of adverse conditions (stress, immobilization or chemical challenge).[2,6] An increase in physical resistance to the toxic effects of noxious chemicals such as alloxan, cyclophosphan, ethymidine and benzo-tepa was observed after intragastric administration of a 33% ethanol extract of the roots to rats (1–5 ml/kg).[4,6] Intragastric administration of a 33% ethanol extract of the roots to mice (10 ml/kg) decreased the toxicity of diethylglycolic acid in mice, but did not reduce the severity of electroshock-induced convulsions.[6] Administration of a 10% decoction of the root to frogs (0.1 ml, ventral lymph sac) protected the frogs against injection of lethal doses of cardiac glycosides.[6] Intra-gastric administration of a 33% ethanol root extract to rats and mice (0.1 or 1.0 ml/kg) for 12 to 14 days increased body temperature after experimentally-induced hypothermia when administered.[6] Intragastric administration of a freeze-dried extract of the roots 80 mg/kg or 320 mg/kg for 3 days decreased blood glucose levels of mice by 35% and 60%, respectively, as compared with placebo.[15] The reduction of blood glucose levels appears to be due to enhancement of the synthesis of glycogen and high-energy phosphate compounds.

Investigations to elucidate the adaptogenic effect on the lymphatic system assessed the ability of the root extracts to inhibit cortisone-induced weight decreases of the thymus and spleen in rats.[4,6] Rats treated for eight days with a 33% ethanol extract of the roots (1.0 ml/kg, intraperitoneal) showed a decrease in spleen and thymus weights due to cortisone administration.[6] Intragastric administration of a 33% ethanol extract of the roots showed normalizing effects on experimentally-induced hypothermia, and had sedative activity in rats and mice.[4,6]

Administration of an aqueous extract of the root (500 mg/kg by gavage) to rats suppressed light-induced decrease in locomotor activity, indicating a reduction in the anxiety levels of the animals.[16] Intragastric administration of an aqueous root extract (500 mg/kg) reduced a stress-induced enlargement of the adrenal gland, normalized a decrease in rectal temperature due to chronic stress, and enhanced sexual behavior in mice;[17] prolonged the swimming times of rats;[18] and suppressed stress-induced (cold and water immersion) gastric ulcer formation in rats.[19]

The *in vivo* adaptogenic effects of Siberian ginseng are produced through the enhancement of energy, nucleic acid, and protein metabolism of the tissues. During stressful conditions, a β-lipoprotein glucocorticoid complex is generated in the blood. This complex inhibits the permeation of cell membranes by sugars, and also inhibits hexokinase activity *in vivo* and *in vitro*.[6] *Eleutherococcus senticosus* extracts increase the formation of glucose-6-phosphate, which in turn decreases the competition between the different pathways of its utilization. In animal tissues, depleted of adenosine triphosphate, glucose-6-phosphate is oxidized via the pentose phosphate shunt, yielding substrates for the biosynthesis of nucleic acids and proteins.[6] The constituents syringin (eleutheroside B), and (−) syringaresinol-4,4"-O-β-D-diglucoside (eleutheroside E) are thought to be responsible for the antistress activities.[4] Intragastric administration of a butanol extract of the roots to mice (170 mg/kg/day for 6 days per week for 6 weeks) enhanced the activities of oxidative enzymes in skeletal muscle and superoxide dismutase resulting in improve aerobic metabolic rates.[20] Administration of an aqueous extract of the roots to mice (0.17 mg/g/day, gavage) for 9 weeks increased the activity of succinate dehydrogenase and malate dehydrogenase in skeletal muscle.[21]

A polysaccharide fraction, isolated from the roots increased the activity of lymphokine-activated killer (LAK) cells (0.01 mg/ml) and enhanced interleukin 2-stimulated LAK cell activities *in vitro*.[22] A 95% ethanol extract of the roots increased phagocytosis of *Candida albicans* by human granulocytes and monocytes by

30–45% *in vitro* at a concentration 0.008 μg/ml.[23] Intraperitoneal administration of a polysaccharide fraction, isolated from a root extract, had immunostimulant activity in mice in a dose of 10 mg/kg, as demonstrated by the clearance of colloidal carbon.[24] A pyrogen-free polysaccharide fraction (designated PES) isolated from the roots stimulated phagocytosis and T-cell dependent B-cell functions in lymphocytes *in vitro*, as determined by plaque forming cell (PFC) stimulation assays, and the production of anti-bovine serum albumin antibodies.[25] Intraperitoneal administration of the polysaccharide fraction-PES to mice (100 mg/kg) daily for 7 days significantly increased PFC counts, anti-BSA antibodies, and the phagocytic activity of lymphocytes.[25]

## Safety Information

### A. Adverse Reactions
In a clinical study involving 64 patients with atherosclerosis, oral administration of 4.5 to 6 ml of a 33% ethanol extract of the root for 6–8 treatments of 25–35 days each, resulted in a few cases of insomnia, shifts in heart rhythm, tachycardia, extrasystole and hypertonia.[6] In another study of 55 patients with rheumatic heart lesions, two patients experienced hypertension, pericardial pain and palpitations and pressure headaches after ingesting 3 ml daily of a 33% ethanol extract of the root for 28 days.[6] A single case report of neonatal androgenization was tentatively associated with the ingestion of Siberian ginseng tablets during pregnancy.[26] However, analysis of the raw materials used in the preparation of the tablets indicated that the plant was most likely was *Periploca sepium*, the Chinese silk vine, a common adulterant of Siberian ginseng.[27] Administration of this plant to rats (1.5 g/kg by gavage) did not produce any androgenic effects, suggesting that the neonatal androgenization was not due to the ingested plant material.[28]

### B. Contraindications
Siberian ginseng preparations should not be used during pregnancy or nursing due to a lack of safety dataingested plant material.[4] Clinical trials have demonstrated that Siberian ginseng may increase blood pressure, therefore it should not be administered to patients with a blood pressure in excess of 180/90 mm Hg.[4]

### C. Drug Interactions
One case report of an increased level of serum digoxin levels due to the concomitant use of Siberian ginseng has been reported.[29] However, the identity of the product as *Eleutherococcus senticosus* was never established, and it is believed that the plant material ingested may have been *Periploca sepium*, as known as the Chinese silk vine, a plant that is known to contain cardiac glycosides.[30]

In one clinical trial, Siberian ginseng was shown to inhibit blood coagulation in athletes after intensive training,[14] therefore it should be used with caution in patients taking anticoagulant or antiplatelet drugs.

## D. Toxicology

A 33% ethanol extract of the root had an oral $LD_{50}$ of 14.5 g to 25 g/kg in various animal models.[4,6] Toxicity studies in rats showed no toxic effects after chronic administration for two months. No teratogenic effects were reported in the offspring from male and female mice given 10 mg/kg of eleutherosides from *Eleutherococcus senticosus* for 16 days.[4]

## E. Dose and Dosage Forms

The recommended daily dose of Siberian ginseng is 2–3 grams of the root, or equivalent preparations in divided doses.[4,8,31] Available dosage forms include powdered crude drug or extracts in capsules, tablets, teas, syrups, and fluid extracts. The 33% ethanol extract of Siberian ginseng that was used in most of the clinical trials is not currently available on the U.S. market.

## References

1    Sonnenborn U, Hänsel R. *Eleutherococcus senticosus*. In: De Smet PAGM, Keller K, Hänsel R, Chandler RF (eds) Adverse effects of herbal drugs, vol 2, Berlin Heidelberg, Springer-Verlag, 1993:pp 159–169.

2    Halstead, BW, Hood, LL. *Eleutherococcus senticosus*, Siberian Ginseng: An Introduction to the Concept of Adaptogenic Medicine. Oriental Healing Arts Institute, Long Beach, CA, 1984.

3    Awang D. Eleuthero. *Canadian Pharm J*. 1996a;129:52–54.

4    Anon. Radix Urticae, *WHO Monographs on Selected Medicinal Plants*, Volume II, World Health Organization, Geneva, Switzerland, 2000, in press.

5    Soejarto DD, Farnsworth NR. *Botanical Museum Leaflets Harvard University*, 1978;26:339.

6    Farnsworth NR, Kinghorn AD, Soejarto DD, Waller DP. Siberian ginseng (*Eleutherococcus senticosus*): Current status as an adaptogen. In: Wagner H, Hikino H, Farnsworth NR (eds) *Economic and Medicinal Plant Research*, vol. 1. London, Academic Press, 1985, pp 217–284.

7    Bladt S, Wagner H, Woo WS. Taiga-wurzel. DC- und HPLC analyse von *Eleutherococcus* – bzw. *Acanthopanax*-extrakten und diese enthaltenden phytopräparaten. *Deutsche Apotheker Zeitung*. 1990;130:1499–1508.

8    Blumenthal M et al. (eds.). The Complete German Commission E Monographs, Boston, American Botanical Council, *Eleutherococcus senticosus* (Eleutherococci radix), 1998.

9    Asano K et al. Effect of *Eleutherococcus senticosus* extracts on human physical working capacity. *Planta Med*. 1986;(4):175–177.

10   Bohn B et al. Flow-cytometric studies with *Eleutherococcus senticosus* extract as an immunomodulatory agent. *Arzneimittelforschung*. 1987;37:1193–1196.

11   Dowling EA et al. Effect of *Eleutherococcus senticosus* on submaximal and maximal exercise performance. *Medicine and Science in Sports and Exercise*. 1995;28:482–489.

12   McNaughton L et al. A comparison of Chinese and Russian ginseng as ergogenic aids to improve various facets of physical fitness. *Int Clin Nutr Rev*. 1989;9:32–35.

13   *Pharmacopoeia of the People's Republic of China* (English ed.), Beijing, Chemical Industry Press, 1997.

14    Azizov AP. Effects of *Eleutherococcus*, elton, Leuzea and leveton on the blood coagulation system during training in athletes. *Eksp Klin Farmakol.* 1997;60:58–60.

15    Medon PJ et al. Hypoglycemic effect and toxicity of *Eleutherococcus senticosus* following acute and chronic administration in mice. *Acta Pharmacol Sinica.* 1981;2:281.

16    Winterhoff H et al. Effects of *Eleutherococcus senticosus* on the pituitary-adrenal system of rats. *Pharmaceutical and Pharmacological Letters.* 1993;3:95–98.

17    Nishiyama N et al. Effect of *Eleutherococcus senticosus* and its components on sex- and learning behaviour and tyrosine hydroxylase activities of adrenal gland and hypothalmic regions in chronic stressed mice. *Shoyakugaku Zasshi.* 1985;39:238–242.

18    Nishibe S et al. Phenolic compounds from the stem bark of *Acanthopanax senticosus* and their pharmacological effect in chronic swimming stressed rats. *Chem Pharm Bull.* 1990;38:1763–1765.

19    Fujikawa T et al. Protective effects of *Acanthopanax senticosus* Harms. From Hokkaido and its components on gastric ulcer in restrained cold water stressed rats. *Biol Pharm Bull.* 1996;19:1227–1230.

20    Sugiura H et al. Effects of *Eleutherococcus* extracts on oxidative enzyme activity in skeletal muscle, superoxide dismutase activity and lipid peroxidation in mice. *Japanese Journal of Fitness and Sports Medicine.* 1992;41:304–312.

21    Sugiura H et al. Effects of aqueous extracts from *Eleutherococcus* on the oxidative enzyme activities in mouse skeletal muscle. *Annual Proceedings of the Gifu Pharmaceutical University.* 1989;38:38–48.

22    Cao GW et al. Influence of four kinds of polysaccharides on the induction of lymphokine-activated killer cells *in vivo. J Med Coll Pla.* 1993;8:5–11.

23    Wildfeuer A et al. Study of the influence of phytopreparations on the cellular function of body defence. *Arzneimittelforschung.* 1994;44:361–366.

24    Wagner H et al. Immunstimulier wirkende Polysaccharide (Heteroglykane) aus höheren Pflanzen. *Arzneimittelforschung.* 1984;35:1069–1075.

25    Shen ML et al. Immunopharmacological effects of polysaccharides from *Acanthopanax senticocus* on experimental animals. *Int J Immunopharmacol.* 1991;13:549–554.

26    Koren G et al. Maternal ginseng use associated with neonatal androgenization. *JAMA.* 1990;264:1866.

27    Awang D. Maternal use of ginseng and neonatal androgenization. *JAMA.* 1991;264:2865.

28    Waller DP et al. Lack of androgenicity of Siberian ginseng. *JAMA.* 1991;265:1826.

29    McRae S. Elevated serum digoxin levels in a patient taking digoxin and Siberian ginseng. *Canadian Med Assoc J.* 1996;155:293–295.

30    Awang D. Siberian ginseng toxicity may be a case of mistaken identity. *Canadian Med Assoc J.* 1996b;155:1237.

31    Collisson RJ. Siberian ginseng (*Eleutherococcus senticosus* Maxim.). *Brit J Phytother.* 1991;2:61–71.

# 23

# Saw Palmetto

## Synopsis

Evidence from over 35 clinical trials supports the use of standardized saw palmetto extracts for the treatment of lower urinary symptoms (LUTS) secondary to benign prostatic hyperplasia (BPH). Treatment with the extract increases peak urine flow, and decreases nocturia, without increasing serum PSA levels. Patients should contact their physician prior to self-medicating with saw palmetto to obtain a diagnosis of BPH, and to rule out urinary tract infections or prostate cancer. The recommended dose of a standardized saw palmetto extract (standardized to contain 70–95% free fatty acids, esters and sterols) is 320 mg per day, either 160 mg twice daily or 320 mg in a single dose. Treatment for 1–3 months may be required before the therapeutic effects are apparent. No drug interactions have been reported, and adverse reactions are limited to gastrointestinal disturbances such as nausea and diarrhea. There are no clinical data to justify the use of saw palmetto extracts in either women or children. In addition, due to possible estrogen and androgen antagonism by saw palmetto extract, administration during pregnancy, nursing or to young children is contraindicated.

## Introduction

Saw palmetto (*Serenoa repens*), also referred to as sabal or American dwarf palm tree, is a small, low-lying scrubby palm, indigenous to the West Indies and the southeastern United States.[1-4] The tree produces a fruit (berry), which is harvested when ripe, dried and used for the production of commercial extracts. The Native American Indians utilized saw palmetto berries to treat genitourinary problems, and more generally as a nutritional tonic. Saw palmetto was used traditionally in medicine as a sedative,[1] for the treatment of prostate inflammation,[2] dysmenorrhea,[3] and sore throats and colds.[4] During the early part of the 20th century, a tea made from the berries was used as a mild diuretic, and as a treatment for prostate enlargement. Saw palmetto was officially listed in the United States National Formulary (NF) until 1950. At this time, use of saw palmetto in the U.S. fell into disrepute due to a lack of substantiating clinical evidence, and it was subsequently deleted from the NF. In

Europe however, scientific evaluation of the herb continued, and numerous clinical trials have demonstrated the efficacy of saw palmetto fruit extracts for the symptomatic treatment of benign prostatic hyperplasia.[5]

## Quality Information

*   The correct Latin name for the tree is *Serenoa repens* (Bartr.) Small. (Arecaceae).[6,7] Botanical synonyms for *Serenoa repens* include*: Brahea serrulata* (Mich.) H. Wendl., *Chamaerops serrulata* Mich., *Corypha repens* Bartr., *Serenoa serrulata* Roem. et Schult, *Sabal serrulata* (Mich.) Nuttall. Ex. Schult., *Serenoa serrulata* Hook., *Serenoa serrulatum* (Michx.) Benth et Hook. Common names include: American dwarf palm tree, dwarf palm tree, dwarf palmetto, fan palm, sabal, sabal fructus, saw palmetto, saw palmetto berries, serenoa.[6–8]
*   Standardized extracts and other commercial preparations of saw palmetto are prepared from the dried ripe fruits (berries).[9]
*   The major chemical constituents in extracts of saw palmetto berries include free fatty acids and their corresponding ethyl esters, sterols and lipids. The main fatty acid constituents include lauric, linoleic, linolenic, myristic, oleic, palmitic and stearic acids,[10–12] and the major sterols include β-sitosterol, stigmasterol and daucosterol.[12]
*   Daily dosage: 1–2 g of the ripe dried fruits or 320 mg (160 mg twice daily or 320 once daily) of a lipidosterolic extract (*n*-hexane, supercritical fluid extract ($CO_2$) or ethanol 90% v/v) standardized to contain between 70–95% free fatty acids and corresponding ethyl esters or equivalent preparations.[9]

## Medical Uses

For the treatment of lower urinary tract symptoms (nocturia, polyuria, and urinary hesitancy) secondary to benign prostatic hyperplasia (BPH), stages I and II.[5,9,13–37] There are no therapeutic indications to justify the use of saw palmetto in women or children.

Although saw palmetto has been used as an aphrodisiac and a nutritional tonic, as well as for the treatment of bronchitis, cystitis, dysmenorrhea and the common cold,[8] there are no clinical data to support these indications.

## Summary of Clinical Evidence

Over 35 controlled and uncontrolled clinical trials have assessed the efficacy of standardized extracts of saw palmetto for the symptomatic treatment of BPH. At least 20 uncontrolled clinical trials have assessed the safety and efficacy of various saw palmetto extracts for the treatment of urinary symptoms associated with BPH.[13–25] All trials reported positive improvements in both objective and subjective variables after treatment with saw palmetto extract.[13-25] Results from the largest uncontrolled study demonstrated an improvement in post-residual urine volume (50% decrease) and a decrease in urinary frequency (37%) and nocturia (54%) in 1334 men after treatment with 320 mg/day of the extract for 6 months.[16] In a 3-year prospective multi-center trial, 435 men with BPH were treated with a saw palmetto extract (320 mg per day).[17] A steady improvement in micturition, characterized by a marked decrease in symptoms and residual urine volume (50% decrease), and an increase in peak urinary flow rate, were observed after treatment.[17] The clinical efficacy of a saw palmetto extract (160 mg, twice daily for 3 months) was assessed in another multi-center open trial involving 505 patients with mild to moderate BPH.[18] Results showed that patients treated with the extract had a 25% increase in maximal and a 27% increase in mean urinary flow rates, respectively, and a 35% improvement in the mean International Prostate Symptom Score (IPSS) after 90 days of treatment.[18] Other clinical trials have shown a reduction in urinary symptoms and the severity of the disease after 1–6 months of treatment with *S. repens* fruit extract (320 mg/day).[19-25] In general, clinical trials involving periodic evaluations of the subjects throughout the course of treatment with saw palmetto extract have shown that subjective and objective improvements were progressive over time.[17,24,25] However, clinical data from uncontrolled trials must be assessed with caution, as the placebo effect may be as high as 60% in the treatment of BPH.

The efficacy of saw palmetto has been assessed in eleven randomized, double-blind, placebo-controlled studies (RCTs),[26-37] and three comparison trials.[38-40] In the placebo-controlled trials, patient numbers ranged from 22 to 205, and the dose of the extract was generally 160 mg, twice daily for 1 to 3 months. All, but one trial,[35] show that the extract was significantly more effective than placebo for reducing the symptoms of mild to moderate BPH. The results from positive trials showed an increase in the urinary flow rate and a decrease in postvoid residual volume.[26-34,36,37] The largest placebo-controlled trial (n = 205) assessed the safety and efficacy of saw palmetto extract in subjects with mild to moderate BPH.[36] Patients were treated with 320 mg/day of the extract or a matching placebo for 3 months. This study concluded that the extract was superior to placebo for reducing the total symptom score (polyuria, nocturia, dysuria, urgency, hesitancy), and improving the quality of life score and urinary volume.[36] A placebo-controlled double blind clinical study was performed with 176 patients who had been unresponsive to placebo in previous clinical studies.[33] After 30 days of treatment (*S. repens* fruit extract 160 mg, twice daily) there was a significant reduction in dysuria, urinary frequency and nocturnal urination in the treated group as compared with the placebo recipients. A significant increase in mean peak urinary flow rate was observed in the treated group, as compared with placebo (28.9 vs 8.5%), and the global efficacy of the extract was rated higher than placebo by both patients and physicians.[33]

A double-blind, placebo-controlled study assessed the efficacy of a lipidosterolic extract in reduction of edematous and congestive processes in human BPH tissue.[37]

Pathohistological analysis of enucleated BPH tissue from 18 patients treated pre-
operatively with the lipidosterolic extract or placebo showed a significant decrease
in glandular congestion and stromal edema in the prostate tissue subjects treated with
the *S. repens* fruit extract.[37]

In the one negative trial, 70 men with mild to moderate BPH were treated with
either the extract or placebo.[35] Symptomatic improvement, including a significant
increase in flow rate was observed, however, there was no statistical difference
between the two treatments.[35] In general, the quality of the randomized, placebo-
controlled clinical trials is low due to small patient numbers, short trial duration, variable inclusion and exclusion criteria and a lack of uniform symptom score analysis.

The efficacy of saw palmetto extracts for the treatment of BPH has been compared
with the prescription drugs finasteride, and $\alpha_1$-receptor antagonists, prazosin and
alfuzosin.[38–42] One large international, double-blind, randomized, clinical trial compared the efficacy of the extract (320 mg daily) with that of finasteride (5 mg daily)
in the treatment of 1098 men with mild to moderate BPH, using the International
Prostate Symptom Score (I-PSS) as the outcome measure.[38] After six months of
treatment, the I-PSS decreased from baseline by 37% in men treated with *S. repens*
fruit extract as compared with 39% among finasteride recipients.[38] Improvements in
patient self-related quality of life scores, and the primary end-point of objective
symptom score were also observed in both groups. Both treatments improved peak
urinary flow rates and reduced size of the prostate. Peak urinary flow rate increased
from 10.6 ml/sec at baseline to 13.3 ml/sec after 6 months with *S. repens* fruit extract
treatment, as compared to 10.8 ml/sec to 14.0 ml/sec with finasteride. Prostate size
was reduced by 6% in men treated with the saw palmetto and by 18% in men treated
with finasteride. Serum prostate specific antigen levels were reduced 41% by finasteride, but remained unchanged in patients treated with *Serenoa repens* fruit
extract.[38] Other smaller (41–63 patients), short-term double-blind randomized comparative trials assessed the efficacy of *S. repens* fruit extract (320 mg daily) with the
$\alpha_1$-receptor antagonists alfuzosin and prazosin.[39,40] In a three week comparative trial
with alfuzosin, using Boyarsky's rating scale, investigators showed that the total
mean symptom score improved by 27 and 39% in patients treated with *S. repens* fruit
extract and alfuzosin, respectively.[39] Improvements in the peak urinary flow rates
were better in the alfuzosin-treated group but did not reach statistical significance. In
a 12-week randomized trial comparing *Serenoa repens* fruit extract (n = 20) and pra-
zocin (n = 21), improvements in urinary frequency, mean urinary flow rate and
postvoid residual urine volume were similar between the groups but no statistical
analysis of the data was provided by the investigators.[40] Larger, well-designed randomized trials, of longer duration, are necessary to adequately compare the clinical
efficacy of saw palmetto and $\alpha_1$-receptor antagonists.

A meta-analysis of 13 of the clinical trials with regard to the efficacy of an *n*-hexane
extract of saw palmetto was performed for the available information regarding peak
urinary flow and nocturia.[41] In all trials, the placebo effect was seen as an average
increase in peak flow of 0.94 mL/s, and an increase of a further 1.87 mL/s was seen in

patients treated with the extract. The frequency of nocturia was decreased by an average of 0.65 for placebo and a further 0.55 for the extract (p < 0.001).[41] A systematic review of 18 randomized controlled trials assessed the efficacy of saw palmetto extract for the treatment of BPH in 2929 men.[42] Treatment was adequately concealed in 9 trials and 16 trials were double-blinded. The mean study duration was 9 weeks (range, 4–48 weeks). As compared with placebo, men treated with saw palmetto extract had decreased urinary tract symptom scores and peak urine flow.[42] Other reviews of the clinical trials have indicated that lipidosterolic extracts of S. repens, improve urinary tract symptoms and flow rates in men with benign prostatic hyperplasia.[14,43]

## Mechanism of Action

The antiandrogenic and antiestrogenic activities of n-hexane extracts of saw palmetto berries have been demonstrated in vitro.[44–49] Dihydrotestosterone (DHT) and testosterone uptake by cytosol receptors of human skin and other tissues was reduced by 40.9% and 41.9% respectively, after treatment with the extract.[44] The binding of [3H]-DHT to both cytosolic and nuclear receptors in cultured human foreskin fibroblasts was inhibited by 90 and 70%, respectively, after treatment with a sterol fraction isolated from the extract ($IC_{50}$ 7.1 units/ml).[45] The extract inhibits androgen binding to the cytosolic androgen receptor of rat prostatic tissue ($IC_{50}$ 330–367.5 $\mu$g/ml), in a specific and competitive manner.[46,47] However, the extract did not compete for [3H]-DHT binding to the androgen receptor in cultured fibroblasts.[48]

The effect of an n-hexane extract of berries was evaluated on two long-term human prostatic cell lines, LNCaP and $PC_3$, responsive and unresponsive to androgen stimulation, respectively.[50] In LNCaP cells, the extract (100 $\mu$g/ml) induced a dual proliferative/differentiated effect that was not observed in $PC_3$ cells. In $PC_3$ cells, cotransfected with wild-type androgen receptor and CAT reporter genes under the control of an androgen responsive element, the extract inhibited androgen-induced CAT transcription (70% at 25 $\mu$g/ml). These results indicate that the androgen receptor plays a role in mediating the effects of saw palmetto in LNCaP cell lines.[50]

Treatment of Chinese hamster ovary cells with an n-hexane extract of the berries (1–10 $\mu$g/ml) completely inhibited the effects of prolactin on potassium conductance, protein kinase C activity, and intracellular concentrations of calcium.[51] These results suggest that the extract inhibited prolactin-induced prostate growth by interfering with signal transduction of prolactin receptor.[51] Lipidosterolic extracts of S. repens noncompetitively inhibited radioligand binding to human prostatic $\alpha_1$-adrenoceptors and agonist-induced [3H]-inositol phosphate formation.[52]

Extracts of saw palmetto berries (n-hexane, ethanol and supercritical fluid extracts with carbon dioxide) inhibit the activity of 5$\alpha$-reductase in vitro.[45,48,53–57] Inhibition of 5$\alpha$-reductase activity, and reduction of testosterone metabolism to DHT was observed in the rat ventral prostate (50% reduction, 100 $\mu$g/ml), and in human foreskin fibroblasts (90%) in vitro after treatment with the extract.[45] The conversion of DHT to 5$\alpha$-androstane-3$\alpha$–17$\beta$-diol by 3$\alpha$-ketosteroid oxidoreductase was also partially inhibited in cultured foreskin fibroblasts.[45] An *n*-hexane extract inhibited both 5α-reductase and 17β-hydroxysteroid dehydrogenase activity in primary cultures of human prostate epithelial and fibroblast cells derived from BPH tissues ($IC_{50}$ = 60 and 40 μg/ml; and 30 and 200 μg/ml, respectively).[54] One study reported that various saw palmetto extracts did not inhibit the activity of 5α-reductase isolated from human prostate tissues or DHT binding to the rat prostatic androgen receptor at concentrations up to 100 μg/ml.[55] The reasons for these conflicting results are unclear, and may be due to the methodology used in the studies.

Results from a recent study have shown that human 5α-reductase has two isoforms, type 1 and type 2, of which finasteride is a selective inhibitor of type 2 isoform (Ki 7.3 nM).[56] This investigation also demonstrated that an *n*-hexane extract of saw palmetto inhibited both isoforms of 5α-reductase, $IC_{50}$'s of 7.2 μg/ml (type 1) and 4.9 μg/ml (type 2).[56] Inhibition was non-competitive for the type 1 isoform and an uncompetitive inhibitor of type 2 isoform of the enzyme.[56] An ethanol extract of the berries inhibited the activity of 5α-reductase in the epithelium (29%) and stroma (45%) of human BPH tissue.[57] When the extract was fractionated into saponifiable, non-saponifiable and hydrophilic subfractions, only the saponifiable subfraction, containing lauric, oleic, myristic and palmitic acids (90% of the total fatty acid content) was active.[58] Lauric acid was the most active, and inhibited epithelium and stromal 5α-reductase activity by 51 and 42% respectively.[58]

Testosterone metabolism in primary cultures of epithelial cells and fibroblasts separated from BPH tissues and prostate cancer tissues was assessed.[54] In all cultures, androst-4-ene-3,12-dione, formed by the oxidation of testosterone by 17β-hydroxysteroid dehydrogenase, represented 80% of all metabolites recovered. The *n*-hexane extract of *S. repens* fruits inhibited the formation of androst-4-ene-3,12-dione in both cell-types, indicating that it inhibited the activity of 17β-hydroxysteroid dehydrogenase, while finasteride was inactive.[54] A lipidosterolic extract of *S. repens* fruits significantly inhibited basic fibroblast growth factor (b-FGF)-induced proliferation of human prostate cell cultures at 30 μg/ml.[59] Lupenone, hexacosanol and the unsaponified fraction of the lipidosterolic extract markedly inhibited the bFGF-induced cell proliferation, whereas a minimal effect on basal cell proliferation was observed.[60]

A supercritical carbon dioxide extract of the fruit inhibited cyclooxygenase ($IC_{50}$ 28.1 μg/ml) and 5-lipoxygenase ($IC_{50}$ 18.0 μg/ml) *in vitro*.[61] A lipidosterolic extract of the fruit inhibited A23187-stimulated production *in vitro* of leukotriene $B_4$ in human polymorphonuclear neutrophils.[62] An ethanol extract of the fruits suppressed A23187-stimulated synthesis of leukotriene $B_4$ ($IC_{50}$ 8.3 μg/ml) and thromboxane $B_2$ ($IC_{50}$ 15.4 μg/ml in rat peritoneal leukocytes *in vitro*.[63]

Intragastric administration of an *n*-hexane berry extract to castrated rats for 60–90 days inhibited both estradiol- and testosterone-induced increases of prostate total weight.[64] Intragastric administration of an ethanol extract to castrated rats reduced the weight of the ventral prostate, seminal vesicles and coagulation glands induced by testosterone treatments.[63] Intragastric administration of an ethanol extract inhibited both estradiol- and DHT-stimulated prostatic growth in nude mice to which

human benign hyperplastic tissue had been transplanted.[65] Oral administration of an
*n*-hexane extract (160 mg/day) inhibited the binding of $[^3H]-17\beta$-estradiol to the
nuclear estrogen receptors in prostatic tissue samples from patients with BPH.[49]
Binding was measured by saturation analysis and enzyme-linked immunosorbent
assay determination of cytosolic and nuclear estrogen and androgen receptors.[49] An
increase in the activity of $3\alpha$-hydroxysteroid-oxidoreductase (3-HSOR), the enzyme
that metabolizes DHT into the inactive androstanediol form, in human benign prostatic hyperplasia tissues was reported following treatment of patients with a hexane
fruit extract (320 mg daily for 3 months).[58] Analysis of enzyme kinetics showed that
the $V_{max}$ of 3-HSOR was significantly enhanced in stroma tissues of treated patients.
Since 3-HSOR also has a strong substrate affinity for prostaglandins, the increase in
the activity of this enzyme may also increase the metabolism of prostaglandins,
thereby accounting for the reduction of prostaglandin-mediated congestion or
intraprostatic edema formation.[58]

Intragastric administration of an ethanol extract of the fruit to rats (5.0 g/kg) inhibited carrageenan-induced pedal edema.[66] Administration of a hexane extract of the
fruit to rats by gastric lavage (10 ml/kg) decreased histamine-, 48/80- and dextran-
induced capillary permeability, and dextran-generalized edema.[67] External application of an ethanol extract of the fruits (500 μg) inhibited croton-oil-induced ear
edema by 42% in mice.[63]

## Pharmacokinetics

Tissue distribution was measured in rats after intragastric administration of a lipidosterolic extract supplemented with radiolabelled oleic or lauric acid or β-sitosterol.[68] The results of this investigation demonstrated that the uptake of the extract
was much higher in the prostate than either the liver or genitourinary tissues.[68] The
pharmacokinetics of a saw palmetto extract was investigated in 12 health male subjects, in a randomized, cross-over study of the bioequivalence of a new oral preparation (320 mg capsules) in comparison to a reference preparation (160 mg).[69] The
subjects received one 320 mg capsule (new formulation) or two 160 mg capsules as
control. The extract was rapidly absorbed with maximum absorbtion at 1.5–1.58 h
and peak plasma levels of 2.54–2.67 μg/ml. The area under the curve ranged from
7.99 to 8.42 μg.h/ml. The plasma concentrations were identical.[69]

## Safety Information

### A.  Adverse Reactions
Both short and long-term clinical studies have found that saw palmetto extracts have
are very well tolerated.[13–36,42] Adverse reactions are mild and infrequent, erectile
dysfunction is more frequent with finasteride (4.9%) than saw palmetto (1.1%, p <
0.001).[42] Occasional nausea, diarrhea and other minor gastrointestinal complaints
have also been reported.[18]

## B.  Contraindications

Due to the reported antiandrogenic and antiestrogenic activities of the fruit, its should not be administered during pregnancy, nursing or to children.

## C.  Drug Interactions

No drug interactions have been reported.

## D.  Toxicology

Documented clinical studies report that saw palmetto extracts are well tolerated in humans.[16–43] Minor gastrointestinal side effects have been reported in most of the clinical trials, but results from standard blood chemistry tests were normal.[43]

## E.  Dose and Dosage Forms

Daily dosage: 320 mg (160 mg twice daily or 320 once daily) of a lipidosterolic extract (*n*-hexane, supercritical fluid extract ($CO_2$) or ethanol 90% v/v) standardized to contain between 70–95% free fatty acids and corresponding ethyl esters or equivalent preparations.[16–42]

## References

1    Dragendorff G. *Die Heilpflanzen der Verschiedenen Volker und Zeiten*, F. Enke, Stuttgart, Germany, 1898.
2    Bombardelli E, Morazzoni P. *Serenoa repens* (Bartram) J.K. Small. *Fitoterapia.* 1997;68:99–113.
3    Novitch M, Schweiker RS. (1982) Orally administered menstrual drug products for over-the-counter human use, establishment of a monograph. *Fed Regist.* 47:55076–55101.
4    Anon. *The Herbalist*, Hammond Book Company, Hammond, Indiana, 1931.
5    Harnischfeger G, Stolze H. *Serenoa repens* – Die Sägezahnpalme. *Zeit Phytotherapie.* 1989;10:71–76.
6    German Commission E Monograph, Sabal fructus (Sägepalmenfrüchte), *Bundesanzeiger*, No 173 (18.09.1986), No. 43 (02.03.1989), No. 50 (13.03.1990), No. 11 (17.01.1991).
7    *British Herbal Pharmacopoeia*. Bournemouth, British Herbal Medicine Association, 1996.
8    Farnsworth NR, ed. *Napralert database*. University of Illinois at Chicago, IL, January 20, 1999 production, an on-line database available directly through the University of Illinois at Chicago or through the Scientific and Technical Network (STN) of Chemical Abstracts Services.
9    Anon. Herba Hyperici. *WHO Monographs on Selected Medicinal Plants*, Vol. II, World Health Organization, Traditional Medicines Programme, Geneva, Switzerland, 2001, in press.
10   De Swaef SI, Vlietinck AJ. Simultaneous quantitation of lauric acid and ethyl laureate in *Sabal serrulata* by capillary gas chromatography and derivatisation with trimethyl sulphonium hydroxide. *J Chromatogr.* 1996;719:479–482.
11   Wajda-Dubois JP et al. Comparative study on the lipid fraction of pulp and seeds of *Serenoa repens* (Palmaceae). *Oleagineux Corps Gras Lipides.* 1996;3:136–139.
12   Hänsel R et al. Eine Dünnschichtchromatographische untersuchung der Säbalfrüchte. *Planta Med.* 1964;12:136–139.

13  Gerber GS, Zagaja GP, Bales GT, Chodak GW, Contreras BA. Saw palmetto (*Serenoa repens*) in men with lower urinary tract symptoms: effects on urodynamic parameters and voiding symptoms. *Urology*. 1998;51:1003–1007.

14  Lowe FC et al. Review of recent placebo-controlled trials utilizing phytotherapeutic agents for treatment of BPH. *Prostate*. 1998;37:187–193.

15  Vahlensieck W et al. Benigne prostatahyperplasie-behandlung mit Säbalfrüchteextrakt. *Fortschritte der Medizin*. 1993;111:323–326.

16  Anderson JT. α1-Blockers vs 5α-reductase inhibitors in benign prostatic hyperplasia: a comparative review. *Drugs and Aging*. 1995;6:388–396.

17  Bach D, Ebeling L. (1996) Long-term treatment of benign prostatic hyperplasia-results of a prospective 3-year multicenter study using Sabal extract IDS 89. *Phytomedicine*. 1996;3:105–111.

18  Braeckman J. The extract of *Serenoa repens* in the treatment of benign prostatic hyperplasia: A multicenter open study. *Current Therapeutic Res*. 1994;55:776–786.

19  Schneider HJ, Uysal A. Internationaler Prostata-Symptomenscore (I-PSS) im klinischen Alltag. *Urologe* (B). 1994;34:443–447.

20  Braekman J et al. Efficacy and safety of the extract of *Serenoa repens* in the treatment of benign prostatic hyperplasia: therapeutic equivalence between twice and once daily dosage forms. *Phytother Res*. 1997;11:558–563.

21  Derakhshani P et al. Beeinflussung des Interationalen Prostata-Symptomen Score unter der therapie mit Sägepalemenfrüchteextrakt bei täglicher Einmalgabe. *Urologie* (B). 1997;37:384–391.

22  Ziegler H, Hölscher U. Wirksamkeit des Spezialextraktes WS 1473 aus Sägepalemenfrüchteextrakt bei Patienten mit benigner Prostatahyperplasie im Stadium I-II nach Alken-offene Multicenter-studie. *Jatros Uro*. 1998;14:34–43.

23  Redecker KD, Funk P. *Sabal*-Extrakt WS 1473 bei benigner Prostatahyperplasie. *Extracta Urologica*. 1998;21:23–25.

24  Hanus M, Matouskova M. Alternativni lecba BPH- Permixon (Capistan). *Rozhledy Chirurgia*. 1993;72:75–79.

25  Romics I et al. Experience in treating benign prostatic hypertrophy with *Sabal serrulata* for one year. *J Int Urol Nephrol*. 1993;25:565–569.

26  Boccafoschi C, Annoacia S. Confronto fra estratto di *Serenoa repens* e placebo mediante prova clinica controllata in pazienti co adenmatosi prostatica. (Italian) *Urologia*. 1983;50:1–14.

27  Champault G et al. A double-blind trial of an extract of the *plant Serenoa repens* in benign prostatic hyperplasia. *Brit J Clin Pharmacol*. 1984;18:461–462.

28  Cukier C et al. *Serenoa repens* extract vs placebo. *Comptes Rendus de Therapeutiques and de Pharmacologie Clinique*. 1985;4:15–21.

29  Descotes JL et al. Placebo-controlled evaluation of the efficacy and tolerability of Permixon® in benign prostatic hyperplasia, after exclusion of placebo responders. *Clinical Drug Investigations*. 1995;9:291–297.

30  Emili E et al. Clinical results on a new drug in the treatment of benign prostatic hyperplasia (Permixon). *Urologia*. 1983;50:1042–1049.

31  Tasca A et al. Treatment of obstructon in prostatic adenoma using an extract of *Serenoa repens*. Double-blind clinical test vs placebo (Italian). *Minerva Urol e Nefrol*. 1985;37:87–91.

32  Mandressi A et al. Treatment of uncomplicated benign prostatic hypertrophy (BPH) by an extract of *Serenoa repens*: clinical results. *J Endocrinol Invest*. 1987;10 (Suppl.2):49.

33  Mattei FM, Capone M, Acconcia A. Impiego dell'estratto di *Serenoa repens* nel trattamento medico della ipertrofia proststica benigna. *Urologia*. 1988;55:547–552.

34    Gabric V, Miskic H. Behandlung des benignen Prostataadenoms und der chronischen Prostatitis. *Therapiewoche.* 1987;37:1775–1788.

35    Reese-Smith H et al. The value of Permixon in benign prostatic hypertrophy. *British J Urol.* 1986;58:36–40.

36    Braeckman J et al. A double-blind, placebo-controlled study of the plant extract *Serenoa repens* in the treatment of benign hyperplasia of the prostate. *Eur J Clini Res.* 1997;9:247–259.

37    Marks LS, Dorey FJ, Macairan ML, Santos PB, Garris JB, Jahner DK, Stonebrook KA, Parkin AW, Tyler VE. Clinical effects of saw palmetto extract in men with symptoms of BPH. *Proceedings of the American Urological Association Annual meeting,* 1999; Dallas, TX.

38    Carraro JC et al. Comparison of phytotherapy (Permixon®) with finasteride in the treatment of benign prostatic hyperplasia: a randomized international study of 1098 patients. *Prostate.* 1996;29:231–240.

39    Grasso M et al. Comparative effects of alfuzosin versus *Serenoa repens* in the treatment of symptomatic benign prostatic hyperplasia. *Archivos Espanoles de Urologia.* 1995;48:97–103.

40    Adriazola Semino M et al. Tratamiento sintomatico de la hipertrofia benigna de prostata. Estudio comparativo entre Prazosin y *Serenoa repens. Archivos Espanoles de Urologia.* 1992;45:211–213.

41    Lowe FC, Fagelman E. Phytotherapy in the treatment of benign prostatic hyperplasia: an Update. *Urology.* 1999;53:671–678.

42    Wilt TJ, Ishani A, Stark G, MacDonald R, Lau J, Mulrow C. Saw palmetto extracts for the treatment of benign prostatic hyperplasia. *JAMA.* 1998;280:1604–1609.

43    Plosker GL, Brogden RN. *Serenoa repens* (Permixon®). A review of its pharmacology and therapeutic efficacy in benign prostatic hyperplasia. *Drugs and Aging.* 1996;9:379–395.

44    El-Sheikh MM, Dakkak MR, Saddique A. The effect of Permixon® on androgen receptors. *Acta Obstet Gynecol Scand.* 1988;67:397–399.

45    Sultan C et al. Inhibition of androgen metabolism and binding by a liposterolic extract of "*Serenoa repens* B" in human foreskin fibroblasts. *Journal of Steroid Biochemistry.* 1984;20:515–519.

46    Briley M, Carilla E, Fauran F. Permixon, a new treatment for benign prostatic hyperplasia, acts directly at the cytosolic androgen receptor in rat prostate. *Brit J Pharmacol.* 1983;79:327.

47    Carilla E et al. Binding of Permixon, a new treatment for prostatic benign hyperplasia, to the cytosolic androgen receptor in rat prostate. *Journal of Steroid Biochemistry.* 1984;20:521–523.

48    Hagenlocher M et al. Specific inhibition of 5-$\alpha$ reductase by a new extract of *Sabal serrulata. Aktuelle Urologie.* 1993;24:146–149.

49    Di Silverio F et al. Evidence that *Serenoa repens* extract displays an antiestrogenic activity in prostatic tissue of benign prostatic hypertrophy patients. *European Urology.* 1992;21:309–314.

50    Ravenna L et al. Effects of the lipidosterolic extract of *Serenoa repens* (Permixon®) on human prostatic cell lines. *Prostate.* 1996;29:219–230.

51    Vacher P et al. The lipidosterolic extract from *Serenoa repens* interferes with prolactin receptor signal transduction. *J Biomed Sci.* 1995;2:357–365.

52    Goepel M et al. Saw palmetto extracts potently and noncompetitively inhibit human $\alpha_1$-adrenoceptors *in vitro. Prostate.* 1999;38:208–215.

53    Niederprüm HJ et al. Testosterone 5$\alpha$-reductase inhibition by free fatty acids from *Sabal serrulata* fruits. *Phytomedicine.* 1994;1:127–133.

54  Delos S et al. Testosterone metabolism in primary cultures of human prostate epithe-lial cells and fibroblasts. *Journal of Steroid Biochemistry and Molecular Biology.* 1995;55:375–383.

55  Rhodes L et al. Comparison of finasteride (Proscar®), a 5α-reductase inhibitor, and various commercial plant extracts in *in vitro* and *in vivo* 5α-reductase inhibition. *Prostate.* 1993;22:43–51.

56  Iehle C et al. Human prostatic steroid 5α-reductase isoforms, a comparative study of selective inhibitors. *Journal of Steroid Biochemistry and Molecular Biology.* 1995;54:273–279.

57  Weisser H et al. Effects of Sabal serrulata extract IDS 89 and its subfractions on 5α-reductase activity in human benign prostatic hyperplasia. *Prostate.* 1996;28:300–306.

58  Weisser H et al. Enzyme activities in tissue of human benign prostatic hyperplasia (BPH) after three months of treatment with *Sabal serrulata* extract IDS 89 (Strogen) or placebo. *European Urology.* 1997;31:97–101.

59  Paubert-Braquet M et al. Effect of the lipidosterolic extract of *Serenoa repens* (Per-mixon) and its major components on basic fibroblast growth factor-induced prolifera-tion of cultures of human prostate biopsies. *European Urology.* 1998;33:340–347.

60  Di Silverio F et al. Effects of long-term treatment with *Serenoa repens* (Permixon®) on the concentrations and regional distribution of androgens and epidermal growth fac-tor in benign prostatic hyperplasia. *Prostate.* 1998;37:77–83.

61  Breu, W et al. Antiphlogistische Wirkung eines mit hyperkritischem Kohlendioxid gewonnenen Sabalfrucht-Extraktes. *Arzneimittelforschung.* 1992;42:547–551.

62  Paubert-Braquet M et al. Effect of the lipidic lipidosterolic extract of *Serenoa repens* (Permixon®) on the ionophore A23187-stimulated production of leukotriene B4 (LTB4) from human polymorphonuclear neutrophils. *Prostaglandins, Leucotrienes and Essential Fatty Acids.* 1997;57:299–304.

63  Koch E. Pharmakologie und Wirkmechanismen von Extrackten aus Sabalfrüchten (*Sabal fructus*), Brennesselwurzeln (*Urticae radix*) und Kürbissamen (*Curcurbitae peponis* semen) bei der Behandlung der benignen Prostatahyperplasie. In: Loew D, Rietbrock N. (Eds.) *Phytopharmaka in Forschung und klinischer Anwendung,* Damstadt, Steinkopff, 1995;57–79. Odenthal KP, Rauwald HW. Lipophilic extract *from Sabal serrulata* inhibits contractions in smooth muscle tissue. *Aktuelle Urol.* 1996;27:152–157.

64  Otto U et al. Transplantation of human benign hyperplastic prostate tissue into nude mice: first results of systemic therapy. *Urol Int.* 1992;48:167–170.

65  Paubert-Braquet M et al. Effect of *Serenoa repens* extract (Permixon®) on estradiol/testosterone-induced experimental prostate enlargement in the rat. *Pharma-cological Research.* 1996;34:171–179.

66  Hiermann A. About the contents of *Sabal* fruits and their anti-inflammatory effect. *Archiv der Pharmazie (Weinheim).* 1989;322:111–114.

67  Tarayre, JP et al. Anti-edematous action of a hexane extract from *Serenoa repens* Bartr. drupes. *Annales Pharmaceutiques Françaises.* 1983;41:559–570.

68  Bernard P, Cousse H Chevalier G. Distribution of radioactivity in rats after oral admin-istration of lipidosterolic extract of *Serenoa repens* (Permixon®) supplemented with [1-14C]-lauric acid, [1-14C] oleic acid or [4-14C] beta-sitosterol. *European Journal of Drug Metabolism and Pharmacokinetics.* 1997;22:73–83.

69  De Bernardi di Valserra M, Tripodi AS, Contos S, Germogli R. *Serenoa repens* cap-sules: a bioequivalence study. *Acta Toxicol Ther.* 1994;15:21–39.

# 24

# St. John's Wort

**Synopsis**

The results from randomized controlled clinical trials support the use of St. John's wort (*Hypericum perforatum*) for the symptomatic treatment of mild to moderate depression. The recommended dose of a standardized St. John's wort extract (containing 0.3% hypericin or 5% hyperforin) is 900 mg per day in three divided doses. As with other antidepressant drugs, observation of the therapeutic effects of St. John's wort may require 2 to 4 weeks of therapy. If a significant antidepressant effect has not been observed by 6 weeks, a physician should be consulted. Adverse reactions range from minor gastrointestinal disturbances and photosensitivity, to acute neuropathy in sensitive patients. St. John's wort has been reported to reduce plasma and serum levels of drugs such as theophylline, digoxin, warfarin, cyclosporin, and protease inhibitors by inducing drug metabolism by cytochrome P450. Due to a lack of safety data, St. John's wort should not be administered to patients allergic to the plant, or during pregnancy or nursing.

**Introduction**

St. John's wort (*Hypericum perforatum*) is a herbaceous aromatic perennial plant in the family Clusiaceae.[1] The common name, St. John's wort, appears to be in reference to John the Baptist, as the plant begins to flower around the 25th of June, the day of St. John feast. Euryphon, a Greek physician in 288 BC, first described the medicinal value of the plant. Hippocrates reportedly employed St. John's wort as a cooling and anti-inflammatory agent.[2] Dioscorides, Paracelsus and Pliney the Elder recommended St. John's wort for the treatment of various ailments such as sciatica, minor burns and wounds, pain, and as an antivenom agent.[2,3] In 1650, Culpepper employed St. John's wort for the treatment of wounds and venomous stings. In the early 19th century, the Eclectic doctors were using oily preparations of the plant for the treatment of ulcers, diarrhea, hysteria and nervous conditions with depression.[2] During the early part of the 20th century Madaus, a German physician employed St. John's wort preparations internally for the treatment of neuralgia, neuroses, neurasthenias, hysteria and insomnia, and externally for the treatment of wounds.[3] More recently, St. John's wort has been employed primarily for the treatment of depression and has been tested as an antiviral agent.[4]

## Quality Information

- The correct Latin name for the plant is *Hypericum perforatum* L. (Clusiaceae).[1] Botanical synonyms that may appear in the scientific literature include *Hypericum officinarum* Crantz, *Hypericum officinale* Gater ex. Steud., *Hypericum vulgare* Lam.[5] The plant family Clusiaceae is also referred to as Guttiferae or Hypericaceae.[1] The vernacular (common) names include: balsana, common St. John's wort, devil's scorge, hard hay, Johanniskraut, Eisenblut, erba di San Giovanni, hardhay, Johanniskraut, John's wort, Lord God's wonder plant, St. John's wort, and witcher's herb.[1]
- Commercial products of St. John's wort are prepared from the dried whole or cut flowering tops, or aerial parts of *Hypericum perforatum*.[1]
- *Hypericum perforatum* is native to Europe, Asia and Northern Africa, South Africa, Australia, New Zealand, South America and is naturalized in the USA.[1–4] The plant material is harvested at flowering time.[1]
- The major chemical constituents include hypericin, pseudohypericin and related naphthodianthrones (0.05–0.3%), flavonoids [hyperoside, quercitrin, isoquercitrin, rutin] (2–4%), acylphloroglucinols [hyperforin, adhyperforin] (2–4%), and catechin tannins (6.5–15%).[1,5,6]

## Medical Uses

For symptomatic treatment of mild to moderate depression (as described by the International Classification of Diseases 10: F32.0 and F32.1).[7–19]

Used externally for the treatment of minor wounds, burns and skin ulcers.[4,20] The antiviral effects of hypericin have been demonstrated in experimental investigations, however, the use of St. John's wort as a potential therapeutic agent for the treatment of human retroviral infections requires clinical evaluation.[4]

## Summary of Clinical Evidence

Over 28 controlled clinical trials (published since 1985) have assessed the efficacy of St. John's wort for the treatment of mild to moderate depression.[14] Twelve of the trials, performed using an ethanol extract and involved a total of 950 patients, while the other sixteen trials used a dried 80% methanol extract and involved a total of 1170 patients.[14] A systematic review and meta-analysis of 23 of the randomized clinical trials including 1757 patients assessed the efficacy of the herb for the symptomatic treatment of mild to moderate depression.[9] Twenty of the clinical trials were double-blinded, one was single blinded and two were uncontrolled clinical trials. Fifteen of the trials involving 1008 patients were placebo-controlled and eight trials involving 749 patients were comparison studies with other antidepressant drugs. All trials, with the exception of two, had follow-up periods of four to eight weeks. The dose administered during the trials ranged from 300 to 1000 mg per day of a standardized extract containing 0.4–2.7 mg of hypericin. Seventeen of the trials used the Hamilton depression scale, measuring somatic symptoms, to determine the effectiveness of treatment. While twelve of the trials used the clinical global impressions, an observer rated instrument with three items (severity of illness, global improvement and an efficacy index). The meta-analysis concluded that St. John's wort was significantly superior to placebo for the treatment of mild to moderate depression, and similarly effective as standard antidepressants such as maprotiline (3 × 25 mg), or imipramine (3 × 25 mg). Side effects were much lower in the herb treated group (19.8%) as compared with the standard antidepressant drugs (52.8%).[9]

A systematic criteria-based review was performed on eighteen of the controlled clinical trials using extracts of the herb as a treatment for depression.[7] Of the 18 trials assessed, twelve of the trials (nine placebo-controlled and three comparison trials) met the methodological inclusion criteria, and were included in the review. The results of the cumulative data show that St. John's wort was superior to placebo for the symptomatic treatment of depression as quantified by the Hamilton Depression Scale. The results of the comparison studies, conducted with standard antidepressant drugs suggest that St. John's wort had a similar therapeutic profile as imipramine (50–75 mg/day) or maprotiline (75 mg/day). Methodological flaws in the clinical trials included: no intention to treat analysis, lack of control over compliance, and a lack of a better description of the extract or placebo involved in the study.[7]

Twelve double blind, placebo-controlled studies and 3 comparison clinical trials assessing the efficacy of St. John's wort for the treatment of mild to moderate depression, were reviewed.[15] The results of this review concluded that the antidepressant activity of a standardized extract of the herb (300 mg standardized to 900 μg of hypericin, three time daily) was sufficiently documented. However, the review further concluded that no dose finding studies have been conducted, and that trials involving inpatients with severe depression and endogenously depressed patients were lacking. In the 3 comparison trials, the activity of St. John's wort was comparable to low doses of amitriptyline (30 mg) or maprotiline (75 mg). However, the review concluded that additional clinical trials, with a longer follow-up and in comparison with higher doses of standard antidepressant drugs, are warranted.[15]

A randomized, double-blind, multicenter clinical trial assessed the efficacy, safety and tolerability of 900 mg/day of a hydroalcoholic extract of St. John's wort in comparison with 75 mg amitriptyline.[17] After a one week placebo run-in phase, 156 subjects were treated with 300 mg of the extract or 25 mg of amitriptyline three times daily for 6 weeks. The subjects were assessed using the Hamilton Depression Scale (HAMD), Montgomert-Asberg Rating Scale for Depression (MADRS), and Clinical Global Impressions Scale (CGI) at baseline and then at 1, 2, 4 and 6 weeks after treatment. Changes in the HAMD from baseline to 6 weeks: 20 to 10 in the extract treated patients, and 21 to 6 in the amitriptyline treated patients (p < 0.05). Changes in the MADRS (baseline to 6 weeks) were 27 to 13 in the extract treated patients, and 26 to 6.5 in the amitriptyline treated patients (p < 0.05). Similar scores in the CGI scale were observed in both groups.[17]

A randomized, double-blind, multicenter, trial compared the effectiveness of a dried 80% methanol extract of St. John's wort with imipramine in 209 patients with severe depressive syndrome (ICD-10 F33.2).[16] The subjects were treated with 1800 mg/day of the standardized extract or 150 mg per day of imipramine for 6 weeks. All subjects were assessed using the Hamilton Depression Scale (HAMD), von Zerssen's Depression Scale (ZDS), and Clinical Global Impressions Scale (CGI) at baseline and then at 1, 2, 4 and 6 weeks after treatment. Changes in the HAMD from baseline to 6 weeks: 25.3 to 14.4 in the extract treated patients, and 26.1 to 13.4 in the imipramine treated patients (p < 0.021). Changes in the ZDS, 28.9 to 13.6 in the extract treated patients, and 26 to 6.5 in the imipramine treated patients (p < 0.05). Results in the CGI scale showed a trend in favor of imipramine. The effectiveness of the extract was not significantly different to that of imipramine, however, analysis of the subgroups showed that the efficacy of the methanol extract was better in moderately severe depression.[16]

A prospective, randomized, double-blind, placebo-controlled trial assessed the safety and efficacy of an ethanol extract of St. John's wort for the treatment of 151 patients with mild to moderate depression (ICD-10; F 32.0 mild; F 32.1 moderate).[12] The patients were administered one 250 mg tablet of the extract twice daily (corresponding to 1 mg hypericin) or a matching placebo for 6 weeks. The primary efficacy variable was Hamilton Depression Scale (HAMD), and secondary variables were the risk-benefit Clinical Global Impressions scales I-III and a validated patient self-assessment on a Visual Analogue Scale. After 6 weeks of therapy, a decrease in the HAMD score was observed and 56% of patients treated with the extract were classified as responders as compared with 15% of patients taking placebo.[12] A randomized, double-blind, placebo-controlled, multicenter study assessed the safety and efficacy of two different extracts of St. John's wort.[8] In this trial, 147 patients suffering from mild to moderate depression as described by DSM-IV criteria were treated with placebo or one of two St. John's wort extracts (900 mg/day for 42 days) differing in their hyperforin content (0.5% or 5% hyperforin).[8] At the end of the treatment period, the patients receiving the extract containing 5% hyperforin exhibited the largest decrease in the HAMD (−10.3 points) (p = 0.004 vs. placebo), followed by −8.5 points with the extract containing 0.5% hyperforin and −7.9 in the placebo group. More severely depressed patients showed a 22.4% and a 53.8% reduction in the HAMD after treatment with the extract containing 0.5% and 5.0% hyperforin, respectively.[8]

A four-week double-blind, placebo-controlled crossover study assessed the effects of St. John's wort on sleep quality in 12 healthy volunteers.[13] Subjects treated with a dried hydromethanolic extract of the herb (300 mg, three times daily) showed improved sleep quality with an increase in deep sleep phases.[13] A randomized, double-blind, placebo-controlled trial was performed measured the pharmacodynamic effects of two St. John's wort extracts with different contents of hyperforin (0.5% or 5.0%) but identical hypericin content in 54 healthy volunteers.[21] In this three-armed trial, volunteers were treated with 900 mg of the extract or placebo for 8 days. A quantitative topographic EEG (qEEG) was performed on days 1 and 8, and was used as an indicator of drug-induced pharmacological activity. In both treatment groups, reproducible central pharmacodynamic effects were observed between 4 and 8 hours after administration, and confirmed on day 8. The extract containing the 5% hyperforin showed a marked tendency to produce higher increases in qEEG baseline power performances than the extract containing 0.5% hyperforin. Higher baseline outputs were observed on day 8 in the delta, theta and alpha-1 frequency values. As compared with placebo, patients treated with the extract containing 5% hyperforin had an increase in qEEG power performance in the delta frequency values after a single dose and in the theta and alpha-1 frequencies after 8 days of treatment.[21]

A randomized, single-blind controlled study evaluated the efficacy of St. John's wort for the treatment of seasonal affective disorders (SAD) in conjunction with light therapy.[22,23] Twenty patients who fulfilled the diagnostic criteria for SAD were treated with 900 mg of a hydroalcoholic extract of St. John's wort in combination with either bright (3000 lux) or dim light ( < 300 lux) conditions. Light therapy was administered for 2 hours daily. A significant reduction of the Hamilton Depression Scale score in both groups but no statistically significant difference between the two groups was observed.[22,23]

A double-blind, placebo-controlled crossover trial assessed the effects of St. John's wort on the EEG.[24] Twelve healthy subjects were treated with 900 mg of a dried hydromethanolic extract for 6 weeks, and the effects on EEG were measured. A reduction in alpha activity and audio-visual latencies in evoked potentials were observed, along with an increase in beta and theta activities.[24] In another randomized, double-blind clinical trial involving 24 healthy subjects, the effects of a dried hydromethanolic extract of St. John's wort (300 mg, 3 times daily for 4 weeks) on the resting EEG, and visually and acoustically evoked potentials, were compared with that of maprotiline (10 mg 3 times daily).[25] An increase in theta and beta-2 activity was observed in those patients treated with the extract, while a decrease in theta activity was noted in those patients treated with maprotiline.[25] The extract increased deep sleep as demonstrated by visual analysis of the sleeping phases and automatic analysis of slow-wave EEG activities. REM sleep was not influenced.[13]

Numerous case reports and drug monitoring studies have also assessed the safety and efficacy of St. John's wort in over 5000 patients.[10,11,19] One uncontrolled drug monitoring study involved 3250 subjects.[19] Evaluation of the patients (mean age 51 years) at the beginning of the trial determined that 49% were mildly depressed, 46% were moderately depressed, and 3% were considered to be severely depressed. The patients were treated with 3 × 300 mg/day of a dried hydroalcoholic extract of St. John's wort, and symptoms were evaluated after 2 and 4 weeks of therapy. At the end of therapy, 80% of patients were improved or symptom-free, while 13 to 16%

remained unchanged or worse. Minor adverse reactions were reported in 2.4% of patients.[19] A post-marketing trial was performed with 2404 patients with mild to moderate depressive symptoms.[26] Patients were treated with 2–4 capsules equivalent to 0.6 to 1.8 mg total hypericin daily for 4–6 weeks. Symptomatic improvement was evaluated as good to very good in 77% of cases and satisfactory in 15% of patients.[26]

The effect of a standardized ethanol extract of St. John's wort (0.5 mg total hypericin or 1.4 g crude drug, 4 weeks) on the electroencephalogram (EEG) was determined following oral administration to 40 patients with symptoms of depression.[27] An increase in theta activity, a decrease in alpha-activity and no change in beta activity were observed, indicating the induction of relaxation.[27] A significant increase in nocturnal melatonin plasma concentrations was observed in 13 healthy subjects treated with a hydroethanolic extract of the herb equivalent to 0.53 mg of total hypericin daily for 3 weeks.[28] A significant increase in the concentration of urinary neurotransmitters was observed two hours after oral administration of a standardized ethanol extract of St. John's wort to six women with symptoms of depression.[29]

The photodynamic effects of hypericin, incorporated into a nonionic hydrophilic ointment base, were assessed after external application to the skin of patients with herpes communis.[30] The infected dermal surface of treated patients recovered rapidly and the effects were lasting in most cases.[30]

## Pharmacokinetics

Plasma levels of hypericin and pseudohypericin were measured after oral administration of a St. John's wort extract.[31,32] Single and multiple dose pharmacokinetics of hypericin and pseudohypericin were determined in 12 healthy men.[31] After a single dose of 300, 900 or 1800 mg of the extract (containing 250, 750 or 1500 $\mu$g of hypericin and 526, 1578, or 3156 $\mu$g of pseudohypericin, respectively), plasma levels were measured by high-performance liquid chromatography for up to 3 days. The median plasma levels were 1.5, 4.1 and 14.2 ng/ml for hypericin and 2.7, 11.7 and 30.6 ng/ml for pseudohypericin, respectively for the three doses given. The median half-life for hypericin was 24.8–26.5 hours as compared with 16.3 to 36.0 hours for pseudohypericin. Median lag-time of absorption was 2.0 to 2.6 hours for hypericin and 0.3 to 1.1 hours for pseudohypericin. During long-term dosing (900 mg/day) a steady state was reached after 4 days. Mean maximal plasma level during the steady-state was 8.5 ng/ml for hypericin and 5.8 ng/ml for pseudohypericin.[31]

A randomized, placebo-controlled clinical trial was performed to evaluate the pharmacokinetics and dermal photosensitivity of hypericin and pseudohypericin in human subjects after single and multiple dose administration.[32] Single and multiple dose pharmacokinetics of hypericin and pseudohypericin were determined in 13 subjects after administration of either a placebo or 900, 1800 mg or 3600 mg of the extract (containing 0, 2.81, 5.62 and 11.25 mg combined of hypericin and pseudohypericin). Maximum total hypericin plasma levels were observed at 4 hours after dosing, and were 0.028, 0.061 and 0.159 mg/L, respectively. Prior to, and 4 hours after drug intake, the subjects were exposed at small areas of their back to increasing doses of solar simulated irradiation. No dose-related increase in light sensitivity was observed. In the multiple dose part of the trial, 50 volunteers received 600 mg of the herb extract three times daily. A slight increase in the solar simulated irradiation sensitivity was observed.[32]

In an open, randomized, four-way crossover study in 6 healthy volunteers, the pharmacokinetics of hyperforin was determined after single doses of 300, 600, 900 or 1200 mg of an alcohol extract containing 5% hyperforin.[33] After ingestion of 300 mg of the extract, the maximum plasma level of hyperforin (150 ng/ml) was reached 3.5 hours after administration. The half-life and mean residence time were 9 and 12 hours respectively. Hyperforin kinetics was linear up to 600 mg of extract. Increasing doses to 900 or 1200 mg of extract resulted in lower $C_{max}$ and area under the curve (AUC) values than those expected from linear extrapolation of data from the lower doses. As part of a double-blind, randomized, placebo-controlled study in 54 subjects, single and multiple dose pharmacokinetics of hyperforin was measured in 9 volunteers, who received 900 mg of an alcohol extract containing 5% hyperforin once daily for eight days. No accumulation of hyperforin in the plasma was observed. Using the AUC values from the repeated dose study, the estimated steady state plasma concentrations of hyperforin after 900 mg of extract per day was approximately 100 ng/ml.[33]

## Mechanism of Action

The mechanism by which St. John's wort exerts is antidepressant effect has been assessed by behavioral studies in rodents, and by measuring the exploratory and locomotor activities of the animals in a foreign environment.[34,35] Administration of a 95% ethanol extract of St. John's wort to male gerbils (2 mg/kg, gavage) suppressed clonidine-induced depression.[35] Administration of the 95% ethanol extract to male mice (5 mg/kg, gavage) enhanced exploratory activity of mice in a foreign environment, significantly prolonged narcotic-induced sleeping time in a dose-dependent manner, and exhibited reserpine antagonism.[35] Intragastric administration of the extract to mice increased the activity in the water wheel test, and decreased the aggressiveness of socially isolated males.[35] Intragastric administration of a 50% ethanol extract to male mice prolonged pentobarbital-induced sleeping time at a dose of 13.25 mg/kg, and had CNS depressant activity at a dose of 25.5 mg/kg.[36] The activity of St. John's wort was comparable to that of 2mg/kg of diazepam.[36] Administration of a methanol extract, containing both hypericin and pseudohypericin, to mice (500 mg/kg, gavage), produced a dose-dependent increase in ketamine-induced sleeping time, and also increased body temperature of the animals.[37]

Administration of a carbon dioxide extract (38.8% hyperforin, 30 mg/kg, gavage) or an ethanol extract (300 mg/kg, gavage) of St. John's wort to rodents exhibited dose-dependant antidepressant activities as measured by the behavioral despair test and the learned helplessness paradigm.[38] The results were comparable to that of imipramine at an intraperitoneal dose of 10 mg/kg.[38] Administration of an ethanol extract containing 4.5% hyperforin (50, 150 and 300 mg/kg/day, gavage) or a $CO_2$ extract, devoid of hypericin, but containing 38.8% hyperforin (5, 15 and 30 mg/kg/day, gavage), exhibited similar antidepressant activity in several behavioral models in rodents.[39,40] The ethanol extract of St. John's wort enhanced the dopaminergic behavioral responses, whereas the $CO_2$ extract enhanced the serotoninergic effects.[39] Measurement of some metabolites of biological amines in the urine of various animal models has established a correlation between the excretion in the urine of 3-methoxy-4-hydroxyphenylglycol, the main metabolite of noradrenaline, with the start of the therapeutic antidepressant activity.[29] Intragastric administration of a methanol extract to mice (500 mg/kg, gavage) had antidepressant effects as measured by a decreased immobility time in the tail suspension test and forced swimming tests.[37]

Topical applications of an aqueous extract of St. John's wort (20% concentration) to the skin of guinea pigs and rabbits accelerated the healing of experimentally-induced wounds.[41,42] Intragastric administration of a 60% ethanol extract of the dried leaves to rats (0.1 ml/animal) accelerated wound healing by enhancing the strength and rate of wound contraction and epithelialization.[43]

To determine the mechanism of action of St. John's wort, initial *in vitro* studies focused on the inhibition of monoamine oxidase (MAO) and catechol-O-methyltransferase inhibition (COMT), the two enzymes responsible for the catabolism of biological amines. Early investigations assessed the inhibition of MAO using a series of xanthones isolated from extracts of St. John's wort.[44,45] In rat brain mitochondria, hypericin was reported to inhibit MAO type A ($IC_{50}$ $6.8 \times 10^{-5}$ mol/L) and type B ($IC_{50}$ $4.2 \times 10^{-5}$ mol/L) *in vitro*.[46] However, analysis hypericin-fraction used in these experiments showed that it was not pure and contained other constituents including some flavonoid derivatives.[4] Xanthone-containing fractions of a hydroalcoholic extract of St. John's wort, free of hypericin and tannins, showed significant inhibition of monoamine oxidase type A (serotonin) *in vitro*.[47] Other studies demonstrated that only the flavone aglycones, quercitrin, and the xanthone derivative, norethyriol, significantly inhibited $MAO_{-A}$.[47–49] Molecular modeling of the constituents of St. John's wort has indicated that the flavonoids are the most likely candidates for inhibition of MAO, as their structures are similar to that of other $MAO_A$ inhibitors, toloxotone and brofaromine.[50]

Inhibition of MAO by six fractions of a St. John's wort extract was assessed *in vitro* and *ex vivo*.[51] *In vitro* inhibition of $MAO_A$ in rat brain homogenates was achieved only at high concentrations (1–10 mM/L) of the crude extract or a flavonoid-rich fraction.[51] In albino rats, neither the crude extract, nor the xanthone-containing fractions inhibited $MAO_A$ or $MAO_B$ after intraperitoneal administration of 300 mg/kg of the extract.[51] In addition, purified hypericin did not inhibit $MAO_A$ in either *in vitro* or *ex vivo* experiments.[51] In another study,[52] hypericin inhibited MAO at a concentration of 1 mM, total extract $10^{-4}$ M*, and one fraction (containing hypericins and flavonols)

$10^{-5}$ M* (*molar concentrations were based on a mean molar mass of 500). Hypericin and a crude methanol extract weakly (1 mM*) inhibited the activity of COMT, while two fractions, containing flavonols and xanthones, inhibited COMT at $10^{-4}$M*. The inhibitory concentrations observed during this study appear to be too high to be of any clinical significance.[52]

Other possible mechanisms to explain the antidepressant effects of St. John's wort include its ability to reduce the levels of interleukin-6. The release of interleukin-6 was suppressed in blood samples obtained from depressed patients after treatment with the extract.[53] Interleukin-6 (IL-6) is involved in the modulation of the hypothalamic-pituitary-adrenal axis (HPA) within the nervous and immune systems, and elevated IL-6 levels activate the HPA axis increasing levels of adrenal hormones that play a role in depression.

Inhibition of the re-uptake of serotonin has also been proposed as a mechanism for the antidepressant effects of St. John's wort. Serotonin re-uptake inhibition ($IC_{50}$ 6.2–25 μg/ml),[54,55] and inhibition of both GABA re-uptake ($IC_{50}$ 1 μg/ml) and GABA$_A$-receptor binding ($IC_{50}$ 3 μg/ml) have been reported in vitro.[56] A hydroalcoholic extract of the fresh flowers and buds (standardized to 0.1% hypericin) was subjected to a battery of assays involving 39 receptor types and two enzyme systems.[57] In this investigation, receptor assays exhibiting at least 50% of radioligand displacement or a 50% inhibition of MAO was considered to be active. The extract exhibited an affinity for the adenosine (non-specific), GABA (A and B), serotonin, benzodiazepine, inositol triphosphate receptors and inhibited monoamine oxidase (A and B). Hypericin lacked any significant MAO (A or B) inhibitory activity in concentrations up to 10 μM, and had an affinity only for N-methyl-D-aspartate (NMDA) receptors measured in rat forebrain membrane.[57] An ethanol extract of the herb inhibited radioligand binding to the NMDA, GABA$_A$ and GABA$_B$ receptors ($IC_{50}$ 7.025, 3.24 and 3.31 μg/ml, respectively).[58] The extract also inhibited synaptosomal GABA and L-glutamate uptake in vitro ($IC_{50}$ 1.11 and 21.25 μg/ml, respectively).[58]

A methanol or carbon dioxide extract of the herb, and one of the active constituents, hyperforin, significantly inhibited synaptosomal re-uptake of serotonin, noradrenalin, dopamine, L-glutamate and GABA in vitro.[59] The carbon dioxide extract, containing 38.8% hyperforin, was more active than a methanol extract containing 4.5% hyperforin. For hyperforin, the inhibition was most pronounced with the following order of affinity: noradrenalin > dopamine > GABA > serotonin >> glutamate ($IC_{50}$ from 0.043–0.445 μg/ml).[59,60] Neither hyperforin nor the carbon dioxide extract inhibited MAO-A or B in concentrations up to 50 μg/ml.[59] Hyperforin (10 mg/kg, i.p.) enhanced the extracellular levels of dopamine, noradrenaline, serotonin and glutamate in the synaptic cleft in the rat locus coeruleus.[61]

A methanol extract of the dried flowers inhibited radiolabeled flumazenil binding to the benzodiazepine sites of the GABA receptor in rat brain preparations in vitro ($IC_{50}$ 6.83 μg/ml).[62] The number of serotonergic 5-HT1-A and 5-HT2-A receptors significantly increased in the brains of rats treated with an ethanol extract of the herb (2700 mg/kg) for 26 weeks, whereas the affinity of both serotonergic receptors remained unaltered.[63] These data suggest that prolonged administration with the extract induced upregulation of the 5-HT$_1$-A and 5-HT$_2$-A receptors.[63] The affinity of hypericin for thirty receptor and re-uptake sites was determined in an *in vitro* study. At 1 µM, hypericin inhibited less than 40% of specific radioligand binding at all sites tested except the acetylcholine and sigma receptors.[64]

Hydroalcoholic extracts of St. John's wort flowers have antiviral activity against the influenza virus A2 (Mannheim 57), herpes virus type 2, poliovirus II and vaccinia virus *in vitro*.[65–66] However, extracts of the dried stem (100 µg/ml) were not active against herpes simplex 1 or 2 or HIV *in vitro*.[67] *In vitro* antiviral activity for hypericin against Friend-MuLV-leukemia virus, hepatitis B virus, murine cytomegalovirus, human cytomegalovirus (Davis strain), para-influenza 3 virus, vaccinia virus, vesicular stomatitis virus and equine infectious anemia virus has been demonstrated.[68–72]

Hypericin and pseudohypericin inhibit the replication of herpes simplex virus (HSV) types 1 and 2, and the human immunodeficiency virus (HIV-1) *in vitro*.[72–75] Hypericin inhibited the activity of HIV reverse transcriptase *in vitro* (IC$_{50}$ 0.77 mM).[71,76] In rodent models, hypericin inhibited HSV, Rauscher murine leukemia and Friend-MuLV-leukemia viruses after intravenous, intraperitoneal or intragastric administration.[77–79] Intraperitoneal administration of an aqueous extract of the herb to mice (5% concentration) had virucidal activity against tick-born encephalitis virus.[80] *In vitro* experiments have demonstrated that hypericin displayed marginal activity against Moloney murine leukemia virus and did not show selective antiviral activity against HSV, influenza A, adenovirus or poliovirus.[79] However, when the virus was incubated with hypericin prior to infecting the cells, the compound had antiviral activity against all enveloped viruses tested (IC$_{50}$ range 1.56 to 25 µg/ml), but did not exhibit activity against non-enveloped viruses.[79] The antiviral activity of hypericin involves a process of photoactivation, involving the formation of a singlet oxygen and inactivating both viral fusion and syncytia formation.[81]

Hypericin is a potent protein kinase C (PKC) inhibitor *in vitro*.[82–84] Treatment of glioma cell lines with hypericin inhibited cell proliferation, and induced cell death due to the inhibition of PKC.[85] Receptor tyrosine kinase activity of epidermal growth factor is also inhibited by hypericin and may be linked to the antiviral and anti-neoplastic effects.[84,86]

## Safety Information

### A. *Adverse Reactions*

Phototoxicity has been reported in animals after ingestion of *Hypericum* during grazing.[87] The dose was estimated at approximately 30–50 times higher than the normal therapeutic doses.[87] Photosensitization in fair-skinned individuals has been demonstrated in a controlled clinical trial involving metered doses of hypericin and exposure to ultraviolet UVA/UVB irradiation.[88] Patients were treated with 600 mg of a hydroalcoholic extract of the herb (0.24–0.32% total hypericin) three times daily for

15 days. A measurable increase in erythema in light-sensitive subjects was observed after UVA irradiation. The plasma concentration of hypericin and pseudohypericin in these subjects was double that seen during normal therapeutic treatment of depression.[88] One case of reversible erythema in light exposed areas has been reported in a 61-year-old female, who had been taking St. John's wort for three years.[89] A single case of acute neuropathy was reported in a patient taking St. John's wort, after exposure to sunlight.[90] Drug monitoring studies indicate that side effects of the herb are rare and mild, and include minor gastrointestinal irritations, allergic reactions, tiredness and restlessness,[9,12,19] however these studies did not exceed 8 weeks. Clinical studies have suggested that the use of St. John's wort does not effect general performance or the ability to drive.[91,92]

## B. Contraindications

St. John's wort should not be administered to patients with an allergy to plants of the Clusiaceae during pregnancy or nursing. See below for drug interaction information. For patients taking protease inhibitors or cyclosporin, risk-benefit of concomitant therapy with St. John's wort should be carefully considered prior to treatment.

## C. Drug Interactions

Although the ingestion of high tyramine-containing foods such as pickled or smoked foods and cheese, and selective serotonin re-uptake inhibitors such as fluoxetine are contraindicated with MAO inhibitors, *in vivo* data linking St. John's wort to MAO inhibition are lacking.[93] The combination of St John's wort with other standard antidepressant drugs, such as tricyclic antidepressants or fluoxetine is not recommended, unless under medical supervision.

There are now numerous reports in the medical literature indicating that St. John's wort extracts induce hepatic enzymes that are responsible for drug metabolism and thereby reduce the serum levels of these drugs.[94-98] Co-adminitration of theophylline with a St. John's wort extract lowered serum level of theophylline in a patient previously stablized, requiring an increase in the theophylline dose.[94] However, the patient's drug regime also included furosemide, potassium, morphine, zolpidem, valproic acid, ibuprofen, amitriptyline, albuterol, prednisone, zafirlukast and inhaled triacinolone acetonide.[94] Co-administration of St. John's wort and digoxin reduced serum digoxin concentrations after ten days of treatment.[95] A decrease in the serum cyclosporin, warfarin and phenprocoumon concentrations has been observed in patients within weeks after the addition of St. John's wort extracts to their therapeutic regimen.[96] Concomitant use of St. John's wort in five patients previously stablized on serotonin-reuptake inhibitors resulted in symptoms of central serotonin excess.[97] The U.S. Food and Drug Administration has publised a report concerning a significant drug interaction between St. John's wort and Indinavir, a protease inhibitor used to treat HIV infections.[98] St. John's wort substantially reduced indinavir plasma concentrations, due to an induction of the cytochrome P450 metabolic pathway. As a consequence the concomitant use of St. John's wort and protease inhibitors or nonnucleoside reverse transcriptase inhibitors is not recommended and may result in suboptimal antiretroviral drug concentrations, leading to a loss of virologic response and the development of resistance.[98]

## D.  Toxicology

The mutagenicity of hydroalcoholic extracts of St. John's wort containing 0.2 to 0.3% of hypericin and 0.35 mg/g quercetin was assessed in various *in vitro* and *in vivo* test systems.[99–103] The *in vitro* studies were performed using the Ames test; the hypoxanthine guanidine phosphoribosyl transferase test (up to 4 μl/ml of extract); the unscheduled DNA synthesis test (up to 1.37 μl/ml of extract); and the cell transformation test in Syrian hamster embryo cells (up to 10 μl/ml of extract). The *in vivo* tests included: the spot test in mice (up to 10 μl/ml of extract), the chromosome aberration test with bone marrow cells of Chinese hamsters (10 ml/kg extract, gavage), and the micronucleus test in rodent bone marrow (2 g/kg extract, gavage). While some positive findings were observed *in vitro* in the Ames test,[100,103] all *in vivo* test results were negative, indicating that the hydroalcoholic extract is not mutagenic in animals. In a 26-week study, intragastric administration of 900 and 2700 mg/kg of the hydroalcoholic extract to rats and dogs had no effect on fertility, development of the embryo and pre- and post-natal development.[104]

## E.  Dose and Dosage Forms

A standardized hydroethanolic or dried hydromethanolic extract up to a daily dose of 900 mg of extract (in three divided doses), equivalent to 0.2 to 2.7 mg of total hypericin daily,[1,7,9,10,15,19] or 5% hyperforin.[8] The average daily dose of the crude drug is two to four grams per day.[20] As with other antidepressant drugs, observation of the therapeutic effects of St. John's wort may require 2 to 4 weeks of therapy. If a significant antidepressant effect has not been observed by 6 weeks, a physician should be consulted.

## References

1    Anon. Herba Hyperici. *WHO Monographs on Selected Medicinal Plants*, Volume II, World Health Organization, Geneva, Switzerland, in press.
2    Hobbs C. St. John's wort, *Hypericum perforatum* L. *Herbal Gram.* 1989;18/19:24–33.
3    Hahn G. *Hypericum perforatum* (St. John's wort) – a medicinal herb used in antiquity and still of interest today. *J Naturopathic Med.* 1992;3:94–96.
4    Bombardelli E, Morazzoni P. *Hypericum perforatum*. *Fitoterapia.* 1995;66:43–68.
5    Hänsel R et al., eds. *Hägers Handbuch der pharmazeutischen Praxis*, 5th ed. Vol. 5. Berlin, Springer-Verlag, 1993:476–495.
6    Nahrstedt A, Butterweck V. Biologically active and other chemical constituents of the herb of *Hypericum perforatum* L. *Pharmacopsychiatry.* 1997;30:129–134.
7    Ernst E. St. John's wort, an antidepressant? A systematic criteria-based review. *Phytomedicine.* 1995;2:47–71.
8    Laakmann G, Dienel A, Kieser M. Clinical significance of hyperforin for the efficacy of *Hypericum* extracts on depressive disorders of different severities. *Phytomedicine.* 1999;5:435–442.

9   Linde K et al. St. John's wort for depression-an overview and meta-analysis of ran-
    domized clinical trials. *Brit Med J.* 1996;313:253–258.
10  Maisenbacher HJ et al. Therapie von Depressionen in der Praxis. Ergebnisse einer
    Anwendungsbeobachtung mit Hyperici herba. *Natura Medica.* 1992;7:394–399.
11  Pieschl D et al. Zur Behandlung von Depressionen. Verblindstudie mit einem pflan-
    zlichen Extrakt Johanniskraut. *Therapiewoche.* 1989;39:2567–2571.
12  Schrader E et al. *Hypericum* treatment of mild-moderate depression in a placebo-con-
    trolled study. A prospective, double-blind, randomized, placebo-controlled, multicen-
    tre study. *Human Psychopharmacology.* 1998;13:163–169.
13  Schultz H, Jobert M. Effects of *Hypericum* extract on the sleep EEG in older volunteers.
    *J Geriatr Psych Neurol.* 1994;7:S39–43.
14  Schultz H et al. Clinical trials with phyto-psychopharmacological agents. *Phytomedi-
    cine.* 1997;4:379–387.
15  Volz HP. Controlled clinical trials of Hypericum extracts in depressed patients – an
    overview. *Pharmacopsychiatry.* 1997;30:72–76.
16  Vorbach et al. Efficacy and tolerability of St. John's wort extract LI 160 versus
    imipramine in patients with severe depressive episodes according to ICD-10. *Pharma-
    copsychiatry.* 1997;30:81–85.
17  Wheatley D. LI 160, an extract of St. John's wort, versus amitriptyline in mildly to
    moderately depressed outpatients – a controlled 6-week clinical trial. *Pharmacopsy-
    chiatry.* 1997;30:77–80.
18  Wheatley D. Hypericum extract – Potential in the treatment of depression. *CNS Drugs.*
    1998;9:431–440.
19  Woelk H et al. Benefits and risks of the *Hypericum* extract LI 160: drug-monitoring study
    with 3250 patients. *J Geriatr Psych Neurol.* 1994;7:S34–38.
20  Blumenthal M et al. (eds). St. John's wort, *The Complete German Commission E
    Monographs.* American Botanical Council, Austin TX, 1998.
21  Schellenberg R et al. Pharmacodynamic effects of two different *Hypericum* extracts in
    healthy volunteers measured by quantitative EEG. *Pharmacopsychiatry.* 1998;31
    (Suppl. 1):44–53.
22  Martinez B et al. Hypericum in the treatment of seasonal affective disorders. *J Geriatr
    Psych Neurol.* 1994;7:S29–33.
23  Kasper S. Treatment of seasonal affective disorder (SAD) with *Hypericum* extract.
    *Pharmacopsychiatry.* 1997;30:89–93.
24  Johnson D. Neurophysiologische Wirkungen von *Hypericum* im Doppelblindversuch
    mit Probanden. *Nervenheilkunde.* 1991;10:316–317.
25  Johnson D et al. Effects of *Hypericum* extract LI 160 compared with maprotiline on
    resting EEG and evoked potentials in 24 volunteers. *J Geriatr Psych Neurol.*
    1994;7:S44–46.
26  Schakau D et al. Risk/benefit profile of St. John's wort extract. *Psychopharmakother-
    apie.* 1996;3:116–122.
27  Kugler J et al. Therapie depressiver Zustände. *Hypericum*-Extrakt Steigerwald als
    Alternative zur Benzodiazepin-Behandlung. *Zeitschrift für Allgemeine Medizin.*
    1990;66:21–29.
28  Demisch L et al. Einfluß einer subchronischen Gabe von Hyperforat auf die nächtliche
    Melatonin – ind Kortisolsekretion bei Probanden. Nünberg, AGNP Symposium
    Abstract 1991.
29  Mülder H, Zöller M. Antidepressive Wirkung eines auf den Wirkstoffkomplex Hyper-
    icin standardisierten *Hypericum* Extrakts. Biochemische und klinische Untersuchun-
    gen. *Arzneimittelforschung.* 1984;34:918–920.

30   Ivan H. Preliminary investigations on the application of *Hypericum perforatum* in herpes therapy. *Gyogyszereszet*. 1979;23:217–218.

31   Staffeldt B et al. Pharmacokinetics of hypericin and pseudohypericin after oral intake of the *Hypericum perforatum* extract LI 160 in healthy volunteers. *J Geriatr Psych Neurol*. 1994;7:S47–53.

32   Brockmöller J et al. Hypericin and pseudohypericin: pharmacokinetics and effects on photosensitivity in humans. *Pharmacopsychiatry*. 1997;30:94–101.

33   Biber A et al. Oral bioavailability of hyperforin from *Hypericum* extracts in rats and human volunteers. *Pharmacopsychiatry*. 1998;31 (Suppl. 1):36–43.

34   Öztürk Y. Testing the antidepressant effects of *Hypericum* species on animal models. *Pharmacopsychiatry*. 1997;30:125–128.

35   Okpanyi SN, Weischer ML. Animal experiments on the psychotropic action of Hypericum extract. *Arzneimittelforschung*. 1987;37:10–13.

36   Girzu M et al. Sedative activity in mice of a hydroalcohol extract of *Hypericum perforatum* L. *Phytother Res*. 1997;2:395–397.

37   Butterweck V et al. Effects of the total extract and fractions of *Hypericum perforatum* in animal assays for antidepressant activity. *Pharmacopsychiatry*. 1997;30:117–124.

38   Chatterjee SS et al. Hyperforin as a possible antidepressant component of *Hypericum* extracts. *Life Sci*. 1998a;63:499–510

39   Bhattacharya SK et al. Activity profiles of two hyperforin-containing *Hypericum* extracts in behavioral models. *Pharmacopsychiatry*. 1998;31 (Suppl. 1):22–29.

40   Dimpfel W et al. Effects of a methanolic extract and a hyperforin-enriched $CO_2$ extract of St. John's wort (*Hypericum perforatum*) on intracerebral field potentials in the freely moving rat. *Pharmacopsychiatry*. 1998;31 (Suppl. 1):30–35.

41   Fedorchuk AM. Effect of *Hypericum perforatum* on experimentally infected wounds. *Mikrobiologichnii Zhurnal (Kiev)*. 1964;26:32.

42   Lazareva KN et al. The results of a study of some drug plants of the Bashkir USSR. *Sbornik Nauchnykh Trudov Bashkir Gosudarstvennogo Meditsinskii Institut*. 1968;17:54.

43   Rao SG et al. *Calendula* and *Hypericum*: two homeopathic drugs promoting wound healing in rats. *Fitoterapia*. 1991,62:508–510.

44   Suzuki O. et al. Inhibition of monoamine oxidase by isogentisin and its 3-O-glucoside. *Biochem Pharmacol*. 1978;27:2075–2078.

45   Suzuki O. et al. Inhibition of type A and type B monoamine oxidase by naturally occurring xanthones. *Planta Med*. 1981;42:17–21.

46   Suzuki O. et al. Inhibition of monoamine oxidase by hypericin. *Planta Med*. 1984;50:272–274.

47   Hölzl J et al. Investigation about antidepressive and mood changing effects of *Hypericum perforatum*. *Planta Med*. 1989;55:643.

48   Demisch L et al. Identification of selective MAO-type-A inhibitors in *Hypericum perforatum* L. (Hyperforat®). *Pharmacopsychiatry*. 1989;22:194.

49   Sparenberg B et al. Untersuchungen über antidepressive Wirkstoffe von Johanniskraut. *Pharmazie Zeitschrift Wissenschaften*. 1993;22:194.

50   Höltje HD, Walper A. Molecular modeling of the antidepressive mechanism of *Hypericum* ingredients. *Nervenheilkunde*. 1993;12:339–340.

51   Bladt S, Wagner H. Inhibition of MAO by fractions and constituents of *Hypericum* extracts. *J Geriatr Psych Neurol*. 1994;7:S57–S59.

52   Thiele HM, Walper A. Inhibition of MAO and COMT by *Hypericum* extracts and hypericin. *J Geriatr Psych Neurol*. 1994;7:S54–56.

53   Thiele B et al. Modulation of cytokine expression by *Hypericum* extracts. *J Geriatr Psych Neurol*. 1994;7(Suppl):S60–S62.

54  Neary JT, Bu YR. *Hypericum* LI 160 inhibits uptake of serotonin and norepinephrine in astrocytes. *Brain Res.* 1999;816:358–363.

55  Perovic S, Müller WEG. Pharmacological profile of *Hypericum* extract: effect on serotonin uptake by postsynaptic receptors. *Arzneimittelforschung.* 1995;45:1145–1148.

56  Müller WE et al. Effects of *Hypericum* extract LI 160 on neurotransmitter uptake systems and adrenergic receptor density, *2nd International Congress on Phytomedicine*, Munich, 1996.

57  Cott JM. *In vitro* receptor binding and enzyme inhibition by *Hypericum perforatum* extract. *Pharmacopsychiatry.* 1997;30(Suppl):108–112.

58  Wonnemann M et al. Effects of *Hypericum* extract on glutamatergic and gabaminergic receptor systems. *Pharmazie.* 1998;53:38.

59  Chatterjee SS et al. Hyperforin inhibits synaptosomal uptake of neurotransmitters *in vitro* and shows antidepressant activity *in vivo. Pharmazie.* 1998b;53:9.

60  Müller WE et al. Hyperforin represents the neurotransmitter reuptake inhibiting constituent of *Hypericum* extract. *Pharmacopsychiatry.* 1998;31(Suppl. 1):16–21.

61  Kaehler ST et al. Hyperforin enhances the extracellular concentrations of catecholamines, serotonin and glutamate in the rat locus coeruleus. *Neurosci Lett.* 1999;262:199–202.

62  Baureithel KH et al. Inhibition of benzodiazepine binding *in vitro* by amentoflavone, a constituent of various species of *Hypericum. Pharmaceutica Acta Helvetiae.* 1997;72:153–157.

63  Teufel-Mayer R, Gleitz J. Effects of long-term administration of *Hypericum* extracts on the affinity and density of the central serotonergic 5-HT$_1$-A and 5-HT$_2$-A receptors. *Pharmacopsychiatry.* 1997;30:113–116.

64  Raffa RB. Screen of receptor and uptake-site activity of hypericin component of St. John's wort reveals sigma receptor binding. *Life Sci.* 1998;62:PL265–270.

65  May G, Willuhn G. Antiviral activity of aqueous extracts from medicinal plants in tissue cultures. *Arzneimittelforschung.* 1978;28:1–7.

66  Mishenkova EL et al. Antiviral properties of St. John's wort and preparations produced from it. *Trudy S'ezda mikrobiologii Ukrainskoi* 4th 1975, 222.

67  Pacheco P et al. Antiviral activity of Chilean medicinal plant extracts. *Phytother Res.* 1993;7:415–418.

68  Anderson DO et al. *In vitro* virucidal activity of selected anthraquinones and anthraquinone derivatives. *Antiviral Res.* 1991;162:185–196.

69  Carpenter S, Kraus GA. Photosensitization is required for inactivation of equine infectious anemia virus by hypericin. *Photochem Photobiol.* 1991;53:169–174.

70  Hudson JB et al. Antiviral assays on phytopharmaceuticals: the influence of reaction parameters. *Planta Med.* 1994;604:329–332.

71  Lavie G et al. Hypericin as an antiretroviral agent. Mode of action and related analogues. *Annals of the New York Academy of Sciences.* 1992;556–562.

72  Wood S et al. Antiviral activity of naturally occurring anthraquinones and anthraquinone derivatives. *Planta Med.* 1990;56:651–652.

73  Cohen PA et al. Antiviral activities of anthraquinones, bianthrones and hypericin derivatives from lichens. *Experientia.* 1996;523:180–183.

74  Degan S et al. Inactivation of the human immunodeficiency virus by hypericin: evidence for photochemical alterations of P24 and A block in uncoating. *Aids Research in Human Retroviruses.* 1992;811:1929–1936.

75  Weber ND et al. The antiviral agent hypericin has *in vitro* activity against HSV-1 through non-specific association with viral and cellular membranes. *Antiviral Chemistry and Chemotherapy.* 1994;5:83–90.

76    Schinazi RF et al. Anthraquinones as a new class of antiviral agents against human immunodeficiency virus. *Antiviral Research.* 1990;135:265–272.

77    Lavie G et al. Studies of the mechanisms of action of the antiretroviral agents hypericin and pseudohypericin. *Proceedings of the National Academy of Sciences (USA).* 1989;8615:5963–5967.

78    Meruelo D et al. Therapeutic agents with dramatic antiretroviral activity and little toxicity at effective doses: aromatic polycyclic diones hypericin and pseudohypericin. *Proceedings of the National Academy of Sciences USA.* 1988;85:5230–5234.

79    Tang J et al. Virucidal activity of hypericin against enveloped and non-enveloped DNA and RNA viruses. *Antiviral Research.* 1990;136:313–325.

80    Fokina GI et al. Experimental phytotherapy of tick-borne encephalitis. *Soviet Progress in Virology.* 1991;1:27–31.

81    Lenard J et al. Photodynamic inactivation of infectivity of human immunodeficiency virus and other enveloped viruses using hypericin and Rose bengal: inhibition of fusion and syncytia formation. *Proceedings of the National Academy of Sciences (USA).* 1993;901:158–162.

82    Agostinis P et al. A comparative analysis of the photosensitized inhibition of growth factor regulated protein kinases by hypericin derivatives. *Biochemical and Biophysical Research Communications.* 1996;2203:613–617.

83    Agostinis P et al. Photosensitized inhibition of growth factor regulated protein kinases by hypericin. *Biochem Pharmacol.* 1995;4911:1615–1622.

84    De Witt PA et al. Inhibition of epidermal growth factor receptor tyrosine kinase activity by hypericin. *Biochem Pharmacol.* 1993;46:1929–1936.

85    Couldwell WT et al. Hypericin: a potential antiglioma therapy. *Neurosurgery.* 1994;35:705–710.

86    Panossian AG et al. Immunosuppressive effects of hypericin on stimulated human leucocytes: inhibition of the arachidonic acid release, leukotriene B4 and interleukin-1 production and activation of nitric oxide formation. *Phytomedicine.* 1996;3:19–28.

87    Siegers CP et al. Zur Frage der Phototoxizitat von *Hypericum. Nervenheilkunde.* 1993;12:320–322.

88    Roots I et al. Evaluation of photosensitization of the skin upon single and multiple dose intake of *Hypericum* extract. *Second International Congress on Phytomedicine.* Munich, 1996.

89    Golsch S et al. Reversible Erhöhung der photosensitivitat im UV-B-Bereich durch Johanniskrautextrakt Präparate. *Hautarzt.* 1997;48:249–252.

90    Bove GM. Acute neuropathy after exposure to sun in a patient treated with St. John's wort. *Lancet.* 1998;352:1121.

91    Herberg KW. Psychotrope Phytopharmaka im Test. Alternative zu synthetischen Psychopharmaka? *Therapiewoche.* 1994;44:704–713.

92    Schmidt U et al. Johanniskraut-Extrakt zur ambulanten Therapie der Depression. Aufmerksamkeit und Reaktionsvermogen bleiben erhalten. *Fortschrift der Medizin.* 1993;111:339–342.

93    Cott J, Misra R. Medicinal plants: a potential source for new psychotherapeutic drugs. In: Kanba S et al., (eds.). *New drug development from herbal medicines in neuropsychopharmacology,* New York, Brunner/Mazel, Inc. 1997.

94    Nebel A et al. Potential metabolic interaction between St. John's wort and theophylline. *Annals of Pharmacother.* 1999;33:502.

95    Johne A, Brockmöller J, Bauer S et al. Interaction of St. John's wort extract with digoxin, Jahreakongress für Klin. Pharmakol., Berlin June 10–12, 1999.

96    Ernst E. Second thoughts about the safety of St. John's wort. *Lancet.* 1999;354:2014–16.

97    Lantz MS, Buchalter E et al. St. John's wort and antidepressant drug interactions in the elderly. *J Geriatr Psych Neurol.* 1999;12:7–10.

98    Piscitelli SC, Burstein AH, Chaitt D et al. Indinavir concentrations and St. John's wort. *Lancet.* 2000;355:547–548.

99    Okpanyi SN et al. Genotoxizität eines standardisierten *Hypericum* Extrakts. *Arzneimittelforschung.* 1990;40:851–855.

100   Poginsky B et al. Johanniskraut (*Hypericum perforatum* L.). Genotoxizitat bedingt durch den Quercetingehalt. *Deutsche Apotheker Zeitung.* 1988;128:1364–1366.

101   Schimmer O et al. The mutagenic potencies of plant extracts containing quercetin in *Salmonella typhimurium* TA 98 and TA 100. *Mut Res.* 1988;206:201–208.

102   Schimmer O et al. An evaluation of 55 commercial plant extracts in the Ames mutagenicity test. *Pharmazie.* 1994;49:448–451.

103   Leuschner J. Preclinical toxicological profile of Hypericum extract LI 160. *Second International Congress on Phytomedicine*, Munich, 1996.

# 25

# Valerian

## Synopsis

The results from controlled clinical trials support the use of valerian (*Valeriana officinalis*) as a mild sedative, and for the symptomatic treatment of insomnia. Numerous valerian-containing products are available on the commercial market, including "standardized extracts", tinctures, capsules and tablets. The considerable variation in the composition of valerian extracts and the instability of some of its chemical constituents makes the standardization of these products difficult. The recommended daily dose of valerian ranges from 200 to 600 mg per day. Results from one controlled clinical trial have demonstrated that 900 mg per day of valerian root is no more effective than 450 mg per day. Adverse reactions include stomach upset, headaches and allergic reactions. Chronic administration is not recommended due to a lack of long-term safety studies. No drug interactions have been reported. While valerian administration has not been shown to impair vigilance, patients should be cautioned that motor reflexes, driving ability and the operation of heavy machinery might be adversely affected. Due to a lack of safety data, valerian should not be administered during pregnancy or nursing, or to children under the age of twelve.

## Introduction

Valerian has a long history of use as a medicinal plant in many parts of the world.[1] Dioscorides recommended valerian for the treatment of digestive tract problems, nausea, as well as liver and urinary tract disorders.[1] Galen, a Greek physician and pharmacist, who used valerian to treat insomnia, first mentioned the use of valerian as a sedative during 2nd century AD. However, the use of valerian for the treatment of nervous disorders was not firmly established until the middle of the 18th century.[1] It is believed that the English physician, John Hill, was the first to use valerian therapeutically as a sedative. During the early 19th century the Eclectic physicians in the United States employed valerian for the treatment of various nervous disorders. By the latter part of the 19th century, the volatile oil of valerian had already been analyzed, and in 1883 valerian was described in standard medical texts such as the *Dispensary of the United States of America*, and its medical uses were described.[2] Currently, valerian is the most well known herbal sedative worldwide, and it is listed in at least 20 different pharmacopoeias from around the globe. Extracts of valerian root, alone and in combination with other plant extracts, are often prescribed in Europe and other parts of the world as a substitute for the benzodiazepines in the treatment of insomnia and anxiety. A wide variety of valerian products are available, including a number of standardized extracts. Valerian products are sold in health food stores and pharmacies, and in the United States these products are regulated as dietary supplements.

## Quality Information

*   The correct Latin name for the plant is *Valeriana officinalis* L.[3] Taxonomic synonyms that may be found in the scientific literature include *Valeriana alternifolia* Ledeb., *Valeriana excelsa* Poir., *Valeriana sylvestris* Grosch.[3] The vernacular or common names for the plant include: all heal, amantilla, Balderbrackenwurzel, baldrian, Baldrianwurzel, cat's love, cat's valerian, fragrant valerian, garden heliotrope, great wild valerian, Katzenwurzel, kesso root, racine de valeriane, St. George's herb, setwall, valerian fragrant, valerian, valeriana, vandal root, and wild valerian.[3]
*   Commercial products of valerian are prepared from the underground parts of *Valeriana officinalis* L. *sensu lato* (Valerianaceae) including the rhizomes, roots, and stolons.[3]
*   *Valeriana officinalis sensu lato* is an extremely polymorphous complex of subspecies with natural populations dispersed over temperate and sub-polar Eurasian zones. The species is common in damp woods, ditches, and along the streams in Europe, and is cultivated as a medicinal plant, especially in Belgium, England, France, Germany, Holland, Eastern Europe, Russia and the USA.[3]
*   The chemical composition of valerian root and rhizome varies considerably depending on the subspecies, variety, age of the plant, growing conditions, type and the age of the extract. The principal chemical constituents of the volatile oil are bornyl acetate and bornyl isovalerate.[3] Other significant constituents include $\beta$-caryophyllene, valeranone, valerenal, valerenic acid, and other sesquiterpenoids and monoterpenes. Another important group of chemical constituents includes a series of non-glycosidic bicyclic iridoid monoterpene epoxy-esters known as the valepotriates. The major valepotriates are valtrate and isovaltrate, while smaller amounts of dihydrovaltrate, isovaleroxy-hydroxy-didrovaltrate, 1-acetyl may also be present. It should be noted that the valepotriates are unstable compounds due to their epoxide structure. Degradation of the valepotriates occurs fairly rapidly during storage or processing, especially if the root has not been carefully dried. However, the degradation products baldrinal, homobaldrinal, and valtroxal, also have sedative activity.[3]

## Medical Uses

Valerian is used clinically as a mild sedative, and for the symptomatic treatment of chronic insomnia.[4–15] One clinical trial has suggested that valerian may not be very useful for acute insomnia.[11]

## Summary of Clinical Evidence

To date at least ten controlled clinical trials have assessed the effects of valerian extracts in healthy volunteers, patients with sleep disorders, and pharmacodynamic studies based on quantitative EEG analysis. Five trials used a freeze-dried aqueous extract,[4–8] three used ethanol extracts,[9–11] and one trial did not specify the type of extract that was employed.[12]

Three placebo-controlled, double-blind in three consecutive trials, involving a total of 165 healthy volunteers, assessed the effects of valerian (450 mg or 900 mg of an aqueous root extract) on sleep latency (using a self rating scale) and sleep EEG.[4–6] In the first study, 128 healthy volunteers were treated with 400 mg of an aqueous valerian extract, or a combination valerian and hops preparation or placebo.[4] Treatment with the valerian extract alone improved subjective ratings for sleep quality and sleep latency, and night awakenings, while dream recall and somnolence the next morning were unaffected. The improvements in sleep latency and sleep quality were better in those patients who considered themselves to be habitually poor sleepers.[4] In the second trial, a group of 8 volunteers suffering from mild insomnia were treated with either 450 mg or 900 mg of an aqueous valerian extract.[6] The patients treated with 450 mg of a valerian extract exhibited a significant decrease in sleep latency ($p < 0.05$) and improved sleep quality as compared with placebo, but had no significant effects on sleep EEG. Treatment with 900 mg per day of valerian did not further reduce sleep latency.[6]

A similar placebo-controlled, double-blind crossover trial involving 10 healthy subjects assessed the effect of an aqueous extract of valerian root (450 mg or 900 mg) on subjective and objective sleep parameters.[8] One group of patients slept at home (n = 10) and one group slept in a sleep laboratory (n = 8). Outcome measures were evaluated on the basis of questionnaires, self-rating scales and nighttime motor activity. In addition, polygraph sleep recordings and spectral analysis of the sleep EEG was performed in the laboratory group. Under home conditions, both doses showed a significant dose-dependent decrease in sleep latency ($p < 0.01$) and awaking time ($p < 0.05$) after treatment with 450 or 900 mg/day of a valerian extract.[8] No changes were observed in sleep stages or sleep EEG's.

The effects of 60 and 120 mg of valerian were assessed in a placebo-controlled, double-blind crossover by computer analysis of sleep stages and psychometric methods in 20 subjects.[13] Both doses showed a decrease of sleep stage 4 and a slight reduction in REM sleep. A slight increase in stages 1, 2 and 3 were also observed. After administration of 120 mg of valerian, the frequency of REM-phases of sleep declined during the first half of the night and increased in the second half. Changes in the beta intensity of the EEG during REM sleep showed a stronger hypnotic effect for the 120 mg dose. The maximum effect was observed 2–3 hours after administration.[13]

Two double-blind, placebo-controlled pharmacodynamic studies, assessed the effects of an ethanol extract in 36 patients with sleep disorders.[9,10] The first trial demonstrated that an extract of valerian root significantly increased sleep latency, in poor and irregular sleepers, but it had no effect on night awakenings or dream recall.[9] Treatment with the valerian extract increased slow-wave sleep in those patients with low baseline values, while REM sleep remained unaltered.[9] In the second study, used a randomized crossover design to compare valerian extract (1200 mg) with diazepam (10 mg), lavender extract (1200 mg), passion flower extract (1200 mg), kava extract (600 mg) or placebo.[10] Unlike diazepam, all herbal extracts caused a relative increase in amplitude in the theta band of the EEG, and no increase in the beta frequency band. Marked increased in the long-wave delta frequency range were observed for the valerian extract.[10]

The effects of valerian were assessed in a placebo-controlled double blind study involving subjects with sleep disorders from geriatric hospitals.[7] The study involved 150 patients treated for 30 days and 80 patients treated with 270 mg of an aqueous valerian extract for 14 days, and assessed using the von Zerssen mood scale and the NOSIE (behavioral disturbance) scale, staying asleep score. After 14 days of treatment, there was a statistically significant improvement ($p < 0.01$) in mood as seen in the von Zerssen score and in the NOSIE scores.[7]

A placebo-controlled double-blind trial involving 121 patients who had experienced significant sleep disturbances for at least four weeks assessed the effects of a 70% ethanol extract of valerian LI 156.[11] The outcomes measured were four rating scales: a physician-rated sleep scale (SS), the Görtelmayer sleep questionaire (G), the von Zersson mood scale (VZ), and the Clinical Global Impressions (CGI). After 28 treatment with 600 mg/day of the extract, there was improvement in the VZ, G and CGI scales. However no significant effects were observed before 14 days of treatment. The results of this trial suggest that unlike typical sleep aids, valerian preparations may take 2 to 4 weeks of treatment to achieve significant improvement.[11]

One comparison, double-blind clinical trial assessed the effects of a valerian and hops extract (150 mg each) with flunitrazepam and a placebo in groups of 20 healthy volunteers.[14] After three doses of the medications, a series of vigilance and reaction tests were administered, and self-assessments were performed using visual analogue scales. No impairments were found in the subjects treated with the valerian/hops combination, but flunitrazepam was associated with multiple impairments due to a decline in vigilance.[14]

The effect of a valerian extract on sleep disturbances was assessed in an uncontrolled post-marketing study involving 11,168 subjects.[15] Treatment with the extract was evaluated as successful in 72.1% of subjects with disturbances of falling asleep, 75.5% of early awaking and 72.2% of restlessness and tension.[15]

Although extracts of valerian have been clearly shown to have CNS depressant activity, the identity of the active constituents remains controversial. Neither the valepotriates, nor the sesquiterpenes-valerenic acid and valeranone, nor the volatile oil alone can account for the overall sedative activity of the plant.[3] It has been suggested that the baldrinals, degradation products of the valepotriates, may be responsible for the sedative action. Currently, it is still not known whether the pharmacological activity of valerian is dependent on one compound, or due a synergistic effect of a group of compounds.[3]

## Pharmacokinetics

No pharmacokinetic data have been published in the scientific literature.

## Mechanism of Action

Numerous *in vitro* and *in vivo* pharmacological studies have demonstrated that *Valeriana officinalis* has sedative activity. *In vitro* studies have demonstrated that valerian extracts bind to GABA ($\gamma$-aminobutyric acid) receptors, adenosine receptors and the barbiturate and benzodiazeopine receptors.[16] Both hydroalcoholic and aqueous total extracts show affinity for the GABA-A receptors,[17] but there is no clear correlation between any of the known chemical components isolated from valerian and GABA-A binding activity.[17] Aqueous extracts of the roots of *Valeriana officinalis* inhibit the reuptake by 50% at concentrations of 1 $\mu$g/ml, and stimulate the release of [$^3$H]-GABA in isolated synaptosomes isolated from rat brain cortex, either in the presence or absence of potassium depolarization.[18,19] A 50% inhibition was observed at 1 $\mu$g/ml and 100% inhibition at 8 $\mu$g/ml. It was concluded that valerian released GABA by reversal of the GABA carrier, which was sodium dependent and calcium independent.[20] This activity may increase the extracellular concentration of GABA in the synaptic cleft, and thereby enhance the biochemical and behavioral effects of GABA. Interestingly, $\gamma$-aminobutyric acid (GABA) has been found in extracts of *Valeriana officinalis*, and appears be responsible for this activity.[20]

Some of the major chemical constituents of valerian include the valepotriates, bornyl acetate (volatile oil) and valerenic acid. The valepotriates and valerenic acids were considered to be the active constituents of valerian, however the pharmacological studies have not been conclusive.[21] Furthermore, the valprotriates hydrolyze quite rapidly and are not present in any significant amount in aqueous or dilute alcohol extracts after a few days. The valepotriates may act as prodrugs, being rapidly transformed in the gastrointestinal tract into baldrinal, homobaldrinal and isovaltral, which have been shown to have sedative activity *in vivo*.[22] Valerenic acid, a sesquiterpene found only in the species *Valeriana officinalis*, inhibits the enzyme-induced metabolism of GABA in the brain, which may contribute to its sedative activity. The valtrates, and in particular dihydrovaltrate, show some affinity for both the barbiturate receptors and the peripheral benzodiazepine receptors.[21] Recently, hydroxypinoresinol, a lignan isolated from valerian, has also been shown to bind to the benzodiazepine receptors.[22]

Numerous pharmacological tests have been used to investigate the sedative effects of valerian in rodents.[2] Various extracts of valerian have been shown to reduce the spontaneous motility of laboratory animals and prolong thiopental sleeping times.[2] A lipid soluble fraction of a valerian extract reduced spontaneous motility in mice by 50% after intragastric administration of 10 mg. The aqueous fraction of the extract was much less active, and reduced the spontaneous motility of mice by 30%, but only after a dose of 100 mg.[16] One *in vivo* study in rodents compared the sedative effects of an ethanol extract of valerian to that of diazepam and haloperidol.[23] Although spontaneous motility, nociception or body temperature was not changed, the extract did reduce picrotoxin-induced convulsions, and increased barbitone-induced sleeping times.[23]

Antidepressant effects of valerian extracts have been demonstrated in rodent models. Intragastric administration of methanol extract of valerian to mice exhibited antidepressant activity.[24] The active constituents were isolated from the extract and identified as α-kessyl alcohol, kessanol and cyclokessyl actate.[24] Other *in vivo* studies have indicated that the CNS depressant activity of the valerian extracts may be due to high concentrations of glutamine in the extract.[20] Glutamine is able to cross the blood-brain barrier, where it is taken up by nerve terminals and subsequently metabolized to GABA. The addition of exogenous glutamine stimulates GABA synthesis in synaptosomes and rat brain slices.[20]

An electroneurophysiological study assessed the effects of valerian extract and valtrate in cats with implanted electrodes on EEG recordings.[25] No changes in the contactial and subcortical EEG were apparent after doses of 5 or 20 mg/kg of valtrate or isovaltrate and 100 or 250 mg/kg of the extract.[25] However, muscular tone was reduced in 30 to 40% of cases, and an increase in the amplitude of the amygdalo-hippocampial stimulation response, indicating a thymoleptic response. The effects of valerian and Valium on surface and deep electro-encephalograms were recorded in rats after intragastric administration.[26] Valerian produced a sedative action, which was not associated with the drowsiness that was observed in animals treated with diazepam.[26]

**Safety Information**

*A. Adverse Reactions*

Clinical trials reported side effects such as stomach upset, headaches and itching were noted occasionally.[15] Acute toxicity of valerian overdose is considered to be very low, however subchronic and chronic toxicity are lacking. Doses up to 20 times the recommended therapeutic dose have been reported to cause only mild symptoms that are resolved within 24 hours.[27] Nonetheless, very large doses may cause bradycardia, arrhythmias, and decreased intestinal motility.[27] Four cases of hepatotoxicity have been reported following the use of preparations containing valerian in combination with other herbs, including skullcap.[28] However, in all cases the patients were taking a combination herbal product containing four different plant species and thus a causal relationship to the intake of valerian is extremely doubtful. In addition, skullcap (or a common adulterant of skullcap, germander) may have been the responsible ingredient. An assessment of the delayed effects associated with an overdose of a preparation containing valerian, hyoscine hydrobromide and cyproheptadine hydrochloride should no evidence of acute liver toxicity and suggested that delayed liver damage was unlikely.[29]

Specific studies concerning valerian's influence on vigilance have demonstrated that there are no sedative side effects that would impair the ability to drive or use machinery.[14] Furthermore, valerian does not produce "hang-over" effects, as seen with the benzodiazepines.[14,30]

## B.  Contraindications

Due to a lack of safety data, valerian preparations should not be used during pregnancy or nursing, or in children under the age of 12 years.[3]

## C.  Drug Interactions and Warnings

Although clinical trials have indicated that ingestion of valerian preparations does not effect vigilance,[14,30] patients should be cautioned that valerian may effect driving capacity or the ability to operating hazardous machinery. Although no drug interactions have been reported, the combination of valerian and alcohol, other sedatives or tranquilizers should be avoided.[3]

## D.  Toxicology

Cytotoxicity of the valepotriates has been demonstrated *in vitro* in cultured rat hepatoma cells (33 mg/ml), while baldrinal was not active.[1] *In vivo*, the valeprotriates were not cytotoxic in rodents when administered orally or intraperitoneally, even in doses at 1350 mg/kg or 65 mg/kg respectively.[29] Some valepotriates have alkylating activity *in vitro*.[32] However, since the valepotriates rapidly decompose upon storage, there is generally very little cause for concern.[1] Furthermore, the *in vivo* tests suggest that valepotriates are so poorly absorbed in the gastrointestinal tract that the *in vivo* toxicity is low.[1] The valepotriates are hydrolyzed to form the baldrinals, which are less toxic than the valepotriates *in vitro*, but *in vivo* they are more cytotoxic, due to the fact that they are readily absorbed in the intestine. Baldrinals have been detected in commercial preparations standardized to valepotriates in levels up to 0.988 mg/dose and may pose some cytotoxic concern.[16] However, many commercial products do not contain the valepotriates or the baldrinals.

No acute toxicity has been reported for valerian extracts in rodents, and prolonged administration of the valprotriates to rats did not exhibit any toxicity to the pregnant animals or their offspring.[21,33]

## E.   Dose and Dosage Forms

*Commercial preparations:*
For treatment of insomnia: 450 mg of valerian extract taken at least one hour before bedtime.[4–6,11]
For symptomatic treatment of anxiety: 200–300 mg of valerian extract in the morning.

*Crude drug:*
Dried Root and Rhizome, 2–3 g drug per cup by oral infusion, 1–5 times per day, up to a total of 10 g and preparations correspondingly.[3]

## References

1    Houghton PJ. The biological activity of valerian and related plants. *J Ethnopharmacol.* 1988;22:121–142.
2    Houghton PJ (ed.) Valerian, the genus *Valeriana.* Hardwood Academic Publishers, Amsterdam, Netherlands, 1997.
3    Anon. Radix Valerianae. *WHO Monographs on Selected Medicinal Plants.* WHO, Geneva, Switzerland: WHO Publications, 1999.
4    Leathwood PD, Chauffard F. Quantifying the effects of mild sedatives. *J Psychol Res.* 1982/1983;17:115.
5    Leathwood PD, Chauffard F, Heck E, Munoz-Box R. Aqueous extract of Valerian root (*Valeriana officinalis* L.) improves sleep quality in man. *Pharmacol Biochem Behav.* 1982;17:65–71.
6    Leathwood PD, Chauffard F. Aqueous extract of valerian reduces latency to fall asleep in man. *Planta Med.* 1985;2:144–148.
7    Kamm-Kohl AV, Jansen W, Brockmann P. Moderne Baldriantherapie gegen nervose Störungen im Selium. *Med Welt.* 1984;35:1450–1453.
8    Balderer G, Borbely A. Effect of valerian on human sleep. *Psychopharmacology.* 1985;87:406–409
9    Schultz H, Stolz C, Muller J. The effect of valerian extract on sleep polygraphy in poor sleepers – a pilot study. *Pharmacopsychiatry.* 1994;27:147–151.
10   Schultz H, Jobert M. Die Darstellung sedierender/tranquilisierender Wirkung von Phytopharmaka im quantifizierten. *Z Phytother.* 1995, Abstract, pg 10.
11   Vorbach EU, Görtelmayer R, Bruning J. Therapie von Insomnien: Wirksamkeit und Vertraglichkeit eines Baldarin-Präparates. *Phytopharmakotherapie.* 1996;3:109–115.
12   Jansen W. Doppelblindstudie mit Balrisedon. *Therapiewoche.* 1977;27:2779–2786.
13   Gessner B, Klasser M, Völp A. Untersuchung uber die Langzeitwirkung von Harmonicum Much auf den Schlaf von schlafgestörten Personen. *Therapiewoche.* 1984;33:5547–5558.
14   Gerhardt U, Linnenbrink N, Georghiadou Ch. Effects of two plant-based sleep remedies on vigilance. *Schweiz Rsch Med.* 1996;85:473–481.
15   Schmidt-Voight J. Treatment of nervous sleep disorders and unrest with a sedative of purely vegetable origin. *Therapiewoche.* 1986;36:663–667.
16   Wagner H, Jurcic K, Schaette R. Comparative studies on the sedative action of *Valeriana* extracts, valepotriates and their degradation products. *Planta Med.* 1980;37:358–362.
17   Cavadas C, Araujo I, Cotrim MD, Amaral T, Cunha AP, Macedo T, Fontes Ribiero C. *In vitro* study on the interaction of *Valeriana officinalis* L. extracts and their amino acids on GABA-A receptor in rat brain. *Arzneimittelforschung.* 1995;45:154–157.
18   Santos MS et al. Synaptosomal GABA release as influenced by valerian root extract, involvement of the GABA carrier. *Arch Int Pharmacodynamics.* 1994;327:220–231.

19    Santos MS et al. An aqueous extract of valerian influences the transport of GABA in synaptosomes. *Planta Med.* 1994;60:278–279.

20    Santos MS et al. The amount of GABA present in the aqueous extracts of valerian is sufficient to account for $^3$H-GABA release in synaptosomes. *Planta Med.* 1994;60:475–476.

21    Morazzoni P, Bombardelli E. *Valeriana officinalis*: traditional use and recent evaluation of activity. *Fitoterapia.* 1995;66:99–112.

22    Houghton PJ. The scientific basis for the reputed activity of valerian. *J. Pharm Pharmacol.* 1999;51:505–512.

23    Hiller KO, Zetler G. Neuropharmacological studies on ethanol extracts of *Valeriana officinalis* L.: behavioural and anticonvulsant activity. *Phytother Res.* 1996;10:145–151.

24    Holm H, Kowallik H, Reinecke A, von Henning GE, Behne F, Scherer HD. *Med Welt.* 1980;31:982.

25    Oshima Y, Matsuoka S, Ohizumi Y. Antidepressant principles of *Valeriana fauriei* roots. *Chem Pharm Bull.* 1995;43:169–170.

26    De Romanis F, Di Tonto U, Renda F, Sopranzi N. Tracciati EEG, di superficie e di profondita di farmaci correlati alla valeriana. *La Clinica Terapeutica.* 1988;126:101–108.

27    Willey LB et al. Valerian overdose – a case report. *Vet Human Toxicol.* 1995;37:364–365.

28    MacGregor FB. Hepatotoxicity of herbal remedies, *Brit Med J.* 1989;299; 1156–57.

29    Chan TYK, Tang CH, Critchley JAJH. Poisoning due to an over-the-counter hypnotic, sleep-qik (hyoscine, cyproheptadine, valerian) *Postgrad Med J.* 1995;71:227–228.

30    Gerhard U et al. Die sedative Akutwirkung eines pflanzlichen Entspannungsdragées im Vergleich zu Bromazepam. *Schweizerische Rundschau für Medizin. (PRAXIS).* 1991;80:1481–1486.

31    Tortarolo M et al. *In vitro* effects of epoxide-bearing valepotriates on mouse early hematopoietic progenitor cells and human T-lymphocytes. *Arch Toxicol.* 1982;51: 37–42.

32    Tufik S. Effects of a prolonged administration of valepotriates in rats on the mothers and their offspring. *J Ethnopharmacol.* 1985;87:39–44.

33    Braun R. Valepotriates with an epoxide structure-oxygenating alkylating agents. *Planta Med.* 1982;41:21–28.

Milton Keynes UK
Ingram Content Group UK Ltd.
UKHW020822141024
449569UK00008B/521